Genetic Analysis
of Complex Traits
Using SAS®

Edited by
Arnold M. Saxton

The correct bibliographic citation for this manual is as follows: Saxton, Arnold, ed. 2004. *Genetic Analysis of Complex Traits Using SAS®*. Cary, NC: SAS Institute Inc.

Genetic Analysis of Complex Traits Using SAS®

Copyright © 2004, SAS Institute Inc., Cary, NC, USA

ISBN 1-59047-507-0

All rights reserved. Produced in the United States of America.

For a hard-copy book: No part of this publication may be reproduced, stored in a retrieval system, or transmitted, in any form or by any means, electronic, mechanical, photocopying, or otherwise, without the prior written permission of the publisher, SAS Institute Inc.

For a Web download or e-book: Your use of this publication shall be governed by the terms established by the vendor at the time you acquire this publication.

U.S. Government Restricted Rights Notice: Use, duplication, or disclosure of this software and related documentation by the U.S. government is subject to the Agreement with SAS Institute and the restrictions set forth in FAR 52.227-19, Commercial Computer Software-Restricted Rights (June 1987).

SAS Institute Inc., SAS Campus Drive, Cary, North Carolina 27513.

1st printing, November 2004

SAS Publishing provides a complete selection of books and electronic products to help customers use SAS software to its fullest potential. For more information about our e-books, e-learning products, CDs, and hard-copy books, visit the SAS Publishing Web site at **support.sas.com/pubs** or call 1-800-727-3228.

SAS® and all other SAS Institute Inc. product or service names are registered trademarks or trademarks of SAS Institute Inc. in the USA and other countries. ® indicates USA registration.

Other brand and product names are registered trademarks or trademarks of their respective companies.

Contents

About the Authors vii
Acknowledgments xi

Chapter 1   Overview   1
   1.1 Introduction   1
   1.2 Book Organization   2
   1.3 SAS Usage   4
       1.3.1 Example of a Basic SAS DATA Step   4
       1.3.2 Example of a Basic Macro   5
   1.4 References   6

PART 1   Classical Quantitative Genetics

Chapter 2   Estimation of Genetic Variances and Covariances by Restricted Maximum Likelihood Using PROC MIXED   11
   2.1 Introduction   11
   2.2 The Nested Half-Sib and Related Designs   13
       2.2.1 Example of Nested Half-Sib Design, One Trait   13
       2.2.2 Example of Nested Half-Sib Design, Two Traits   19
       2.2.3 Example of Clonal Design, Two Environments   23
   2.3 The Diallel and Related Designs   27
       2.3.1 Diallel Example   28
       2.3.2 Extension to the NC2 Design   33
   2.4 Pedigreed Populations   34
   2.5 References   34

Chapter 3   More Estimation of Genetic Parameters   35
   3.1 Introduction   35
   3.2 Genetic Parameters Estimated with Regression   35
   3.3 Genetic Gain and Realized Heritability   41
   3.4 Inbreeding and Relationship   44
   3.5 Heterosis, or Hybrid Vigor   49
   3.6 References   54

Chapter 4 Genetic Selection 55

  4.1 Introduction 55
  4.2 Single-Trait Selection 56
   4.2.1 Individual Selection 56
   4.2.2 Selection on Relatives 57
   4.2.3 Indirect Selection 58
  4.3 Independent Culling 59
  4.4 Selection Index 62
  4.5 Selection on BLUP 65
  4.6 References 66

Chapter 5 Genotype-by-Environment Interaction 69

  5.1 Introduction 69
  5.2 Modeling Genotype-by-Environment Interaction 72
   5.2.1 ANOVA Model with Fixed GEI 72
   5.2.2 Linear-Bilinear Models with Fixed GEI 76
   5.2.3 Linear Mixed Model Approach to GEI Analysis 82
   5.2.4 Generalized Linear Models to Explore GEI 86
  5.3 Smoothing Spline Genotype Analysis (SSGA) 90
  5.4 References 94

Chapter 6 Growth and Lactation Curves 97

  6.1 Introduction 97
  6.2 Modeling Lactation and Growth Curves 98
  6.3 Using PROC REG to Fit Lactation and Growth Data 101
  6.4 Using PROC NLIN to Fit Non-linear Models 110
  6.5 Repeated Measures Theory for Lactation and Growth Curves 123
  6.6 PROC MIXED for Test Day Models 125
  6.7 Prediction of Individual Test Day Data 138
  6.8 The SPECTRA and ARIMA Procedures for Time Series Prediction 139
  6.9 References 146

Chapter 7 Empirical Bayes Approaches to Mixed Model Inference in Quantitative Genetics 149

  7.1 Introduction 149
   7.1.1 Hierarchical Models 151
  7.2 An Example of Linear Mixed Model Inference 153
  7.3 Empirical Bayes Inference for Normally Distributed Data 157
   7.3.1 Empirical Bayes Analysis in the Linear Mixed Model 158
  7.4 Generalized Linear Mixed Models 160
   7.4.1 Inference on Fixed and Random Effects 162
   7.4.2 Using the %GLIMMIX Macro for Quantitative Genetic Inference 163
   7.4.3 Empirical Bayes Inference in GLMM 166

7.5 Empirical Bayes vs. MCMC 169
7.6 Final Comments 174
7.7 References 175

PART 2   Molecular Genetics

Chapter 8   Gene Frequencies and Linkage Disequilibrium 179

8.1 Introduction 179
8.2 Single-Locus Frequencies 180
8.3 Hardy-Weinberg Proportions 184
8.4 Multiple-Locus Frequencies 188
8.5 Marker-Trait Association Tests 191
    8.5.1 Analyzing Samples of Unrelated Individuals 192
    8.5.2 Analyzing Family Data 195
8.6 References 200

Chapter 9   The Extended Sib-Pair Method for Mapping QTL 201

9.1 Introduction 202
9.2 Statistical Model 203
9.3 Inferring the Proportion of Genes IBD Shared by Sibs at QTL 204
    9.3.1 Proportion of Alleles IBD at Marker Loci—Parental Genotypes Known 204
    9.3.2 Proportion of Alleles IBD at Marker Loci—Parental Genotypes Unknown 206
    9.3.3 Multipoint Estimation of Proportion of Genes IBD at QTL 207
9.4 Maximum Likelihood Estimation 208
    9.4.1 Likelihood Ratio Test 209
    9.4.2 Approximate Critical Value for Significance Test 209
9.5 Single Marker Analysis and Interval Mapping of QTL 209
9.6 Program Implementation 210
    9.6.1 Choosing the Scanning Increment 210
    9.6.2 Implementing the Multipoint Method 212
    9.6.3 Implementing the Mixed Model Methodology through PROC IML 214
    9.6.4 Implementing the Mixed Model Methodology through PROC MIXED 216
9.7 Executing the Program 216
    9.7.1 Data Preparation 217
    9.7.2 Running the Main Program 218
    9.7.3 Application Notes 219
    9.7.4 Example Demo 220
9.8 References 222

Chapter 10 Bayesian Mapping Methodology 225

    10.1 Introduction 225
    10.2 Two-Point Linkage Analysis 226
        10.2.1 Inference Regarding Recombination Rates 227
        10.2.2 Example 228
        10.2.3 Testing for Linkage 230
    10.3 Three-Point Analysis 231
    10.4 Genetic Map Construction 238
    10.5 QTL Analysis 240
        10.5.1 Example 242
        10.5.2 Results 246
    10.6 Final Comments 249
    10.7 References 249

Chapter 11 Gene Expression Profiling Using Mixed Models 251

    11.1 Introduction 251
    11.2 Theory of Mixed Model Analysis of Microarray Data 252
    11.3 Mixed Model Analysis of Two-Color Microarray Data Using PROC MIXED 254
        11.3.1 Data Preparation 254
        11.3.2 Computation of Relative Fluorescence Intensities 255
        11.3.3 Computation of Gene-Specific Significance Models 257
        11.3.4 Explore and Interpret the Results 258
    11.4 ANOVA-Based Analysis of Affymetrix Data Using SAS Microarray Solution 260
        11.4.1 Download Data 260
        11.4.2 Examine Data 262
        11.4.3 Submit the Analysis 263
        11.4.4 Specifications for Mixed Model Analysis 264
        11.4.5 Submit the Analysis 266
    11.5 Discussion 276
    11.6 References 277

Additional Reading 279
Index 281

About the Authors

Mónica G. Balzarini is a professor of statistics in the Agricultural College at the National University of Córdoba, Argentina. She earned a B.Sc. in agriculture from the National University of Córdoba, an M.Sc. in biometry from the University of Buenos Aires, and a Ph.D. from Louisiana State University. Dr. Balzarini has trained many scientists in using SAS and has served as a statistical consultant to numerous agricultural researchers around the world. She is the author/editor of the first statistical software developed in Argentina, as well as a pioneer investigator in statistical genomics in her country (CONICET investigator). Dr. Balzarini has been honored as president of the Argentinean Region of the International Biometric Society.

Aldo Cappio-Borlino earned a *laurea* degree in physics from the University of Turin, Italy. An associate professor of animal breeding and genetics at the University of Sassari, Italy, he has used SAS since 1996. Cappio-Borlino's scientific interests involve the mathematical modeling of biological phenomena of interest in animal science, such as lactation curves, growth curves, and feed utilization.

Wendy Czika is a research statistician at SAS Institute and the principal developer of SAS/Genetics procedures. She has been using SAS software since beginning graduate school in 1996 at North Carolina State University, where she recently received a Ph.D. in statistics under the direction of Dr. Bruce Weir. The focus of Dr. Czika's research has been statistical methods for analyzing genetic marker data.

James D. Fry, assistant professor of biology at the University of Rochester, received a Ph.D. in biology from the University of Michigan in 1988 and started using SAS software shortly thereafter. He uses computer models to study the genetics of complex traits in mites, morning glories, and *Drosophila* flies.

Greg Gibson is an associate professor of genetics at North Carolina State University. He has been working with Dr. Russell Wolfinger since 1990 on applications for mixed model analysis of microarray data, and has consulted on the development of SAS/Genetics procedures in quantitative genetics. Dr. Gibson has a B.Sc. from Sydney University in genetics, a Ph.D. in developmental genetics from the University of Basel, and post-doctoral training in population and quantitative genetics from Stanford University.

Jose L. L. Guerra earned a Ph.D. from Louisiana State University with a Ph.D. minor in applied statistics. During the last five years he has served as a consultant in applied statistics in the Department of Molecular Biology at the Federal University of Ceara in Brazil. He learned applied statistics with SAS and has used the software since 1997. Dr. Guerra is currently a visiting professor in the Department of Animal Science at the Federal University of Fortaleza and has a research position at the Brazilian National Center of Research (CNPq), where he develops computer programs for animal breeding and production.

Manjit S. Kang is a professor of quantitative genetics in the Department of Agronomy at Louisiana State University. He earned a B.Sc. in agriculture and animal husbandry from the Punjab Agricultural University in India, an M.S. in biological sciences from Southern Illinois University at Edwardsville, an M.A. in botany from Southern Illinois University at Carbondale, and a Ph.D. in crop science from the University of Missouri at Columbia. The author/editor and publisher of numerous volumes in genetics and plant breeding, Dr. Kang began using SAS in 1974. He published his first SAS program on stability analyses, written in the SAS Matrix Programming Language, in the *Journal of Heredity* in 1985.

Nicolò P. P. Macciotta earned a *laurea* degree in agricultural science from the University of Sassari, Italy. He is a researcher of animal breeding and genetics at the University of Sassari, where his main areas of interest are the mathematical modeling of lactation curves, dairy sheep, and cattle and goat breeding. Macciotta began using SAS software in 1996.

Giuseppe Pulina, Ph.D., graduated in agricultural science from the University of Sassari, Italy. He is a professor of animal science as well as head of the Department of Animal Science at the University of Sassari. A SAS user since 1996, Dr. Pulina is particularly interested in the mathematical modeling of milk production.

Guilherme J. M. Rosa is an assistant professor of statistical genetics in the Department of Animal Science and Department of Fisheries and Wildlife at Michigan State University. He has used SAS since 1995. Rosa earned a Ph.D. in biostatistics from the University of São Paulo in 1998. He was a post-doctoral fellow in statistical genetics at the University of Wisconsin. In his 10 years at the São Paulo State University, Dr. Rosa was a lecturer and later an assistant professor of biostatistics and experimental design.

Arnold M. Saxton earned an M.S. in fish genetics at the University of Washington at Seattle before continuing Ph.D. studies in animal breeding at North Carolina State University. At NCSU in 1980 he ran his first SAS program, and he has actively used SAS ever since for statistical and genetic analyses. After serving on the faculty of the Experimental Statistics Department at Louisiana State University for several years, Dr. Saxton moved to the University of Tennessee, Knoxville, where he is now a professor of animal science. A long-term goal is to develop realistic models of complex traits, with direct benefits to agriculture and human medicine.

Kenneth J. Stalder is an assistant professor in the Animal Science Department at Iowa State University. He has used SAS software since 1990. Dr. Stalder holds a Ph.D. in animal breeding and genetics from Iowa State University. He and the students he works with routinely use SAS in the analysis of complex data sets related to genetics and animal production.

Robert J. Tempelman is an associate professor in the Department of Animal Science and an adjunct associate professor in the Department of Statistics and Probability at Michigan State University. He was formerly an assistant professor in the Department of Experimental Statistics at Louisiana State University. He earned a master's degree in animal breeding from the University of Guelph, Canada, and a Ph.D. in quantitative genetics from the University of Wisconsin. Dr. Tempelman has published numerous papers on mixed model and Bayesian inference, with emphases on quantitative genetics applications in animal science. He began using SAS software in 1985 as an undergraduate majoring in animal science at the University of Guelph.

Russ Wolfinger is Director of Genomics at SAS Institute. He earned a Ph.D. in statistics from North Carolina State University in 1989 and has been at SAS ever since, developing procedures (MIXED, MULTTEST, KDE, and NLMIXED) and conducting research. Currently he is leading the development of solutions that integrate high-end statistics, warehousing, and visualization in the areas of genetics, transcriptomics, and proteomics.

Chenwu Xu is a post-doctoral research associate in statistical genetics at the University of California at Riverside. He received a Ph.D. in genetics in 1998 from Nanjing Agricultural University, China, and has been using SAS for more than 10 years.

Shizhong Xu is a professor of genetics in the Department of Botany and Plant Sciences as well as an adjunct professor of statistics at the University of California at Riverside, where he teaches quantitative genetics and statistical genomics. He received a Ph.D. in quantitative genetics from Purdue University in 1989. Dr. Xu's current research work is focused on the development of statistical methods for QTL mapping and microarray gene expression data analysis. He began using SAS in 1987.

Xiang Yu has a bachelor's degree in bioscience from the University of Science and Technology of China and a Ph.D. in bioinformatics from North Carolina State University. He is currently a biometrician at Merck Research Laboratory. Dr. Yu's primary work interest is pharmacogenomics, or the identification of genetic components that affect drug efficacy and efficiency in clinical trial studies. He has used SAS since 2000.

Acknowledgments

We writers of science toil alone before our computers, trying to simultaneously be clear, accurate, and complete. But to make this book a reality, the assistance of many people has been needed. To them we want to express our gratitude. The staff at SAS Books by Users Press has been most helpful. Our acquisitions editor, Patsy Poole, has given expert advice and a steady hand throughout the process. Copyediting by Ed Huddleston corrected many flaws in early drafts and has given the book a more consistent appearance. We thank Patrice Cherry for the cover design, Joan Stout for index editing, and Candy Farrell for production.

Our goals of being clear and accurate are closer to reality because of the efforts of two technical reviewers at SAS, Jill Tao and Jon Trexler. Thanks also go to Mike Stear at Glasgow University and Lauren McIntyre at Purdue University for additional technical reviews. And of course, we must acknowledge the SAS programmers, whose software assists with such a wide variety of scientific inquiry.

Finally, thanks to our employers and significant others who at least tolerated our work on this project!

. . .

James D. Fry, the author of Chapter 2, thanks S. Durham, R. Miller, and A. Saxton for their SAS insights, and the reviewers for helpful comments on the manuscript. This work was supported by NSF grant DEB-0108730 to J.D.F.

Wendy Czika and Xiang Yu, the authors of Chapter 8, used GAW data in this chapter that are supported by the GAW grant GM31575.

Chenwu Xu and Shizhong Xu, the authors of Chapter 9, thank the Autism Genetic Resource Exchange (AGRE) for permission to access genotypic and phenotypic data. This research was supported by the National Institutes of Health Grant R01-GM55321 and the USDA National Research Initiative Competitive Grants Program 00-35300-9245 to S.X.

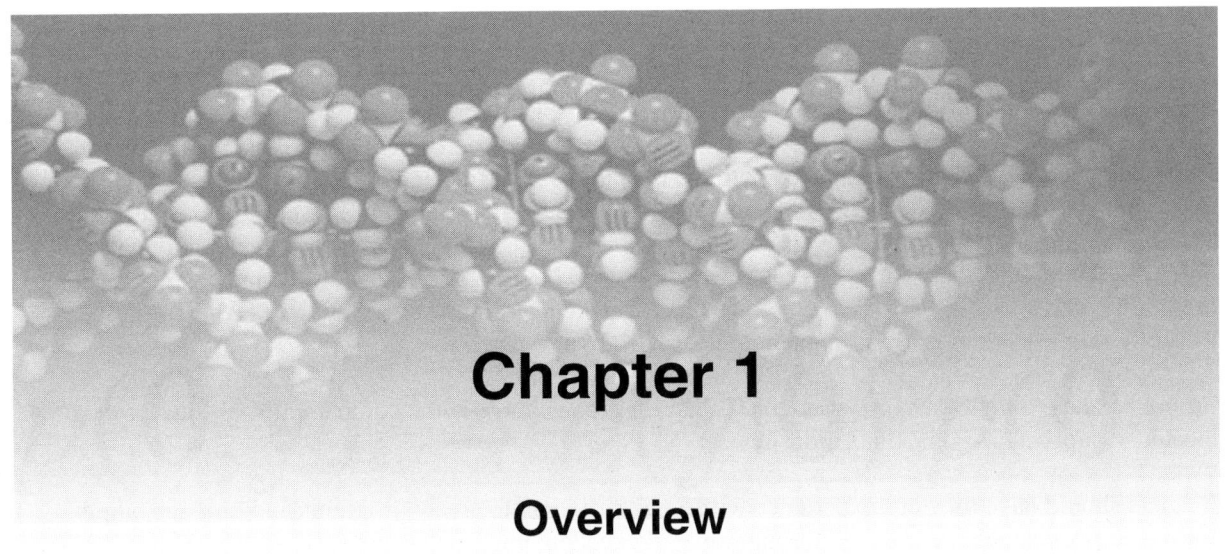

Chapter 1

Overview

1.1 Introduction 1
1.2 Book Organization 2
1.3 SAS Usage 4
   1.3.1 Example of a Basic SAS DATA Step 4
   1.3.2 Example of a Basic Macro 5
1.4 References 6

1.1 Introduction

Some 100 years after the rediscovery of Gregor Mendel's proposed genetic mechanisms, the science of genetics is undergoing explosive growth in theoretical knowledge and in applications. The ability of researchers today to collect enormous amounts of genetic data, along with the increasing sophistication of scientific questions, makes the analysis of these data even more important. One specialty within the genetics field is particularly dependent on the statistical analysis of genetic data. This is *quantitative genetics*, the study of complex traits controlled by many genes. Unlike simple Mendelian genetics, where the large influence of genes on phenotype makes genotype directly observable, complex traits have contributions from many genes of small effect producing the observed trait. Add the uncertainty from environmental contributions to the trait, and complex statistical methods are needed to make genetic inference.

Presentation of software for complex genetic data analysis in textbooks has been limited, however. Becker (1984) provides thorough coverage of data analysis for basic quantitative genetic methods, but provided no computer applications. Other texts that cover analysis of genetic data mention the use of computer programs, but stress theory over applications (Balding et al., 2001; Lynch and Walsh, 1998; Narain, 1990; Liu, 1998). Weir (1996) provides excellent coverage of discrete genetic data. Certainly genetic analysis currently would be done with the aid of computer software. But many of the software tools available are designed for just one type of data. The advantages of general purpose software, like SAS, for genetic analysis are having to learn just one interface, availability of preprogrammed, thoroughly tested statistical procedures, a large user community for support, and compatibility across most computing platforms. The limitation is of course that general purpose software may lack the unique capabilities and efficiency that specialized software has.

This book's objective is to demonstrate the use of SAS software for the wide variety of genetic analyses associated with quantitative genetics. SAS programs will be presented, and used to analyze real experimental data. Actual program code will be explained in sufficient detail that modification for other experiments can be done. However, several large programs will by necessity have to be presented as "black boxes," with little discussion of the internal code and algorithms. In all cases, use of the output to make genetic conclusions will be covered. Genetic theory will be briefly reviewed, and references to the literature will provide access to more detail. The example later in this overview will give a brief illustration of how SAS programming and results will be presented.

Programs and example data discussed in all chapters are available at the companion Web site for the book, located at support.sas.com/companionsites.

Readers should have previous experience with SAS, sufficient to create programs and produce output. Such experience could be obtained through several excellent introductory texts (Cody and Smith, 1997; Schlotzhauer and Littell, 1987; Gilmore, 1999). Familiarity with the statistical analyses is not needed, but understanding of the genetic questions being addressed would be helpful. The book is designed for researchers who know what genetic question they need to answer, and want to use SAS for the analysis. But this book should be of interest to graduate students, bioinformaticians, statisticians, and any other SAS user with interest in joining the highly active field of genetic analysis. We would be very pleased if the book were used as a teaching companion for texts lacking computer applications.

You should realize that much previous work has been done on using SAS for genetic analyses. "Additional Reading" in the back of this book lists references found by a routine literature search where the focus was on a SAS program. Of course this does not reflect the thousands of researchers who have used SAS for genetic analyses but did not present the SAS code. If you have a favorite SAS application, with an example data set you are willing to put into the public domain, please send these to Arnold Saxton; this will give the scientific community an even wider variety of applications than are represented in this book.

1.2 Book Organization

This book is divided into two parts, classical quantitative genetics (Chapters 2–7) and molecular genetics (Chapters 8–11). Until the recent advances in molecular techniques, scientists had no way of identifying the genotypes underlying complex traits. Thus classical methods used pedigree relationships and phenotype in order to obtain genetic inferences. These methods are statistical in nature, relying on standard genetic model breakdown of the observed phenotype,

$$P = A + D + I + PE + M + E + A*E + ...,$$

where Additive, Dominance, Interaction (epistasis), Permanent Environment, Maternal, temporary Environment, Additive by Environment interaction, and so forth are potentially of interest. Additive effects are of fundamental interest, as these are the genetic effects that can be transmitted to the next generation.

Chapter 2 describes how experiments that produce individuals with defined relationships can be used to estimate genetic variances, such as how much of the differences among individuals is produced by additive effects. From there, the key parameter *heritability* can be estimated, which indicates the fraction of variation due to additive effects. Generally more than one phenotypic trait is of interest, and it then is useful to estimate what fraction of genetic variation is shared by two traits, through the genetic correlation. This chapter focuses on the statistical method called analysis of variance and on the SAS MIXED procedure.

Chapter 3 explores similar issues to those in Chapter 2, but uses statistical regression methods. This chapter also examines how pedigree information can be used to produce relationship information. Genetic selection selects the "best" individuals to use as parents with the goal of increasing gene frequencies of the "better" genes. Experiments that do this can estimate heritability from the rate at which the phenotype changes due to selection. Also covered are methods that address the dominance model component, important when crossing genetically different breeds or lines.

Chapter 4 examines methods used for genetic selection, in particular those that address the simultaneous improvement of several traits. For plant and animal breeders, this chapter addresses the key objective of increasing productivity through genetic improvement of agricultural populations. These methods are also used on experimental populations to further the theoretical understanding of complex traits.

Chapter 5 addresses genotype-by-environment interaction, an issue of great importance in plant breeding, where it is often observed that a crop variety will perform differently in different years or locations. Some breeders attempt to find genotypes that are "stable," meaning they perform similarly under a wide variety of environmental conditions. This has been found to be less important in animals, which because of behavior and physiology are less affected by environmental differences.

Chapter 6 presents a wide variety of analysis methodology for experiments where individuals are measured several times during their life cycle. Common examples are dairy cattle whose milk production is measured every year, and any species where multiple measures of size are used to follow growth over time. It is easy to imagine situations where it is important to measure genetic contributions to growth rates, or where the goal is to genetically change the pattern of growth, the growth curve. For example, beef cattle that grow rapidly, but are small at birth, often suit farm management conditions better.

Chapter 7 reexamines many of the issues above, but from a Bayesian perspective. Bayesian statistics is gaining popularity within the genetics community, so this book would be incomplete without providing some coverage from this viewpoint.

This book would also be incomplete if it did not address research questions about complex traits that are now possible with molecular genetic techniques. One new capability is to identify genetic regions, so-called QTL, or quantitative trait loci, along chromosomes that affect observed traits. This is a first step toward being able to observe the genes that underlie complex traits. Chapter 8 presents the SAS/Genetics procedures that provide basic genetic information about QTL, while Chapter 9 gives a detailed description of an experimental approach to identify QTL. Chapter 10 gives the Bayesian perspective, again with the thought that Bayesian methods could one day become the standard approach. Finally, Chapter 11 discusses analysis of microarray data. Microarray experiments produce vast quantities of measurements on levels of gene expression, in some cases approaching the point of measuring expression of every gene in the organism. This information may eventually lead to understanding the "genetic architecture of complex traits," a holy grail where all genes involved in a complex trait will be known and can be identified in individuals, and where the way that these genes function and interact to produce the observed trait will be understood (Mackay 2001). This might be considered the biological equivalent of the physicist's "Theory of Everything."

1.3 SAS Usage

Software used in this book includes the products Base SAS (utility functions), SAS/STAT (statistics and inbreeding), SAS/IML (matrix algebra programming), SAS/GRAPH (graphic displays), SAS/QC (PROC CAPABILITY distribution analysis), SAS/ETS (time series, forecasting, and econometric techniques), and the new SAS/Genetics product. JMP is recommended in Chapter 11 for easy data visualization, and capabilities of the new SAS Microarray product are presented. Access to all of these products depends on your site license and may entail additional costs. If you are new to SAS, you can explore the functionality of many of the SAS products used in this book by purchasing SAS Learning Edition, a low-cost, renewable version of SAS. See the SAS Learning Edition Web site, support.sas.com/le, for details.

1.3.1 Example of a Basic SAS DATA Step

SAS uses a two-step approach to data analysis, consisting of a DATA step where data are made available, and then procedures (PROCs) to process the data. The classic genetic data of Mendel (1866) are used as an example.

```
data mendel;  ❶
  ** data from classic 1865 monograph;
  input experiment NumParents DomRec$ Trait$ Count;  ❷
datalines;
1   253  d RoundSeed     5474
1   253  r WrinkledSeed  1850
2   258  d YellowSeed    6022
2   258  r GreenSeed     2001
3    .   d VioletCoat    705
3    .   r GrayCoat      224
4    .   d InflatedPod   882
4    .   r ConstrictPod  299
5    .   d GreenPod      428
5    .   r YellowPod     152
6    .   d AxialFlower   651
6    .   r TerminalFlower 207
7    .   d LongStem      787
7    .   r ShortStem     277
;
options ls=77;
proc freq data=mendel; weight count; by experiment;  ❸
  tables DomRec / testp=(.75 .25);
  run;
```

❶ The DATA step is entered, with MENDEL assigned as the name for the data set. Comments can be inserted, starting with an asterisk, and ending with a semicolon as all SAS statements do.

❷ Data will be read in for five variables, named in the INPUT statement. Character value variables have a $ sign after their names. Actual data follow the DATALINES statement, with one column per variable. Missing data are represented by a period.

❸ With data MENDEL available, analysis can be done with any of the many procedures in SAS. Here the FREQ procedure is used to test if the observed ratios follow the expected 3:1 ratio for a single dominant locus. Use of data MENDEL (DATA=MENDEL) is explicitly specified, but by default the most recent data set would be used. Analysis is done by the variable EXPERIMENT, meaning each of the seven experiments in MENDEL will be analyzed and reported separately.

Data can be read into SAS in a variety of ways. Perhaps the most convenient way is to keep the data externally in a spreadsheet, and use the IMPORT procedure to create the SAS data set. The following statements show how this might be done, with the file name in quotes giving the exact location of the external data on the computer.

```
proc import datafile="c:\mydata\mendel.xls" out=mendel replace;
run;
proc freq data=mendel; weight count; by experiment;
 tables DomRec / testp=(.75 .25);
 run;
```

1.3.2 Example of a Basic Macro

Some programs will be presented as *macros*, blocks of code that can be used for different experiments with no user modification. Typical macro code looks like this:

```
%macro runthis(dataset,treat,percent=.50);
%let percent1=%sysevalf(1-&percent);
proc freq data=&dataset; weight count; by experiment;
 tables &treat /testp=(&percent &percent1);
run;
%mend;
```

This defines a macro called %RUNTHIS, which takes two required values, DATASET and TREAT, and an optional value PERCENT. These values are substituted into the PROC FREQ code (signified by &VARIABLE), producing an analysis similar to the Mendel example above. To run macro code, users must first define the macro and then call it. Defining a macro can easily be accomplished by opening and submitting it in the SAS editor. This only needs to be done once during the current SAS session. Then the macro can be used (multiple times) by submitting a statement like this, where user-specific values are given for the macro variables:

```
%runthis(mendel,domrec,percent=.75)
```

Alternatively, the macro can be defined by reading it from an external file, using %INCLUDE as here:

```
%include 'c:\sasmacros\runthis.sas';
%runthis(mendel,domrec,percent=.75)
```

The file name in quotes should give the exact location of the macro file on the computer, and this file contains the SAS macro code. Using the %INCLUDE statement is equivalent to opening and submitting the specified file.

Results of this Mendelian example are in Output 1.1. Test results are nonsignificant (P>.40) for all experiments, indicating the data are consistent with the hypothesized 3:1 ratio. In fact, in no case did the observed percentages deviate from theory by more than 1.25 percentage points, sparking a lively debate in the literature on whether Mendel's data are "too good to be true." All of these experiments contain data on F1 crosses, genetically expected to be crosses of heterozygous individuals, Aa by Aa. Offspring can then be symbolically represented as $(A/2 + a/2)*(A/2 + a/2) = AA/4 + Aa/2 + aa/4$, and if A is completely dominant, the phenotypic ratio will be three A? to one aa. If experimental data do not conform to this ratio, then the trait may be the result of more than one genetic locus, or different dominance mechanisms may be involved.

But consider a situation where the observed phenotype takes on more than just "green" and "yellow" values, where in fact the phenotype is a continuous measurement. Further, the measured phenotype "10.23" does not mean the genotype is AaBB, like "yellow" is aa. In fact, since environmental effects can be as large as genetic effects, "10.23" will likely represent many genotypes. But even worse, we do not know how many and which genes are involved. Welcome to the difficulties of complex trait analysis!

Output 1.1 Mendel's seven experiments on single traits in peas.

```
------------------------------ experiment=1 ------------------------------
                       The FREQ Procedure
                 Chi-Square Test for Specified Proportions

                      Chi-Square             0.2629
                      DF                          1
                      Pr > ChiSq             0.6081
------------------------------ experiment=2 ------------------------------
                 Chi-Square Test for Specified Proportions

                      Chi-Square             0.0150
                      DF                          1
                      Pr > ChiSq             0.9025
------------------------------ experiment=3 ------------------------------
                 Chi-Square Test for Specified Proportions

                      Chi-Square             0.3907
                      DF                          1
                      Pr > ChiSq             0.5319
------------------------------ experiment=4 ------------------------------
                 Chi-Square Test for Specified Proportions

                      Chi-Square             0.0635
                      DF                          1
                      Pr > ChiSq             0.8010
------------------------------ experiment=5 ------------------------------
                 Chi-Square Test for Specified Proportions

                      Chi-Square             0.4506
                      DF                          1
                      Pr > ChiSq             0.5021
------------------------------ experiment=6 ------------------------------
                 Chi-Square Test for Specified Proportions

                      Chi-Square             0.3497
                      DF                          1
                      Pr > ChiSq             0.5543
------------------------------ experiment=7 ------------------------------
                 Chi-Square Test for Specified Proportions

                      Chi-Square             0.6065
                      DF                          1
                      Pr > ChiSq             0.4361
```

1.4 References

Balding, D. J., M. Bishop, and C. Cannings, eds. 2001. *Handbook of Statistical Genetics*. New York: John Wiley & Sons.

Becker, W. A. 1984. *Manual of Quantitative Genetics*. 4th ed. Pullman, WA: Academic Enterprises.

Cody, R. P., and J. K. Smith. 1997. *Applied Statistics and the SAS Programming Language*. 4th ed. Upper Saddle River, NJ: Prentice-Hall.

Gilmore, J. 1999. *Painless Windows: A Handbook for SAS Users*. 2d ed. Cary, NC: SAS Institute Inc.

Liu, B.-H. 1998. *Statistical Genomics: Linkage, Mapping, and QTL Analysis*. Boca Raton, FL: CRC Press.

Lynch, M., and B. Walsh. 1998. *Genetics and Analysis of Quantitative Traits*. Sunderland, MA: Sinauer Associates.

Mackay, T. F. C. 2001. The genetic architecture of quantitative traits. *Annual Review of Genetics* 35:303-339.

Mendel, Gregor. 1866. "Versuche über Pflanzen-hybriden" ["Experiments in Plant Hybridization"]. Verhandlungen des naturforschenden Vereines in Brünn, Bd. IV für das Jahr 1865, Abhandlungen, 3–47. (English translation accessed at www.mendelweb.org.)

Narain, P. 1990. *Statistical Genetics*. New York: John Wiley & Sons.

Schlotzhauer, S. D., and R. C. Littell. 1987. *SAS System for Elementary Statistical Analysis*. Cary, NC: SAS Institute Inc.

Weir, B. S. 1996. *Genetic Data Analysis II: Methods for Discrete Population Genetic Data*. Sunderland, MA: Sinauer Associates.

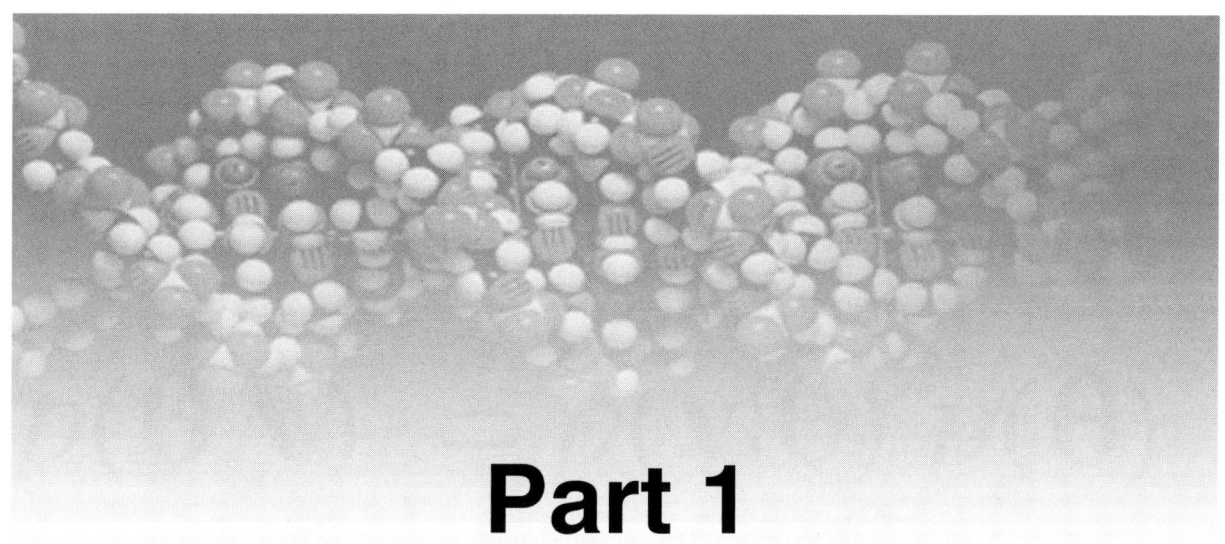

Part 1

Classical Quantitative Genetics

Chapter 2 Estimation of Genetic Variances and Covariances by Restricted Maximum Likelihood Using PROC MIXED **11**

Chapter 3 More Estimation of Genetic Parameters **35**

Chapter 4 Genetic Selection **55**

Chapter 5 Genotype-by-Environment Interaction **69**

Chapter 6 Growth and Lactation Curves **97**

Chapter 7 Empirical Bayes Approaches to Mixed Model Inference in Quantitative Genetics **149**

10

Chapter 2

Estimation of Genetic Variances and Covariances by Restricted Maximum Likelihood Using PROC MIXED

James D. Fry

2.1 Introduction 11
2.2 The Nested Half-Sib and Related Designs 13
    2.2.1 Example of Nested Half-Sib Design, One Trait 13
    2.2.2 Example of Nested Half-Sib Design, Two Traits 19
    2.2.3 Example of Clonal Design, Two Environments 23
2.3 The Diallel and Related Designs 27
    2.3.1 Diallel Example 28
    2.3.2 Extension to the NC2 Design 33
2.4 Pedigreed Populations 34
2.5 References 34

2.1 Introduction

This chapter describes how the MIXED procedure can be used to obtain estimates and significance tests of genetic variances, covariances, and correlations by restricted maximum likelihood (REML). Clonal, nested half-sib, and diallel designs are considered.

One of the main goals of quantitative genetics is to determine the relative contributions of genetic and environmental influences to variation and covariation in quantitative traits (Falconer and Mackay, 1996). In the simplest formulation, the phenotypic variance (V_P) of a population is viewed as the sum of separate genetic (V_G) and environmental (V_E) variances. The genetic variance is in turn made up of additive (V_A), dominance (V_D), and epistatic (V_I) components. The additive variance and the associated narrow-sense heritability, V_A/V_P, play a key role in predicting the response to selection (Falconer and Mackay, 1996).

If two or more traits are considered, the covariances between traits can be partitioned in a similar manner. Thus the phenotypic covariance, COV_P, is the sum of genetic (COV_G) and environmental (COV_E) covariances; COV_G is in turn made up of additive and non-additive components. The additive genetic covariance, COV_A, and the additive genetic correlation, r_A (see Section 2.2.2), are important for predicting the evolutionary response of one trait to selection on a second trait. The concepts of genetic covariance and correlation can be extended to a single trait measured in two or more discrete environments or treatments. Such cross-environment genetic covariances and correlations enable prediction of how a population's mean as measured in one environment would respond to selection in a different environment (Falconer and Mackay, 1996).

Figure 2.1 The nested half-sib design. In this example, *n* males (sires) are mated to *a* females (dams) each. From each dam, *b* offspring (o) are measured for the trait or traits of interest.

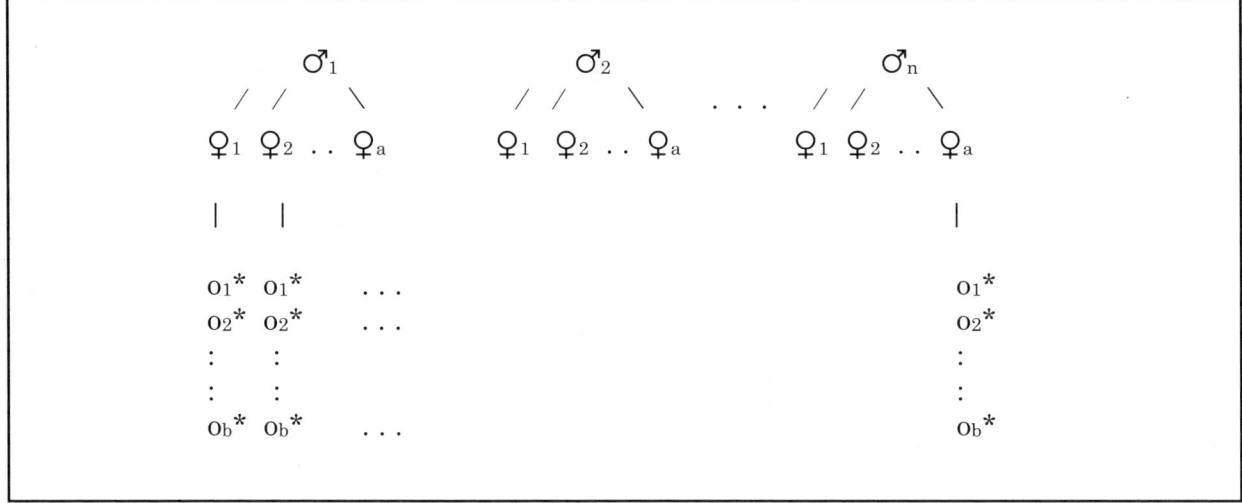

A variety of designs for estimating additive genetic variances and covariances are available. The focus of this chapter is two-generation designs in which measurements are made on only the offspring generation. Perhaps the most widely used design of this kind is the nested half-sib design (Figure 2.1). In this design, each of a number of sires (males) is mated to two or more dams (females). One or more offspring of each dam are then measured for the trait or traits of interest. If cross-environment genetic covariances are being estimated, different offspring of each sire are reared in the different environments. The nested half-sib design has the twin advantages of being relatively easy to implement and providing estimates of genetic variances and covariances that are uncontaminated by common-environment and dominance variances (in contrast, for example, to the full-sib design). For estimating the dominance variance or non-nuclear (maternal and paternal) components of variance, more complicated designs are necessary. These include North Carolina Design 2 and the diallel (see Section 2.3).

Traditionally, V_A and other components of variance have been estimated from these and other designs by the method of moments (Becker, 1992). For the case of a single trait, this approach involves first performing a traditional analysis of variance (ANOVA). The desired variance components, such as that among sires, are then estimated by equating the calculated mean squares to their theoretical expectations and solving. If the data are balanced or approximately so, F-tests involving ratios of mean squares can be used to test hypotheses of interest (e.g., that the variance among sires, and hence V_A, is greater than zero). The method of moments can also be used to estimate genetic covariances between traits; in this case, sums of cross-products are used instead of sums of squares.

The method of moments, however, has important limitations. First, the method runs into well-known problems when the design is unbalanced, as is often the case. Second, even with balanced data sets, only certain hypotheses can be tested by *F*-tests. For example, while the ratio of the sire mean square to the dam mean square provides an *F*-test for the variance among sires in a nested half-sib design, there is no corresponding *F*-test for the covariance among sires when two traits are being considered. Third, in complex designs, those *F*-tests that can be performed often require restrictive assumptions. For example, in a nested half-sib design with two environments, a significant sire × environment interaction indicates that the cross-environment genetic correlation is less than one, but only if the assumption of equal genetic variances among environments is met (Cockerham, 1963).

Restricted maximum likelihood (REML) provides an excellent alternative to the method of moments for estimating quantitative genetic parameters. In contrast to the method of moments, REML does not require balanced designs, and even missing cells present no special problem. Furthermore, REML allows a virtually unlimited number of hypotheses to be tested. The computational limits that formerly made REML impractical for large data sets have mostly been overcome. For these reasons, REML is now the method of choice for estimating quantitative genetic parameters from most designs.

This chapter gives an introduction to using the MIXED procedure to obtain REML estimates and hypothesis tests for half-sib, clonal, and diallel designs. Emphasis will be on practical examples rather than theory. Shaw (1987) and Lynch and Walsh (1998) give excellent introductions to the theoretical foundations of REML that the reader is strongly urged to consult. The book by Littell et al. (1996) describes many of the other uses to which PROC MIXED can be put, and contains an informative comparison between PROC MIXED and PROC GLM.

2.2 The Nested Half-Sib and Related Designs

We begin by considering examples of the nested half-sib design, first with a single trait, and then with two traits. Then we turn to an example of a clonal design with two environments, which serves to illustrate calculation of cross-environment correlations and comparison of genetic variances between environments. The clonal design is for all practical purposes identical to a full-sib design.

2.2.1 Example of Nested Half-Sib Design, One Trait

The use of PROC MIXED for a half-sib design with one trait is illustrated by a data set on seed beetles (data courtesy of Frank Messina, Utah State University). The data come from one of the six blocks in the "seeds present" treatment of Messina and Fry (2003). In this experiment, each of 24 sires was mated to 4 or 5 dams (different for each sire), and 5 female progeny from each dam were measured for two traits, mass at eclosion and lifetime fecundity. This data set will also be used in Section 2.2.2 to illustrate calculation of genetic correlations using PROC MIXED. Here, we will use just the fecundity data.

The following program reads in the data:

```
data beetle;
input sire dam progeny value trait $;
datalines;
1    1    1    58    fec
1    1    1    4.6   mass
1    1    2    64    fec
1    1    2    4     mass
...more datalines...
24   4    4    65    fec
24   4    4    3.9   mass
24   4    5    80    fec
24   4    5    4.6   mass
;
```

The variable VALUE contains the observations for both fecundity and mass; therefore each individual is represented by two lines in the data set, one for fecundity and one for mass. The variable TRAIT identifies the trait, and the variable PROGENY, in conjunction with SIRE and DAM, identifies the individual. Note that in contrast to PROC GLM, when using PROC MIXED, different traits should not be coded as different SAS variables if multiple-trait analyses are to be conducted.

A traditional variance component model (Sokal and Rohlf, 1981) applied to the fecundity data could be written as

$$Y_{ijk} = \mu + S_i + D_{j(i)} + W_{k(ij)} .$$

Here, Y_{ijk} is an observation, μ is the population mean (a fixed effect), S_i is the random effect of the ith sire, $D_{j(i)}$ is the random effect of the jth dam within the ith sire, and $W_{k(ij)}$ is the random effect of the kth individual within the ijth family. The three random effects are assumed to be independent and normally distributed with means zero and variances σ_S^2, σ_D^2, and σ_W^2, respectively. (Note that "D" here stands for "dam," not "dominance.") In quantitative genetics, these three parameters are sometimes called "observational" components of variance, because they can be directly estimated by a conventional ANOVA. In contrast, we are interested in estimating and making inferences about three "causal" components of variance—namely, the additive genetic variance V_A, the common environment or maternal-effect variance V_M, and the within-family environmental variance V_E. From quantitative genetics theory, if the sires and dams are randomly sampled from a random-mating population, and if certain other assumptions are met, then the observational components have the following interpretations in terms of the causal components (Falconer and Mackay, 1996):

$$\sigma_S^2 = \tfrac{1}{4} V_A$$
$$\sigma_D^2 = \tfrac{1}{4} V_A + \tfrac{1}{4} V_D + V_M$$
$$\sigma_W^2 = \tfrac{1}{2} V_A + \tfrac{3}{4} V_D + V_E$$

For simplicity, we will assume that the dominance variance, V_D, equals 0. In this case, we can solve for the causal components to obtain

$$V_A = 4\sigma_S^2$$
$$V_M = \sigma_D^2 - \sigma_S^2$$
$$V_E = \sigma_W^2 - 2\sigma_S^2$$

If dominance variance is present, the first relationship still holds, but $\sigma_D^2 - \sigma_S^2 = V_M + \tfrac{1}{4} V_D$, and $\sigma_W^2 - 2\sigma_S^2 = V_E + \tfrac{3}{4} V_D$. In this case it is not possible to separately estimate V_M, V_D, and V_E. In practice V_A is usually of the most interest. Note that $V_A + V_M + V_E = \sigma_S^2 + \sigma_D^2 + \sigma_W^2 = \sigma_T^2 = V_P$, the total or phenotypic variance.

The traditional approach is to estimate the observational components of variance by ANOVA, and convert the estimates to estimates of the causal components using the relationships above. For balanced or slightly unbalanced data sets, the traditional approach works well for estimating V_A and testing whether it differs from zero. The F ratio MS(sires)/MS(dams) tests the null hypothesis $\sigma_S^2 = 0$, and hence $V_A = 0$. In fact, with balanced data, this test is likely to be more powerful than the REML test of the same hypothesis (Shaw, 1987).

For severely unbalanced data, however, or for making inferences about V_M and/or V_E, REML has important advantages over ANOVA. First, with unbalanced data, REML point estimates of parameters are more efficient (have a lower variance) than those from ANOVA, and hypothesis tests based on REML have a stronger theoretical justification than approximate F-tests (e.g., those provided by PROC GLM). Second, in contrast to ANOVA, REML provides a means to test the null hypotheses $V_M = 0$ and $V_E = 0$ (although the latter test is seldom of interest). It should be kept in mind, however, that these advantages depend on the assumption of normality being met (in formal terms, the vector of observations must come from a multivariate normal distribution). If this assumption is violated, REML and ANOVA point estimates are still unbiased if non-negativity constraints are disregarded, but the REML estimates are not necessarily more efficient. Hypothesis tests under both methods will also be affected by non-normality.

We begin with a PROC MIXED program that provides REML estimates of the observational variance components for fecundity (FEC):

```
proc mixed data = beetle covtest asycov;
class sire dam;
where trait = 'fec';
model value = /solution;
random sire dam(sire);
run;
```

Output 2.1 Nested half-sib design, one trait (fecundity in a seed beetle).

```
                        The Mixed Procedure
                          Model Information
         Data Set                    WORK.BEETLE
         Dependent Variable          value
         Covariance Structure        Variance Components
         Estimation Method           REML
         Residual Variance Method    Profile
         Fixed Effects SE Method     Model-Based
         Degrees of Freedom Method   Containment

                       Class Level Information
         Class     Levels    Values
         SIRE         24     1 2 3 4 5 6 7 8 9 10 11 12 13
                             14 15 16 17 18 19 20 21 22 23
                             24
         DAM           5     1 2 3 4 5

                              Dimensions
                  Covariance Parameters             3
                  Columns in X                      1
                  Columns in Z                    134
                  Subjects                          1
                  Max Obs Per Subject             549
                  Observations Used               549
                  Observations Not Used             0
                  Total Observations              549

                           Iteration History
    Iteration    Evaluations    -2 Res Log Like        Criterion
        0             1          4725.58171669
        1             2          4572.06799839         0.00000000
                       Convergence criteria met.

                     Covariance Parameter Estimates
                                 Standard       Z
         Cov Parm    Estimate      Error      Value      Pr Z
         SIRE         31.7912    19.8135      1.60      0.0543
         DAM(SIRE)   114.61     23.0327       4.98     <.0001
         Residual    177.49     11.9787      14.82     <.0001
```

(continued on next page)

Output 2.1 *(continued)*

```
            Asymptotic Covariance Matrix of Estimates
    Row   Cov Parm        CovP1        CovP2        CovP3
     1    SIRE            392.58      -116.52      -0.01690
     2    DAM(SIRE)      -116.52       530.50     -28.6296
     3    Residual        -0.01690    -28.6296     143.49

                     Fit Statistics
           -2 Res Log Likelihood            4572.1
           AIC  (smaller is better)         4578.1
           AICC (smaller is better)         4578.1
           BIC  (smaller is better)         4581.6

                   Solution for Fixed Effects
                        Standard
    Effect      Estimate     Error      DF    t Value    Pr > |t|
    Intercept    55.4665    1.6425      23     33.77      <.0001
```

(Note that in contrast to PROC GLM and other SAS procedures, only fixed effects should appear in the MODEL statement; here there are no fixed effects other than the intercept.) The results appear in Output 2.1. σ_S^2, σ_D^2, and σ_W^2 are estimated to be about 32, 115, and 177, respectively (see the "Estimate" column under "Covariance Parameter Estimates"). Approximate (asymptotic or large-sample) standard errors for each estimate are also given, along with corresponding z-tests; both are requested by the COVTEST option. The z-tests are generally not very sensitive and should be taken with a grain of salt (Littell et al., 1996); the preferred method for testing hypotheses about the parameters is the likelihood ratio test, described later in the chapter. The SOLUTION option in the MODEL statement causes an estimate of the mean or intercept to be printed (see "Solution for Fixed Effects"); mean fecundity is about 55.

Converting observational components to causal components gives $V_A = 4 \times 31.79 = 127.16$, $V_M = 114.61 - 31.79 = 82.82$, and $V_E = 177.49 - 2 \times 31.79 = 113.91$. The standard error of V_A is simply four times the standard error of σ_S^2, or 79.24. Standard errors for V_M and V_E can be calculated from the asymptotic covariance matrix of the estimates (requested by the "ASYCOV" option), using the fact that for two variables X and Y, $\text{var}(X - Y) = \text{var}(X) + \text{var}(Y) - 2\text{cov}(X,Y)$. The diagonal elements of the matrix are the variances of the three causal component estimates; the order of the variables is the same as given under "Covariance Parameter Estimates." The off-diagonal elements give the covariances between the pairs of estimates. Thus the variance of $V_M = \text{var}(\sigma_D^2 - \sigma_S^2) = 530.50 + 392.58 - 2(-116.52) = 1156.12$; taking the square root gives a standard error of 34.00. Because the ratio of estimated V_M to its standard error is 2.44, a z-test would reject the null hypothesis $V_M = 0$, a conclusion that we will confirm later by a likelihood ratio test. Similarly, $\text{var}(V_E) = \text{var}(\sigma_W^2 - 2\sigma_S^2) = \text{var}(\sigma_W^2) + 4\text{var}(\sigma_S^2) - 4\text{cov}(\sigma_W^2, \sigma_S^2) \cong 1713.7$, for a standard error of 41.4 (here we have used the facts that if c is a constant, $\text{var}(cX) = c^2\text{var}(X)$, and $\text{cov}(X,cY) = \text{cov}(cX,Y) = c\text{cov}(X,Y)$. We can also estimate V_P and its standard error; V_P is the sum of the three observational variance components, 323.89, and $\text{var}(V_P)$ is given by the sum of all the elements of the asymptotic covariance matrix, 776.24 (standard error = 27.86). Finally, we can estimate the narrow-sense heritability, V_A/V_P, which gives 0.39. If you were analyzing many data sets or many traits, it would be worthwhile to write a program to automate these calculations, by first saving the estimates of the causal components and their asymptotic covariance matrix in new data sets using the SAS Output Delivery System (ODS).

We now turn to testing the null hypotheses $V_A = 0$ (equivalently, $\sigma_S^2 = 0$) by a likelihood ratio test. Notice that under "Fit Statistics," the quantity 4572.1 is given for "–2 Res[tricted] Log Likelihood"; the same quantity is given to more decimal places under "Iteration History." The restricted likelihood is a function that depends on both the parameter estimates and the data; it measures the probability of observing the data, given a set of parameter estimates and the assumptions of the model (e.g., normality, etc.). The restricted maximum likelihood parameter estimates provided by SAS are those that, as the name implies, maximize the value of the restricted likelihood, and hence minimize the value of −2Log(Restricted Likelihood). (You can also obtain estimates that maximize the ordinary or unrestricted likelihood by using the TYPE = ML option in the PROC MIXED statement, but using the restricted likelihood is preferable for most purposes and is the default; see Shaw (1987) and Lynch and Walsh (1998) for comparisons of the two types of likelihoods.) To perform the likelihood ratio test, you rerun PROC MIXED, but this time without a SIRE effect in the RANDOM statement:

```
proc mixed data = beetle;
class sire dam;
where trait = 'fec';
model value =;
random dam(sire);
run;
```

The output is not shown; the only piece of information that we need from it is −2 Res Log Likelihood, which is 4576.7. This is greater than before, meaning that the likelihood is smaller than before. This makes sense; one less parameter has been estimated from the data, so the fit of the model is less good. If the null hypothesis $\sigma_S^2 = 0$ is true, and the other model assumptions are met, then twice the difference in log likelihoods between the models with and without the sire effect should have an approximate chi-square distribution with one degree of freedom (the d.f. is one in this case because the models differ by one parameter). Twice the difference in log likelihoods is 4576.7 − 4572.1 = 4.6. Using the PROBCHI function, we can find the probability of getting a chi square this large under the null hypothesis. The SAS code is as follows:

```
data prob;
chiprob = 1 - probchi(4.6 , 1);
proc print;
run;
```

The output (not shown) gives CHIPROB = 0.032, so we can reject the null hypothesis (in fact, because we are testing the hypothesis $\sigma_S^2 = 0$ against the one-sided alternative $\sigma_S^2 > 0$, we are entitled to halve this probability to obtain the significance level). This test is termed a likelihood ratio test because the difference in log likelihoods is equivalent to the log of the ratio of the likelihoods.

Likelihood ratio tests are one of the most useful features of REML, because they permit testing hypotheses that cannot be tested by ANOVA and related methods. A case in point is the hypothesis $V_M = 0$. (Note that this is not equivalent to the hypothesis $\sigma_D^2 = 0$, for which the ratio of the dam mean square to the residual mean square from an ANOVA provides an F-test.) To conduct the test, we need to obtain the log likelihood of a model in which V_M has been constrained to zero. Fortunately, SAS provides a ready means to introduce this constraint. From the relationships above between the causal and observational variance components, note that if $V_M = 0$, then $\sigma_S^2 = \sigma_D^2$. We can introduce this constraint by adding the TYPE = TOEP(1) option to the RANDOM statement. The program is as follows:

```
proc mixed data = beetle covtest;
class sire dam;
where trait = 'fec';
model value = ;
random sire dam(sire)/type = toep(1);
run;
```

Output 2.2 Constraining sire and dam variance components to be equal.

```
                      The Mixed Procedure

                       Model Information
         Data Set                    WORK.BEETLE
         Dependent Variable          VALUE
         Covariance Structure        Banded Toeplitz
         Estimation Method           REML
         Residual Variance Method    Profile
         Fixed Effects SE Method     Model-Based


                     Class Level Information
         Class    Levels    Values
         SIRE       24      1 2 3 4 5 6 7 8 9 10 11 12 13
                            14 15 16 17 18 19 20 21 22 23
                            24
         DAM         5      1 2 3 4 5

                            Dimensions
                  Covariance Parameters            2
                  Columns in X                     1
                  Columns in Z                   134
                  Subjects                         1
                  Max Obs Per Subject            549
                  Observations Used              549
                  Observations Not Used            0
                  Total Observations             549

                          Iteration History
       Iteration    Evaluations    -2 Res Log Like     Criterion
           0              1          4725.58171669
           1              2          4580.19146805     0.00142579
           2              1          4577.17683146     0.00017859
           3              1          4576.83050107     0.00000375
           4              1          4576.82371634     0.00000000
                        Convergence criteria met.

                    Covariance Parameter Estimates
                              Standard        Z
       Cov Parm    Estimate     Error       Value     Pr Z
       Variance     93.4052    17.0977       5.46    <.0001
       Residual    178.77      12.1505      14.71    <.0001

                           Fit Statistics
              -2 Res Log Likelihood            4576.8
              AIC (smaller is better)          4580.8
              AICC (smaller is better)         4580.8
              BIC (smaller is better)          4583.2

                 Null Model Likelihood Ratio Test
                 DF     Chi-Square      Pr > ChiSq
                  1        148.76         <.0001
```

The results appear in Output 2.2. Once again, the relevant part of the output is −2 Res Log Likelihood, which is seen to be 4576.8. This differs from −2 Res Log Likelihood from the full model (4572.1) by 4.7. Using the PROBCHI function, we can determine that a chi square of 4.7 or larger with 1 d.f. (because we have reduced the number of parameters from three to two) has a probability of 0.030. We are again entitled to halve this probability, because the test of $V_M = 0$ is one-tailed (i.e., we expect and indeed observe that $\sigma_D^2 > \sigma_S^2$; we do not expect $\sigma_S^2 > \sigma_D^2$). We can therefore reject the hypothesis $V_M = 0$; maternal or common-environment effects appear to be present. This conclusion is subject to the caveat that we have assumed that dominance variance is absent; our estimate of V_M could be inflated by ¼V_D. It is most unlikely, however, that our V_M estimate of 82.8 comes entirely from dominance variance; this would give an estimate of $V_D = 4 \times 82.8 = 331.2$, which is greater than the observed phenotypic variance. If we had accepted the hypothesis $\sigma_S^2 = \sigma_D^2$, then four times 93.4052, the REML estimate of σ_S^2 and σ_D^2 under the assumption that the two are equal (see "Covariance Parameter Estimates" in Output 2.2), would have been an appropriate estimate of V_A.

2.2.2 Example of Nested Half-Sib Design, Two Traits

If two traits are measured in a nested half-sib design, there are a total of nine observational parameters; these are the three components of variance for each trait, and three components of covariance between traits. Referring to the model equation given in the previous section, the covariance between the effect of a sire on trait 1, S_{1i}, and its effect on trait 2, S_{2i}, will be denoted by $\sigma_{S1,2}$. Similarly, the between-trait covariances of dam and residual (individual) effects will be denoted by $\sigma_{D1,2}$ and $\sigma_{W1,2}$, respectively. These observational components can be interpreted in terms of causal components, just as was the case for variances:

$$\sigma_{S1,2} = \tfrac{1}{4} COV_A$$
$$\sigma_{D1,2} = \tfrac{1}{4} COV_A + \tfrac{1}{4} COV_D + COV_M$$
$$\sigma_{W1,2} = \tfrac{1}{2} COV_A + \tfrac{3}{4} COV_D + COV_E$$

COV_A is the covariance between an individual's breeding value for one trait and its breeding value for the other trait. COV_D, COV_M, and COV_E are the similar covariances for dominance deviations, maternal/common-environment effects, and environmental deviations, respectively. If dominance is ignored, solving for the causal covariances yields completely analogous solutions to those for variances:

$$COV_A = 4\sigma_{S1,2}$$
$$COV_M = \sigma_{D1,2} - \sigma_{S1,2}$$
$$COV_E = \sigma_{W1,2} - 2\sigma_{S1,2}$$

For each causal covariance, we can also define a corresponding correlation. Thus the (additive) genetic correlation between traits, r_A, is defined as $COV_A/(V_{A1}V_{A2})^{1/2}$; we can likewise define the maternal and environmental correlations between traits, r_M and r_E.

We will use the seed beetle data set to illustrate estimating and making inferences about the three causal components of covariance and the corresponding correlations. The two traits are fecundity, considered above, and body mass at eclosion. The following PROC MIXED program provides estimates of the observational components of variance and covariance, from which the causal components can be calculated:

```
proc mixed data = beetle covtest asycov;
class sire dam progeny trait;
model value = trait;
random trait/subject = sire type = un;
random trait/subject = dam(sire) type = un;
repeated trait/subject = progeny(sire*dam) type = un;
run;
```

Several aspects of this program require comment. The MODEL statement tells PROC MIXED to fit TRAIT as a fixed effect; in other words, the two levels of TRAIT are allowed to have different means. The first and second RANDOM statements tell PROC MIXED to estimate the elements of the 2 × 2 covariance matrices of SIRE and DAM effects, respectively. For example, the matrix for SIRE effects is

$$\begin{bmatrix} \sigma_{S1}^2 & \sigma_{S1,2} \\ \sigma_{S1,2} & \sigma_{S2}^2 \end{bmatrix}.$$

The syntax RANDOM TRAIT/SUBJECT = SIRE in this context tells SAS that the rows and columns of the covariance matrix correspond to the levels of TRAIT, and that the variances and covariances to be estimated are those among sires; it does *not* identify TRAIT as a random effect. TYPE = UN indicates that the covariance matrix is "unstructured," meaning that no constraints are to be put on it beyond non-negativity constraints on the variances. As we will see later, other covariance structures can be specified. For the covariance matrix of within-family effects, a REPEATED statement is used instead of a RANDOM statement, but the syntax is otherwise the same; in this case SUBJECT = PROGENY(SIRE*DAM) is used because we want to estimate the variances and covariances among individuals within full-sib families.

Output 2.3 Nested half-sib design, two traits (fecundity and body mass at eclosion in a seed beetle).

```
                      The Mixed Procedure
                       Model Information
         Data Set                    WORK.BEETLE
         Dependent Variable          VALUE
         Covariance Structure        Unstructured
         Subject Effects             SIRE, DAM(SIRE),
                                     PROGENY(SIRE*DAM)
         Estimation Method           REML
         Residual Variance Method    None
         Fixed Effects SE Method     Model-Based
         Degrees of Freedom Method   Containment

                    Class Level Information
         Class      Levels    Values
         SIRE         24      1  2  3  4  5  6  7  8  9 10 11 12 13
                              14 15 16 17 18 19 20 21 22 23
                              24
         DAM           5      1 2 3 4 5
         PROGENY       5      1 2 3 4 5
         TRAIT         2      fec mass

                           Dimensions
              Covariance Parameters            9
              Columns in X                     3
              Columns in Z Per Subject        12
              Subjects                        24
              Max Obs Per Subject             50
              Observations Used             1098
              Observations Not Used            0
              Total Observations            1098

                        Iteration History
     Iteration    Evaluations    -2 Res Log Like      Criterion
          0            1            8693.14459169
          1            2            5558.08421381      0.00000004
          2            1            5558.08415058      0.00000000
                     Convergence criteria met.
```

(*continued on next page*)

Output 2.3 *(continued)*

```
                   Covariance Parameter Estimates
                                    Standard          Z
Cov Parm    Subject         Estimate    Error      Value      Pr Z
UN(1,1)     SIRE             32.0885    19.8835     1.61     0.0533
UN(2,1)     SIRE              0.8115     0.6296     1.29     0.1974
UN(2,2)     SIRE              0.07616    0.03243    2.35     0.0094
UN(1,1)     DAM(SIRE)       114.46      22.9888     4.98    <.0001
UN(2,1)     DAM(SIRE)         1.4373     0.5558     2.59     0.0097
UN(2,2)     DAM(SIRE)         0.07634    0.02262    3.38     0.0004
UN(1,1)     PROGENY(SIRE*DAM) 177.48    11.9784    14.82    <.0001
UN(2,1)     PROGENY(SIRE*DAM)   3.0707   0.4000     7.68    <.0001
UN(2,2)     PROGENY(SIRE*DAM)   0.3428   0.02314   14.82    <.0001

               Asymptotic Covariance Matrix of Estimates
Row  Cov Parm    CovP1       CovP2       CovP3       CovP4       CovP5       CovP6
 1   UN(1,1)    395.35       7.7279      0.1497    -115.29      -1.6506     -0.02366
 2   UN(2,1)      7.7279     0.3964      0.01251    -1.6948     -0.06896    -0.00163
 3   UN(2,2)      0.1497     0.01251     0.001052   -0.02490    -0.00167    -0.00011
 4   UN(1,1)   -115.29      -1.6948     -0.02490   528.48        7.2900      0.1007
 5   UN(2,1)     -1.6506    -0.06896    -0.00167     7.2900      0.3090      0.007155
 6   UN(2,2)     -0.02366   -0.00163    -0.00011     0.1007      0.007155    0.000512
 7   UN(1,1)     -0.01675    0.000135    9.379E-6  -28.6377     -0.4952     -0.00856
 8   UN(2,1)      0.000993  -0.00002    -3.09E-7    -0.4953     -0.03196    -0.00096
 9   UN(2,2)      0.000040   1.262E-6   -1.73E-7    -0.00857    -0.00096    -0.00011

                    Asymptotic Covariance
                     Matrix of Estimates
              Row       CovP7       CovP8       CovP9
               1      -0.01675     0.000993    0.000040
               2       0.000135   -0.00002     1.262E-6
               3       9.379E-6   -3.09E-7    -1.73E-7
               4     -28.6377     -0.4953     -0.00857
               5      -0.4952     -0.03196    -0.00096
               6      -0.00856    -0.00096    -0.00011
               7     143.48        2.4821      0.04294
               8       2.4821      0.1600      0.004794
               9       0.04294     0.004794    0.000535

                       Fit Statistics
           -2 Res Log Likelihood          5558.1
           AIC (smaller is better)        5576.1
           AICC (smaller is better)       5576.2
           BIC (smaller is better)        5586.7

              Null Model Likelihood Ratio Test
              DF    Chi-Square      Pr > ChiSq
               8      3135.06        <.0001

               Type 3 Tests of Fixed Effects
                     Num    Den
           Effect    DF     DF    F Value    Pr > F
           TRAIT      1     46     989.55    <.0001
```

The results are shown in Output 2.3. Parameter estimates are given in the "Estimate" column under "Covariance Parameter Estimates." The covariance matrix in question is identified under the heading "Subject," and UN(1,1), UN(2,1), etc., identify the row and column of the respective matrix. Thus the first estimate, 32.09, is the variance among sires for trait 1, fecundity (unsurprisingly, this is about the same as the estimate from the analysis of this trait alone; see Output 2.1); the second estimate, 0.81, is the among-sire covariance between fecundity and mass; and the third estimate, 0.076, is the variance among sires for trait 2, mass. The equations given earlier can be used to estimate the causal covariances from the observational covariances; thus our estimates are $COV_A = 4(0.8115) = 3.246$, $COV_M = 1.4373 - 0.8115 = 0.6258$, and $COV_E = 3.0707 - 2(0.8115) = 1.4477$. Standard errors could be put on these estimates by the identical logic as given in the previous section. The reader can confirm that the corresponding correlations are $r_A = 0.52$, $r_M = 5.14$, and $r_E = 0.31$. (For example, $r_A = 4\sigma_{S1,2}/(4\sigma_{S1}^2 4\sigma_{S2}^2)^{1/2} = \sigma_{S1,2}/(\sigma_{S1}\sigma_{S2})$; but we will see later that we can get SAS to produce an estimate for r_A directly.) The absurdly high estimate for r_M results from the fact that for mass, the V_M estimate is almost zero ($0.07634 - 0.07616$); in practice there would be little point in reporting an estimate of r_M for this data set.

Two hypotheses about the genetic correlation that we may want to test are $r_A = 0$ and $r_A = 1$. The former hypothesis is of course equivalent to $COV_A = 0$. A convenient way to test this hypothesis is to rerun the PROC MIXED program above, replacing TYPE = UN in the first RANDOM statement with TYPE = UN(1). The UN(1) covariance structure, called banded main diagonal, imposes the constraint that all off-diagonal elements are zero. Doing this yields -2 Res Log Likelihood = 5560.4, compared to 5558.1 with TYPE = UN. The difference is 2.3, which with 1 d.f. has a probability of 0.13 (in this case the test is two-tailed, because COV_A could in theory be either < 0 or > 0). We therefore cannot reject the hypothesis that $r_A = 0$. To test $r_A = 1$, we need to be able to tell PROC MIXED to constrain r_A to this value. Luckily, MIXED provides a way to do this with the TYPE = UNR covariance structure. This structure is identical to the TYPE = UN structure, except that the off-diagonal elements are parameterized in terms of the correlations and variances; i.e., instead of $\sigma_{S1,2}$, the off-diagonal elements are $r_S\sigma_{S1}\sigma_{S2}$, where r_S is the correlation of sire effects, and is equivalent to r_A. Rerunning the program above with this structure gives Output 2.4 (only the relevant part of the output is shown). The first two lines under "Covariance Parameter Estimates" give estimates of σ_{S1}^2 and σ_{S2}^2 (i.e., the variances of sire effects for fecundity and mass), while the third line ("Corr(2,1)") gives an estimate for r_A. This is 0.5191, the same as calculated above. Note that -2 Res Log Likelihood is the same as for Output 2.3; this is because we have not changed the assumptions of the model, only how it is parameterized. We are now in a position to constrain r_A to one by adding the following PARMS statement:

```
parms 32 0.08 0.99999 114 1.4 0.08 177 3 0.3/hold = 3;
```

Output 2.4 Estimation of the genetic correlation between fecundity and body mass at eclosion using the TYPE=UNR covariance structure.

```
                          Iteration History
          Iteration    Evaluations    -2 Res Log Like      Criterion
              0             1           8693.14459169
              1             3           5560.47506002      0.00106283
              2             2           5558.29519517      0.00010932
              3             1           5558.08711278      0.00000165
              4             1           5558.08415136      0.00000000
                          Convergence criteria met.

                    Covariance Parameter Estimates
                                                  Standard         Z
Cov Parm       Subject              Estimate        Error        Value      Pr Z
Var(1)         SIRE                  32.0900       19.8835        1.61     0.0533
Var(2)         SIRE                   0.07616       0.03243       2.35     0.0094
Corr(2,1)      SIRE                   0.5191        0.2721        1.91     0.0564
UN(1,1)        DAM(SIRE)            114.46         22.9869        4.98     <.0001
UN(2,1)        DAM(SIRE)              1.4375        0.5558        2.59     0.0097
UN(2,2)        DAM(SIRE)              0.07634       0.02262       3.38     0.0004
UN(1,1)        PROGENY(SIRE*DAM)    177.48         11.9784       14.82     <.0001
UN(2,1)        PROGENY(SIRE*DAM)      3.0707        0.4000        7.68     <.0001
UN(2,2)        PROGENY(SIRE*DAM)      0.3428        0.02314      14.82     <.0001

                         Fit Statistics
                  -2 Res Log Likelihood        5558.1
                  AIC (smaller is better)      5576.1
                  AICC (smaller is better)     5576.2
                  BIC (smaller is better)      5586.7
```

For all parameters except the third, initial values here are taken from Output 2.4; for constraining r_A to one, we must use 0.99999 rather than 1, since the latter results in an error message. Rerunning PROC MIXED with the line above added gives −2 Res Log Likelihood = 5562.1, which exceeds that from the full model by 4.0. The corresponding chi-square probability with 1 d.f. is 0.046; we are again entitled to halve this probability, since r_A cannot logically exceed one. Therefore we conclude that $0 \leq r_A < 1$.

PARMS statements can also be used to constrain COV_M and COV_E to zero, although the procedure is a little more complicated. The constraints $COV_M = 0$ and $COV_E = 0$ are equivalent to the constraints $\sigma_{D1,2} = \sigma_{S1,2}$ and $\sigma_{W1,2} = 2\sigma_{S1,2}$, respectively. To introduce the latter constraint, for example, we could run PROC MIXED with TYPE=UN in the first RANDOM statement, and a PARMS statement of the following form, where "X" is a number:

```
parms 32 X 0.08 114 1.4 0.08 177 2X 0.3/hold = 2,8;
```

The program should be rerun with various values of X to find, by trial and error, the value that minimizes −2 Res Log Likelihood. −2 Res Log Likelihood from this best-fitting constrained model can then be compared with that from Output 2.3; under the null hypothesis that $COV_E = 0$, the difference should have a chi-square distribution with one degree of freedom.

2.2.3 Example of Clonal Design, Two Environments

We turn now to a design in which one trait is measured in each of two environments or treatments. The example comes from a data set of Fry and Heinsohn (2002). Thirty-five second chromosome lines of *Drosophila melanogaster* were measured for larval viability at each of two temperatures, 18°C and 25°C (denoted "LT" and "R," respectively). The lines were genetically identical except for spontaneous mutations accumulated over 31 generations. The authors were interested in whether mutations had different effects on viability at different temperatures; therefore one of our main goals is to estimate and make inferences about the cross-environment genetic correlation. Although each line was originally tested in six blocks, with one replicate measurement per temperature and block, for illustrative purposes we have combined the adjacent pairs of blocks into three new blocks with two replicates each; this allows the three-way interaction to be estimated in the model below. The commands for reading the data are as follows:

```
data mutants;
input treat $ line block relvia;
datalines;
lt    1      1       0.713615023
lt    1      1       0.707964602
r     1      1       0.643086817
r     1      1       0.937759336
...more datalines...
lt    46     3       0.964539007
lt    46     3       1.097826087
r     46     3       0.762589928
r     46     3       0.824
;
```

The data set will first be analyzed using a traditional variance component model. As we will see, there are better ways to analyze this data set, but the traditional approach is considered first because of its historical importance. In the model, temperature treatment (TREAT) is a fixed effect, while LINE and BLOCK are random effects. All possible interactions among these three crossed factors are included, and considered as random effects. The PROC MIXED program is as follows:

```
proc mixed covtest data = mutants;
class treat block line;
model relvia = treat;
random block line treat*block block*line treat*line treat*block*line;
run;
```

Output 2.5 Traditional variance component model applied to clonal design with two environments.

```
                        The Mixed Procedure

                         Model Information
            Data Set                     WORK.MUTANTS
            Dependent Variable           RELVIA
            Covariance Structure         Variance Components
            Estimation Method            REML
            Residual Variance Method     Profile
            Fixed Effects SE Method      Model-Based
            Degrees of Freedom Method    Containment

                       Class Level Information
         Class      Levels    Values
         TREAT         2      lt r
         BLOCK         3      1 2 3
         LINE         35      1 10 11 12 13 14 15 16 17 19 2
                              22 23 24 25 26 27 28 29 3 30
                              31 32 33 35 37 38 41 42 43 46
                              5 6 7 9

                             Dimensions
            Covariance Parameters              7
            Columns in X                       3
            Columns in Z                     429
            Subjects                           1
            Max Obs Per Subject              415
            Observations Used                415
            Observations Not Used              0
            Total Observations               415

                           Iteration History
       Iteration    Evaluations    -2 Res Log Like      Criterion
           0             1         -280.92923945
           1             3         -329.21676100        0.00024113
           2             2         -329.35419589        0.00001075
           3             1         -329.36037784        0.00000010
           4             1         -329.36043165        0.00000000
                         Convergence criteria met.

                     Covariance Parameter Estimates
                                  Standard        Z
Cov Parm              Estimate      Error       Value     Pr Z
BLOCK                    0            .            .        .
LINE                  0.003076     0.001759      1.75     0.0402
TREAT*BLOCK           0.001809     0.001510      1.20     0.1155
BLOCK*LINE            0.002870     0.001431      2.01     0.0225
TREAT*LINE            0.001548     0.001235      1.25     0.1050
TREAT*BLOCK*LINE         0            .            .        .
Residual              0.02037      0.001740     11.71    <.0001

                            Fit Statistics
              -2 Res Log Likelihood            -329.4
              AIC (smaller is better)          -319.4
              AICC (smaller is better)         -319.2
              BIC (smaller is better)          -323.9

                      Type 3 Tests of Fixed Effects
                          Num      Den
             Effect        DF       DF    F Value    Pr > F
             TREAT          1        2      0.82     0.4603
```

The results are shown in Output 2.5. Variance component estimates of zero are given for BLOCK and the three-way interaction. The other four variance component estimates are positive; one-tailed likelihood ratio tests show that each is significantly different from zero at $P < 0.025$ or lower, except for the TREAT × LINE interaction, which is marginally significant ($P = 0.063$).

The significant LINE effect indicates that the genetic covariance across environments is greater than zero (Fry, 1992). The nearly significant TREAT × LINE interaction suggests that the correlation may be less than one, because one cause of genotype by environment interaction is imperfect correlation of genotype means across environments (Cockerham, 1963). Genotype-by-environment interaction, however, can also be caused by unequal among-line variances in the two environments. Furthermore, while it is possible to estimate the genetic correlation from the magnitude of the LINE and TREAT × LINE variance components (Fry, 1992), the estimate so obtained will be biased downward if among-line variances differ between treatments. It would clearly be desirable to estimate the genetic correlation in a way that is not subject to the assumption of equal among-line variances across environments. Furthermore, the variance component model makes another undesirable assumption, namely that residual variances are the same in the two environments. Violation of either of these assumptions could bias estimates of the parameters.

To fit a model that allows among-line and within-line variances to differ between environments, we must modify the PROC MIXED code in two ways. First, we must use two RANDOM statements instead of one. The first RANDOM statement below, by using the TYPE = UNR covariance structure, tells SAS to estimate separate among-line variances for each level of TREAT, and to estimate the genetic correlation across the two treatments. The second RANDOM statement retains the traditional variance component formulation for the other random effects (the three-way interaction is not included since it was found to be zero). The second modification is the addition of a REPEATED statement with the GROUP= option. This tells SAS to estimate separate residual variances in the two treatments.

```
proc mixed covtest data = mutants;
class treat block line;
model relvia = treat;
random treat/subject = line type = unr;
random block treat*block block*line;
repeated/group = treat;
run;
```

Output 2.6 Clonal design with two treatments: allowing among-line and within-line variances to differ between treatments.

```
                      The Mixed Procedure

                       Model Information
      Data Set                     WORK.MUTANTS
      Dependent Variable           RELVIA
      Covariance Structures        Unstructured using
                                   Correlations, Variance
                                   Components
      Subject Effect               LINE
      Group Effect                 TREAT
      Estimation Method            REML
      Residual Variance Method     None
      Fixed Effects SE Method      Model-Based
      Degrees of Freedom Method    Containment
```

(continued on next page)

Output 2.6 *(continued)*

```
                        Class Level Information
             Class    Levels    Values
             TREAT        2     lt r
             BLOCK        3     1 2 3
             LINE        35     1 10 11 12 13 14 15 16 17 19 2
                                22 23 24 25 26 27 28 29 3 30
                                31 32 33 35 37 38 41 42 43 46
                                5 6 7 9

                             Dimensions
                  Covariance Parameters            8
                  Columns in X                     3
                  Columns in Z                   184
                  Subjects                         1
                  Max Obs Per Subject            415
                  Observations Used              415
                  Observations Not Used            0
                  Total Observations             415

                          Iteration History
         Iteration    Evaluations    -2 Res Log Like      Criterion
             0             1           -280.92923945
             1             3           -339.22068717      593.63135128
             2             2           -339.70880926       16.09133050
             3             1           -340.03261806        0.18714428
             4             2           -343.12633994        0.00287047
             5             2           -344.33560535        0.00013037
             6             2           -344.40924353        0.00000041
             7             1           -344.40947025        0.00000000
                       Convergence criteria met.
```

Covariance Parameter Estimates

Cov Parm	Subject	Group	Estimate	Standard Error	Z Value	Pr Z
Var(1)	LINE		0.006805	0.002965	2.30	0.0109
Var(2)	LINE		0.002507	0.001540	1.63	0.0518
Corr(2,1)	LINE		0.7559	0.2550	2.96	0.0030
BLOCK			0	.	.	.
TREAT*BLOCK			0.001888	0.001537	1.23	0.1096
BLOCK*LINE			0.002723	0.001347	2.02	0.0216
Residual		TREAT lt	0.02552	0.002943	8.67	<.0001
Residual		TREAT r	0.01524	0.001838	8.29	<.0001

```
                         Fit Statistics
            -2 Res Log Likelihood          -344.4
            AIC (smaller is better)        -330.4
            AICC (smaller is better)       -330.1
            BIC (smaller is better)        -339.6
```

Type 3 Tests of Fixed Effects

Effect	Num DF	Den DF	F Value	Pr > F
TREAT	1	2	0.80	0.4664

The results are shown in Output 2.6. This model has two more parameters than the variance component model (this can be verified by counting the number of covariance parameter estimates, ignoring parameters set to zero). −2 Res Log Likelihood for this model is −344.4, much lower than that for the variance component model, which was −329.4 (see Output 2.5). To test whether the more complicated model fits better, we compare −329.4 − (−344.4) = 15.0 to a chi-square distribution with two degrees of freedom; the result is highly significant ($P = 0.0006$). Clearly, one or both of the equal variance assumptions of the variance component model are violated.

Output 2.6 shows estimates of the among-line variances in the low-temperature (LT) and regular (R) treatments to be 0.0068 and 0.0025, respectively; the corresponding residual variance estimates are 0.0255 and 0.0152. While both pairs of estimates suggest higher variances in the LT treatment, the differences need to be tested individually for significance. This can be done by rerunning PROC MIXED, first to constrain among-line variances to be equal, and then to constrain within-line variances. The first constraint can be achieved by changing the covariance structure option to TYPE = CS, for compound symmetry; this covariance structure assumes equal among-line variances in all treatments (i.e., the diagonal elements of the covariance matrix are all the same). The resulting value of −2 Res Log Likelihood is −341.9, which differs from that of the full model by 2.5; with 1 d.f., this gives a two-tailed probability of 0.11. Therefore we cannot reject the hypothesis of equal among-line variances in the two treatments, even though the point estimates differ considerably. Adding the constraint of equal within-line variances can be achieved simply by deleting the REPEATED statement above, restoring the default assumption of equal variances; this gives −2 Res Log Likelihood = −334.5, for a chi square of 9.9, $P < 0.002$. Within-line variances are thus significantly higher in the low-temperature treatment.

Output 2.6 also shows that the best estimate of the cross-environment genetic correlation (Corr(2,1)) is 0.76, with a standard error of 0.26. Using the TYPE = UN(1) covariance structure and the PARMS statement, we can test whether the correlation differs significantly from zero and one, respectively, as described in the previous section. These tests give probabilities of 0.03 and 0.16, respectively; therefore the correlation is significantly greater than zero, but not significantly less than one.

2.3 The Diallel and Related Designs

While the nested half-sib design has the advantage of being applicable to a wide variety of organisms, it has two disadvantages. First, as we have seen, dominance variance and maternal/common-environment effects are confounded. Second, a tacit assumption of the design is that paternal effects are absent; any such effects would be confounded with V_A. More complex designs are needed to provide separate estimates of V_D, V_M, and the paternal-effects variance. Notable among these are the diallel and the North Carolina Design 2 (or NC2 design) with reciprocals. For these designs, it is necessary either to have a hermaphroditic species or to have a set of inbred or isogenic lines. As illustrated in Table 2.1, in each design, individuals or lines are crossed as both males and females. In the diallel without selfs illustrated, all possible crosses among 10 individuals are performed, except for crosses of an individual to itself. The NC2 design with reciprocals differs from the diallel only in lacking certain crosses.

Table 2.1 Illustration of the diallel and NC2 designs.

Dams → Sires ↓	1	2	3	4	5	6	7	8	9	10
1		D 1	D 2	D 3	D 4	D, N 5	D, N 6	D, N 7	D, N 8	D, N 9
2	D 10		D 11	D 12	D 13	D, N 14	D, N 15	D, N 16	D, N 17	D, N 18
3	D 19	D 20		D 21	D 22	D, N 23	D, N 24	D, N 25	D, N 26	D, N 27
4	D 28	D 29	D 30		D 31	D, N 32	D, N 33	D, N 34	D, N 35	D, N 36
5	D 37	D 38	D 39	D 40		D, N 41	D, N 42	D, N 43	D, N 44	D, N 45
6	D, N 46	D, N 47	D, N 48	D, N 49	D, N 50		D 51	D 52	D 53	D 54
7	D, N 55	D, N 56	D, N 57	D, N 58	D, N 59	D 60		D 61	D 62	D 63
8	D, N 64	D, N 65	D, N 66	D, N 67	D, N 68	D 69	D 70		D 71	D 72
9	D, N 73	D, N 74	D, N 75	D, N 76	D, N 77	D 78	D 79	D 80		D 81
10	D, N 82	D, N 83	D, N 84	D, N 85	D, N 86	D 87	D 88	D 89	D 90	

Note: Crosses performed in the diallel and NC2 designs are denoted by "D" and "N," respectively. In both designs, hermaphroditic individuals or inbred lines are mated as both males and females. In the diallel with reciprocals illustrated here, all possible crosses are performed except those within lines ("selfs"). In the NC2 design with reciprocals, dams 1–5 are mated to sires 6–10, and the reciprocal crosses (dams 6–10 × sires 1–5) are also performed. Both designs can also be performed without reciprocals. Pairs of numbered families illustrating various types of relationships are indicated: reciprocal full sibs, 1 and 10; paternal half sibs, 11 and 12; maternal half sibs, 20 and 29; reciprocal half sibs, 31 and 39. For illustrative purposes, families are numbered from 1 to 90.

Historically, the two designs have been treated separately, because estimating variance components from them by the method of moments requires different (and messy) formulae (Lynch and Walsh, 1998; Cockerham and Weir, 1977). The designs are obviously closely related, however, and as we will see, the identical PROC MIXED program can be used to analyze data from either one.

2.3.1 Diallel Example

The data set for illustrating the diallel was kindly provided by S. Good-Avila (Good-Avila and Stephenson, 2003). Ten plants of the hermaphroditic, herbaceous perennial *Campanula rapunculoides* were collected from a natural population and used as parents in crosses like those illustrated in Table 2.1. Offspring were grown in a greenhouse and measured for a variety of traits, of which we will use flower number for our example. For this trait, number of offspring measured per family ranged from three to ten. Data were square-root transformed for normality. The commands for reading the data set are as follows:

```
data diallel;
input sire dam sqrnumb;
datalines;
1    2       10.9545
1    2       12.49
1    2       6.6332
1    2       13.9642
...more datalines...
10   9       11.619
10   9       8.4853
10   9       10.9545
10   9       8.1854
;
```

There are a variety of assumptions we could make about these data. We will use the model of Cockerham and Weir (1977; Lynch and Walsh, 1998). Under this model, the phenotype of the kth offspring of the cross between parent i as sire and parent j as dam is written as

$$Y_{ijk} = \mu + N_i + N_j + T_{ij} + P_i + M_j + K_{ij} + W_{k(ij)}.$$

Here, N_i and N_j are the additive effects of nuclear genes contributed by i and j; T_{ij} is the interaction of the nuclear contributions; P_i and M_j are the paternal and maternal extranuclear contributions, respectively, of sire i and dam j; K_{ij} is the interaction effect of the extranuclear contributions (possibly confounded by a common-environment effect specific to the cross); and $W_{k(ij)}$ is the unique effect of individual k. We assume that the effects are independent and normally distributed, with means zero and respective variances σ_N^2, σ_T^2, σ_P^2, σ_M^2, σ_K^2, and σ_W^2. Our goal is to estimate and make inferences about these six variances.

We can use PROC MIXED to obtain direct estimates of the variances because covariances between different types of relatives in the diallel can be written as linear combinations of the variances, and the TYPE = LIN covariance structure in MIXED provides a flexible way to model covariances as linear combinations of unknowns. There are six possible relationships between any two individuals in the diallel (see Table 2.1 for examples): individuals can be full sibs (FS), reciprocal full sibs (RFS), maternal half sibs (MHS), paternal half sibs (PHS), reciprocal half sibs (RHS), or unrelated. The covariances between the various types of relatives are

$$COV_{FS} = 2\sigma_N^2 + \sigma_T^2 + \sigma_M^2 + \sigma_P^2 + \sigma_K^2$$
$$COV_{RFS} = 2\sigma_N^2 + \sigma_T^2$$
$$COV_{MHS} = \sigma_N^2 + \sigma_M^2$$
$$COV_{PHS} = \sigma_N^2 + \sigma_P^2$$
$$COV_{RHS} = \sigma_N^2,$$

while the covariance between unrelated individuals is zero. Note that the covariances are given by the shared terms in the model equation; e.g., paternal half sibs share N_i and P_i through the common sire, but not other terms. In general, we may write the covariance between members of two families i and j as $a_{ij}\sigma_N^2 + b_{ij}\sigma_T^2 + c_{ij}\sigma_M^2 + d_{ij}\sigma_P^2 + e_{ij}\sigma_K^2$, where the values of a_{ij} through e_{ij} depend on the relationship between i and j. By making use of the TYPE = LIN(5) covariance structure in a RANDOM statement, we can give SAS the values of a_{ij} through e_{ij} for every possible pair of families, and SAS will estimate the five variances σ_N^2 through σ_K^2.

Before running PROC MIXED, we must first create an auxiliary file containing the values of a_{ij} through e_{ij}. This is accomplished by the following code:

```
data dummy;
 do row = 1 to 90;
  output;
   end;

data fam1;
 do sire1 = 1 to 10;
  do dam1 = 1 to 10;
   output;
    end;
     end;

data fam1;
 set fam1;
  if sire1 = dam1 then delete;

data fam1;
 merge dummy fam1;

data fam1;
 set fam1;
  do sire2 = 1 to 10;
   do dam2 = 1 to 10;
    output;
     end;
      end;

data fam1;
 set fam1;
  if sire2 = dam2 then delete;

data dummy;
set dummy;
 do col = 1 to 90;
  output;
   end;

data ii;
 merge fam1 dummy;
 if row < col then delete;
  if sire1 = dam1 or sire2 = dam2 or
   (sire1 ^= sire2 and dam1 ^= dam2 and
    sire1 ^= dam2 and dam1 ^= sire2)
     then delete;
if sire1 = sire2 and dam1 = dam2 then rel = 'fulls';
 else if sire1 = dam2 and dam1 = sire2 then rel = 'rfs';
 else if sire1 = sire2 and dam1 ^= dam2 then rel = 'phs';
 else if sire1 ^= sire2 and dam1 = dam2 then rel = 'mhs';
 else rel = 'rhs';
  do parm = 1 to 5;
   output;
    end;
```

```
data ii;
  set ii;
    if parm = 1 then do;
      if rel = 'fulls' or rel = 'rfs' then value = 2;
        else value = 1;
      end;
    if parm = 2 then do;
      if rel = 'fulls' or rel = 'rfs' then value = 1;
        else delete;
      end;
    if parm = 3 then do;
      if rel = 'fulls' or rel = 'mhs' then value = 1;
      else delete;
      end;
    if parm = 4 then do;
      if rel = 'fulls' or rel = 'phs' then value = 1;
      else delete;
      end;
    if parm = 5 then do;
      if rel = 'fulls' then value = 1;
      else delete;
      end;
    keep parm row col value;
```

The resulting data set, termed II, has four variables: PARM (range 1–5), ROW (1–90), COL (1–90), and VALUE (1 or 2). ROW and COL indicate the row and column of the 90 × 90 covariance matrix of families (i.e., the pair of families, using the numbering scheme for families given in Table 2.1); PARM specifies the parameter, in the order σ_N^2, σ_T^2, σ_M^2, σ_P^2, and σ_K^2; and VALUE gives the coefficient for that parameter (i.e., the value of a_{ij}, b_{ij}, c_{ij}, d_{ij}, or e_{ij}). Only nonzero coefficients need to be specified, because SAS assumes any missing coefficients are zero; furthermore, since covariance matrices are symmetric, it is not necessary to include elements below the diagonal. As a result, the II data set has only 2700 observations, instead of the 90 × 90 × 5 = 40,500 that would be necessary to specify every coefficient in the full matrix.

To make use of the II data set, we run the following PROC MIXED program:

```
proc mixed data = diallel covtest asycov;
class sire dam;
model sqrnumb = /solution;
random sire*dam/ type = lin(5) ldata = ii;
parms/ lowerb = 0,0,0,0,0,0;
run;
```

In the RANDOM statement, SIRE*DAM indicates that the rows and columns of the covariance matrix being estimated correspond to the 90 combinations of sire and dam, and LDATA = II indicates that SAS should look in data set II for the coefficients of the five parameters. The PARMS statement with the LOWERB (lower bound) option is used to tell SAS to constrain each parameter to non-negative values. The order in which effects are given in the CLASS statement is critical, because it determines how SAS will order families; because SIRE is given first, the order will correspond to that in Table 2.1, with families 1–9 having sire 1, etc.

Output 2.7 Diallel with reciprocals but not selfs (flower number in a perennial herb).

```
                        The Mixed Procedure

                         Model Information
           Data Set                    WORK.DIALLEL
           Dependent Variable          SQRNUMB
           Covariance Structures       Linear, Variance
                                       Components
           Subject Effect              Intercept
           Estimation Method           REML
           Residual Variance Method    Parameter
           Fixed Effects SE Method     Model-Based
           Degrees of Freedom Method   Containment

                       Class Level Information
             Class    Levels    Values
             DAM        10      1 2 3 4 5 6 7 8 9 10
             SIRE       10      1 2 3 4 5 6 7 8 9 10

                             Dimensions
                    Covariance Parameters          6
                    Columns in X                   1
                    Columns in Z Per Subject      90
                    Subjects                       1
                    Max Obs Per Subject          667
                    Observations Used            667
                    Observations Not Used          0
                    Total Observations           667

                          Iteration History
         Iteration    Evaluations    -2 Res Log Like    Criterion
              0            1          3196.35212861
              1            4          3140.13836334     0.00084122
              2            2          3139.72100070     0.00009940
              3            2          3139.63775941     0.00000307
              4            1          3139.63473346     0.00000001
                       Convergence criteria met.


                     Covariance Parameter Estimates
                                     Standard      Z
         Cov Parm    Subject    Estimate    Error     Value    Pr Z
         LIN(1)      Intercept    0.4223    0.2468     1.71    0.0435
         LIN(2)      Intercept    2.53E-20      .         .        .
         LIN(3)      Intercept    0.04321   0.1265     0.34    0.3664
         LIN(4)      Intercept         0       .         .        .
         LIN(5)      Intercept    0.3249    0.1946     1.67    0.0475
         Residual                 6.0105    0.3534    17.01   <.0001

                  Asymptotic Covariance Matrix of Estimates
   Row   Cov Parm    CovP1      CovP2      CovP3     CovP4     CovP5      CovP6
    1    LIN(1)      0.06090              -0.00705            -0.00131    0.000952
    2    LIN(2)
    3    LIN(3)     -0.00705               0.01601            -0.00420   -0.00069
    4    LIN(4)
    5    LIN(5)     -0.00131              -0.00420             0.03786   -0.01684
    6    Residual    0.000952             -0.00069            -0.01684    0.1249
```

(continued on next page)

Output 2.7 *(continued)*

```
                        Fit Statistics
            -2 Res Log Likelihood              3139.6
            AIC (smaller is better)            3147.6
            AICC (smaller is better)           3147.7
            BIC (smaller is better)            3157.6

                PARMS Model Likelihood Ratio Test
                  DF    Chi-Square      Pr > ChiSq
                   4       56.72          <.0001

                    Solution for Fixed Effects
                          Standard
    Effect       Estimate    Error      DF    t Value   Pr > |t|
    Intercept    11.4653    0.4316      89     26.57     <.0001
```

The results are given in Output 2.7. Under "Covariance Parameter Estimates," the rows labeled LIN(1) through LIN(5) give estimates of σ_N^2, σ_T^2, σ_M^2, σ_P^2, and σ_K^2, respectively. σ_T^2 and σ_P^2 are both estimated as zero. We can perform likelihood ratio tests of the other three parameters by constraining each one to zero using the PARMS statement. For example, to constrain σ_N^2 to zero, we rerun PROC MIXED, replacing the PARMS statement above with

```
parms 0,0,0.04,0,0.3,6/ lowerb = 0,0,0,0,0,0 hold = 1,2,4;
```

(Note that in addition to constraining σ_N^2, we also constrain the parameters previously estimated as zero). The results of the tests are that σ_N^2 and σ_K^2 are significantly different from zero at $P < 0.0001$ and $P = 0.02$, respectively, while σ_M^2 is not significantly different from zero ($P = 0.33$).

The correct interpretation of the observational components σ_N^2, etc. in terms of causal components depends on the nature of the parents for the diallel. If the parents are hermaphroditic individuals randomly sampled from a random-mating population, and epistasis is ignored, then σ_N^2 and σ_T^2 can be interpreted as ¼V_A and ¼V_D, respectively, and the residual variance (6.01 in this example) estimates ½V_A + ¾V_D + V_E. In contrast, if the parents are fully inbred lines derived from a random-mating base population, then σ_N^2 and σ_T^2 can be interpreted as ½V_A and V_D in the base population, again ignoring epistasis, and assuming that the inbred lines contain a random sample of the variation in the base population.

2.3.2 Extension to the NC2 Design

The diallel program above can also be used to analyze data from an NC2 design like that diagrammed in Table 2.1. No modifications to the program need to be made. Because the II data set tells SAS to expect 90 families, the "missing" families need to be represented in the data set by at least one observation, with a missing value for the dependent variable ("SQRNUMB" in the example above). If this is not done, then SAS will treat family 5, the first family in the NC2 data set, as if it were family 1, and so on; as a result, the relationships assigned to pairs of families by the II data set will be incorrect. Of course, it would also be possible to create an II data set specifically for the case of the NC2 design.

2.4 Pedigreed Populations

All of the examples in this chapter involve manipulated populations, where crosses were performed by the investigator. In many situations, an investigator may want to estimate genetic variance components from a natural or agricultural population where he or she has had no control over which matings occurred. If phenotypic data are available and relationships among individuals are known, PROC MIXED can be used, in conjunction with PROC INBREED, to estimate genetic variances. Just a brief outline of the procedure is given here. Relationship data first need to be entered in a form suitable for PROC INBREED (see SAS Help and Documentation for details). Running PROC INBREED with the MATRIX and COVARIANCE options generates a matrix containing the coefficients of the additive genetic variance (analogous to the a_{ij} in the diallel example) for each pair of individuals. This matrix can be saved using the OUTCOV= option, and later modified for use as an auxiliary file analogous to the II file in the diallel example. This file should be specified by the LDATA= option in a REPEATED statement with the TYPE = LIN(q) covariance structure. As in the diallel example, lines can be added to the auxiliary file to account for common environment, maternal, and/or paternal effects. More details from a Bayesian approach are given in Chapter 7, "Empirical Bayes Approaches to Mixed Model Inference in Quantitative Genetics."

2.5 References

Becker, W. A. 1992. *Manual of Quantitative Genetics*. Pullman, WA: Academic Enterprises.

Cockerham, C. C. 1963. Estimation of genetic variances. In *Statistical Genetics and Plant Breeding*, ed. W. D. Hanson and H. F. Robertson, 53-94. Washington: National Academy of Sciences and National Research Council.

Cockerham, C. C., and B. S. Weir. 1977. Quadratic analyses of reciprocal crosses. *Biometrics* 33:187-203.

Falconer, D. S., and T. F. C. Mackay. 1996. *Introduction to Quantitative Genetics*. Essex, UK: Longman.

Fry, J. D. 1992. The mixed-model analysis of variance applied to quantitative genetics: Biological meaning of the parameters. *Evolution* 46:540-550.

Fry, J. D., and S. L. Heinsohn. 2002. Environment dependence of mutational parameters for viability in *Drosophila melanogaster*. *Genetics* 161:1155-1167.

Good-Avila, S. V., and A. G. Stephenson. 2003. Parental effects in a partially self-incompatible herb *Campanula rapunculoides* L. (Campanulaceae): Influence of variation in the strength of self-incompatibility on seed set and progeny performance. *Amer. Naturalist* 161:615-630.

Littell, R. C., G. A. Milliken, W. W. Stroup, and R. D. Wolfinger. 1996. *SAS System for Mixed Models*. Cary, NC: SAS Institute Inc.

Lynch, M., and B. Walsh. 1998. *Genetics and Analysis of Quantitative Traits*. Sunderland, MA: Sinauer Associates.

Messina, F. J., and J. D. Fry. 2003. Environment-dependent reversal of a life history trade-off in the seed beetle *Callosobruchus maculatus*. *J. Evol. Biol.* 16:501-509.

Shaw, R. G. 1987. Maximum-likelihood approaches applied to quantitative genetics of natural populations. *Evolution* 41:812-826.

Sokal, R. R., and F. J. Rohlf. 1981. *Biometry*. New York: W. H. Freeman.

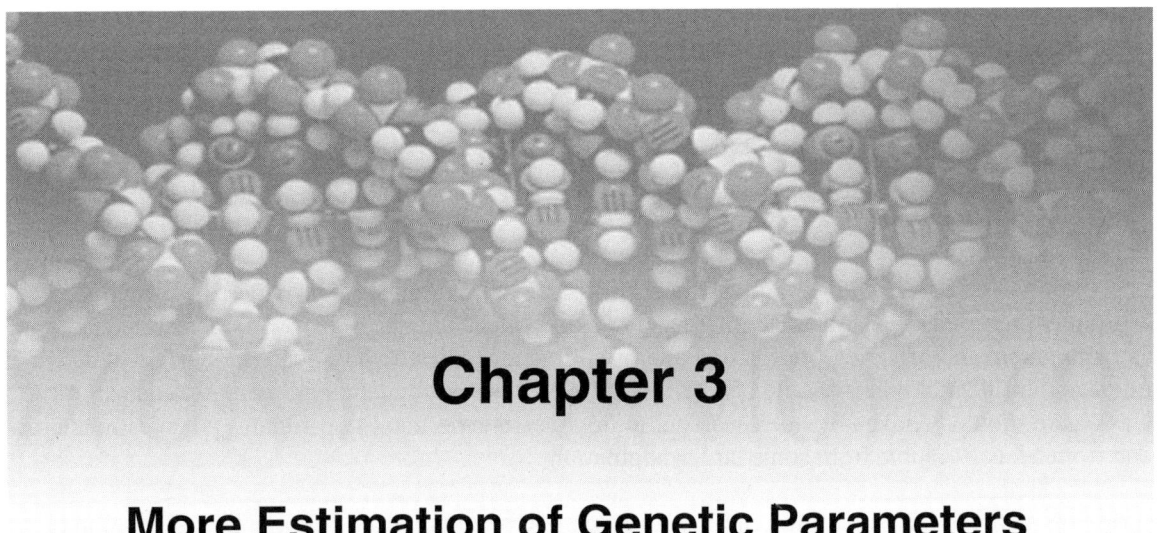

Chapter 3

More Estimation of Genetic Parameters

Kenneth J. Stalder and Arnold M. Saxton

3.1 Introduction 35
3.2 Genetic Parameters Estimated with Regression 35
3.3 Genetic Gain and Realized Heritability 41
3.4 Inbreeding and Relationship 44
3.5 Heterosis, or Hybrid Vigor 49
3.6 References 54

3.1 Introduction

This chapter addresses genetic parameters and estimation methods not covered in other chapters. Topics include regression designs for estimating heritability and genetic correlations, inbreeding, heterosis or crossbreeding, and realized heritability estimated from selection experiments. These techniques are from classic quantitative genetics, providing information on the genetics of populations using only pedigree and phenotypic information. The primary objective is to demonstrate how SAS software can be used to obtain this information, with minimal genetic background provided. More details are available in excellent texts such as Falconer and Mackay (1996) and Lynch and Walsh (1998).

3.2 Genetic Parameters Estimated with Regression

Regression was one of the first methods used to estimate heritability and genetic correlations, producing a direct measure of resemblance of relatives. Francis Galton used this technique when he collected statistical information from parents and their offspring (Hartl and Clark, 1989). The true relationship between any two variables may or may not be of a linear nature. Regardless of the "true" association, linear regression can serve as a method of approximation (Lynch and Walsh, 1998). The general form of a linear regression equation is as follows:

$$y = \alpha + \beta x + e,$$

where y is the dependent or response variable, α is the y intercept value, β is the slope of the line or the regression coefficient, x is the independent or explanatory variable, and e is the residual error.

The goal of linear regression is to find estimates of intercept and slope that provide a best fit to the data. In using linear regression to estimate heritability, the independent and dependent variables are phenotypic data from parent and offspring, mid-parent and offspring, etc. As with any linear model, existence of outliers and influential points should be checked, as these can bias genetic parameter estimates. As always, care should be taken by the researcher to not overstate heritability estimates when estimates are made from a small sample from some larger population.

Regression methods for estimating genetic parameters are often preferred because the association of phenotypic records and genetic relationship among offspring and parents is easily attainable from field data. Additionally, this method of estimating genetic parameters is unbiased by parental selection. Lastly, least squares techniques used to estimate regressions are not as computationally demanding as other estimation procedures. Lynch and Walsh (1998) outline other useful properties of least squares regression analysis.

Biologically, the degree of resemblance of relatives depends on a variety of factors: the rearing environment of individuals, genetic relationships, etc. There are a variety of relationships among members of an extended family. It is reasonable to assume that closer relationship might lead to more phenotypic resemblance among relatives compared to more distant relationships. If there is no strong genetic relationship or no resemblance among relatives, then phenotype of one relative will not help predict the other.

There are a variety of regression designs that could be used to estimate genetic parameters. These include one offspring on dam, one offspring on sire, one offspring on mid-parent, mean offspring on dam, mean offspring on sire, individual offspring records on copied dam records, etc. This section will cover the more common methods of using regression to estimate genetic parameters.

The formula (Falconer and Mackay, 1996) for calculating regression of offspring on parent is

$$b_{OP} = \frac{COV_{OP}}{\sigma_P^2},$$

where COV_{OP} is the covariance of offspring on parents and σ^2_P is the parental variance.

For genetic interpretation of this statistical quantity, theory states (Hartl and Clark, 1989) that the offspring–parent slope is

$$b_{OP} = 1/2 \frac{V_A}{V_P} = \frac{1}{2} h^2,$$

where b_{OP} is the regression of offspring on parent, ½ is used because the regression involves only a single parent, and V_A/V_P or h^2 is heritability.

An example of regression of offspring on a single parent is milk traits from daughters and dams in a dairy herd, since no records exist for the male parent. To conduct this type of study, phenotypic information needs to be collected from the parent and from the offspring. Example data for somatic cell count (SCC) from the University of Tennessee Dairy Experiment Station are read into SAS with this program:

```
data one;
    input CowName$ DamID CurrSCC CurrMilk LactNum DamSCC DamMilk DamLact;
datalines;
   TRST11    3871716   100   47.1   3    650    .      5
   ZUKR02    3878083   152   54.4   2    162    38.1   5
   GENE01    3924135    62   52.6   3     54    34.4   5
   ANCH01    3933356    38   43.5   1    162    34.4   5
   LXUS01    3933356    41   49.0   2    162    34.4   5
   BUCK01    3953108   141   32.6   2    200    54.4   5
   VIEW10    3953973   162    .     2     29    54.4   5
   DAN15     3973832    87   43.5   2     29    58.0   4
   MONT09    3973868    38   38.1   1    100    47.1   3
   HRDL03    3986622   650   36.3   2    214    49.0   4
   DCLO04    4024311   162   61.7   1   1715    23.5   3
   MONT05    4024314   348   50.8   1     31    56.2   3
   FLAG01  110128090    13   54.4   1     13    68.9   3
   JOUR02  110128317   246   38.1   1     62    .      3
   AVRY12  110128438    81    .     1    746    49.0   3
   BRTA77  110128456   152    .     1   1131    41.7   3
   DCLO01  110409807    22   38.1   1     13    47.1   3
;
proc mixed;
    model currscc = damscc /solution outp=rrr influence;
    estimate 'Heritability' damscc 2;
run;
proc univariate plot normal data=rrr;
 var resid;
run;
```

Calculation of the slope of the regression line is easily obtained with any of the linear model procedures in SAS. Here PROC MIXED is used, with the advantage of being able to estimate twice the slope, something PROC REG will not do. Also shown in the MODEL statement is creation of a data set named RRR that contains the residuals, and a request for influential diagnostics. Residuals are processed by PROC UNIVARIATE to check normality and identify outliers.

Thus, heritability can be estimated as twice the regression slope, and the standard error of the estimate is automatically provided by the software. Output 3.1 provides a heritability estimate of 1% with a standard error of 17%. Low heritability is expected for this disease-related trait, and the large standard error reflects the small experiment. Care must be taken in interpretation, as the estimates should be made from a reasonable number of parents. If too few sires or dams are used, the genetic parameter estimates obtained are likely to be biased.

Not shown in Output 3.1 are the influential diagnostics, which did not find any potential problems, and the PROC UNIVARIATE results, which did flag observation 10, with offspring SCC of 650, as a potential outlier. This point is easily seen in Output 3.2. A decision should be made to delete this observation if scientifically justified, because it increases variability and contributes to the standard error of 17%.

Output 3.1 SAS output for heritability estimation using regression of offspring on parent.

```
                    The Mixed Procedure

               Covariance Parameter Estimates
                  Cov Parm       Estimate
                  Residual         25975

                      Fit Statistics
               -2 Res Log Likelihood        213.0
               AIC  (smaller is better)     215.0
               AICC (smaller is better)     215.3
               BIC  (smaller is better)     215.7

                   Solution for Fixed Effects
                            Standard
   Effect       Estimate      Error       DF    t Value    Pr > |t|
   Intercept     145.21      47.6585      15      3.05      0.0082
   DamSCC       0.004835     0.08469      15      0.06      0.9552

                   Type 3 Tests of Fixed Effects
                        Num      Den
           Effect        DF       DF     F Value    Pr > F
           DamSCC         1       15       0.00     0.9552

                           Estimates
                            Standard
   Label         Estimate     Error      DF    t Value    Pr > |t|
   Heritability  0.009670    0.1694      15      0.06      0.9552
```

Output 3.2 Regression of somatic cell count in dairy cows on corresponding values for their dam.

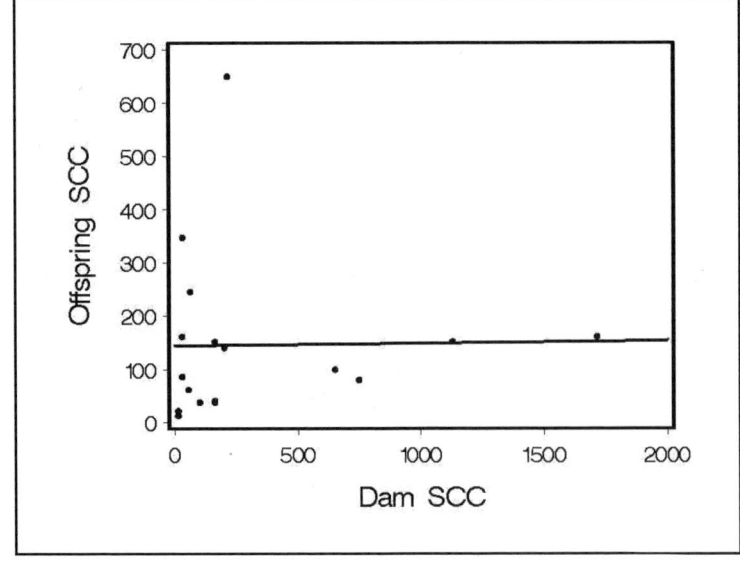

Typical SAS code and options for plotting the offspring on parent regression are shown here; these produce Output 3.2.

```
proc gplot;
  goptions ftext=swissb hpos=30 vpos=30;
  axis1 minor=none w=3 label=(a=90 'Offspring SCC');
  axis2 minor=none w=3 label=( 'Dam SCC');
  symbol1 v=dot i=rl w=3;
  plot currscc*damscc/haxis=axis2 vaxis=axis1;
run; quit;
```

Regression of offspring on mid-parent average is another method that can be used to estimate genetic parameters. This is commonly used when data are available from several offspring from a set of parents, as is the case with litter-bearing species like swine. This method assumes that one of the parents is mated to a single parent of the opposite sex. In most cases this will mean that one female is mated to a single male, but each male may be mated to several females. Assumptions are equal means and variance among the population of males and females, and autosomal inheritance of the trait (so equal resemblance between offspring and sire and dam occurs). If male and female parental means and variances are not equal, then it will be better to calculate a regression of each sex of offspring performance on each individual parent rather than on the midpoint of both parents. Methods that do handle sexes with unequal means or variances are outlined very well in Falconer and Mackay (1996).

Falconer and Mackay (1996) demonstrate that the covariance of offspring and both parents (mid-parent) can be calculated in the following manner:

$$b_{O,MP} = \frac{\text{cov}(O,[P1+P2]/2)}{V_{MP}} = \frac{\text{cov}(O,P1)/2 + \text{cov}(O,P2)/2}{[\sigma^2_{P1} + \sigma^2_{P2}]/4}$$

$$= \frac{V_A/4 + V_A/4}{V_P/2} = h^2$$

These calculations use basic statistical facts for variances and covariances and assume both parents have the same phenotypic variance. Thus the regression of offspring on mid-parent value is two times the regression of offspring on a single parent, and is a direct measure of heritability. Standard error of the heritability estimate will be the standard error of the regression coefficient.

When looking at the calculation of heritability by regression of offspring on single parent or offspring on mid-parent average, biological meaning can be slightly different. If the means and variances are equal between males and females, regression of offspring on mid-parent is likely the best estimate of "effective" heritability. It is best because it factors in both parents when estimating heritability while the heritability estimate obtained from regression of offspring on a single parent is made from either parent, but not both. There are cases such as sex-limited traits where an estimate can be obtained only from one parent and one sex of offspring.

It should be noted that when estimating heritability for a given trait from a data set, several methods can be used. These methods include the regression methods described in this chapter or other methods outlined in this book. If more than one method is used to estimate heritability from the same data set, it is likely that the estimates will differ. Falconer and Mackay (1996) estimated the heritability of abdominal bristle number in *Drosophila melanogaster* by three different methods and arrived at heritability estimates differing by .05. However, all were within range of the standard errors of the estimates. Additionally, heritability estimates are not static and as more records are added to a data set, heritability estimates can change. Similarly, heritability estimates can and often do differ depending on the population being evaluated.

Offspring–parent relationships can be used to estimate genetic correlations. To do this using the regression formula previously outlined in this section, one phenotypic character of interest must be measured in the offspring and the other phenotypic trait of interest must be measured in the parents. For example, protein content might be estimated in corn from the offspring while starch content is measured in the parental lines. The opposite could also be done (starch content measured in the offspring while protein content is measured in the parental line). If both measures are available the arithmetic mean should be used (Falconer and Mackay, 1996). The covariances for offspring and parents are needed for both traits, in this case protein and starch content. The genetic correlation then can be given as

$$r_A = \frac{COV_{XY}}{\sqrt{(COV_{XX} COV_{YY})}}.$$

Users should be aware that calculation of genetic correlations often has some undesirable characteristics. The genetic correlations frequently have large sampling errors. Because of their large sampling errors, the precision of genetic correlations is often less than desired. Additionally, genetic correlations are often population dependent because of differing gene frequencies in various populations (Falconer and Mackay, 1996) and should not be compared across different populations.

Using the somatic cell count example above, all possible regressions of offspring traits on parent traits are done with the following code, producing Output 3.3:

```
proc glm;
  model currscc currmilk=dammilk;
  estimate '2*DamMilk slope' dammilk 2;
run; quit;
proc glm;
  model currscc currmilk=damscc;
  estimate '2*DamSCC slope' damscc 2;
run; quit;
```

Slopes are multiplied by two to estimate genetic correlation for single offspring regressed on single parent, with different multipliers needed for other types of data (Falconer and Mackay, 1996). Estimates of additive genetic correlation are 2.57 and .018, clearly unstable for this small amount of data. Standard errors are produced automatically. The GLM procedure is used here, as it allows multiple dependent variables to be analyzed, whereas PROC MIXED does not. For regression models with no random effects as in these examples, PROC GLM and PROC MIXED will produce identical results.

Output 3.3 All possible regressions of two offspring traits on parent traits.

```
                        The GLM Procedure
                   Number of observations    17
                    Dependent Variables With
                 Equivalent Missing Value Patterns
                                        Dependent
             Pattern          Obs        Variables
                1              15        CurrSCC
                2              12        CurrMilk
NOTE: Variables in each group are consistent with respect to the presence or
      absence of missing values.

Dependent Variable: CurrSCC
                                       Standard
    Parameter              Estimate       Error      t Value     Pr > |t|
    2*DamMilk slope      2.57122313   7.73543873       0.33        0.7449

Dependent Variable: CurrMilk
                                       Standard
    Parameter              Estimate       Error      t Value     Pr > |t|
    2*DamMilk slope     -0.49150166   0.40922263      -1.20        0.2574

Dependent Variable: CurrSCC
                                       Standard
    Parameter              Estimate       Error      t Value     Pr > |t|
    2*DamSCC slope       0.00966995   0.16937735       0.06        0.9552

Dependent Variable: CurrMilk
                                       Standard
    Parameter              Estimate       Error      t Value     Pr > |t|
    2*DamSCC slope       0.01853753   0.00947190       1.96        0.0740
```

3.3 Genetic Gain and Realized Heritability

Prediction of response to selection is described in Chapter 4. In this section, observed response to selection is used to estimate genetic gain and realized heritability. This type of information is useful to assess the effectiveness of genetic selection. If progress is too slow, changes in the selection program must be considered.

Selection experiments for a variety of traits have been and continue to be conducted. These generally have an unselected control line, used to monitor environmental changes. Selected lines may be selected in one direction or selected divergently. Falconer and Mackay (1996) provide an introduction to advantages of various designs. Provided that the selected and unselected individuals were derived from the same original population and that performance for a given trait diverges over time, realized heritability can be calculated. Realized heritability is the ratio of change in population mean per unit selection differential and can be calculated by (Van Vleck et al., 1987)

$$h^2_{realized} = \frac{\overline{P}_{PSP} - \overline{P}_{PRP}}{\overline{P}_S - \overline{P}},$$

where $\overline{P}_{PSP}$ is the performance of the progeny from selected parents, $\overline{P}_{PRP}$ is the performance of the progeny from random parents (if progeny from random parents does not exist, the population mean, P, can be used), and $\overline{P}_S - \overline{P}$ is the selection differential, or selected parent average minus parental population average.

A small example was used by Muir (1986) to illustrate statistical issues, in which *Tribolium* was selected for low body weight. This code shows the data and SAS analysis, and produces Output 3.4:

```
data one;
  input generation bw1 bw2 control;
  bw=bw1; rep=1; diff=bw-control; output;
  bw=bw2; rep=2; diff=bw-control; output;
datalines;
1  216.9  212.1  207.1
2  215.9  212.0  214.4
3  198.0  201.7  215.9
4  193.4  167.8  223.1
5  177.1  161.0  224.3
6  190.2  177.5  213.0
7  171.4  168.6  215.2
8  150.5  131.8  230.4
9  136.7  126.0  233.2
;

proc reg;      ❶
  model bw diff=generation;
  model bw = generation control;
run;

data one; set one;
  classgen=generation;
run;
proc mixed;    ❷
  class classgen rep ;
  model bw = control generation classgen /htype=1;
  random rep rep*generation;
run;

proc mixed;    ❸
  class classgen rep ;
  model bw = control generation  /htype=1 solution;
  random rep rep*generation;
run;
```

❶ A simple linear regression of response, or deviation from the control line, over generation number will estimate selection response per generation if the selection differential is constant. However, to get realized heritability, the slope must be divided by the selection differential value. Alternatively, the regression can be done using cumulative selection differential as the X variable. Results from these analyses in Output 3.4 show a selection response of –10 mg per generation if the control information is not used. Selection response is –12 mg when deviated from the control, or –6.8 mg when the control is used as a covariate, as suggested by Muir (1986). If a constant selection differential of 4 mg is assumed (i.e., the parent's body weight is 4 mg lower than the population average), then realized heritability would be 6.8/4 = 1.7, with standard error similarly calculated from the output. The 170% heritability simply reflects that for each mg that the parent body weight is lower than the average, the progeny body weight is 1.7 mg lower, something that genetically is theoretically impossible.

❷ The experiment has two replicate selected lines, and Muir (1986) suggests a more appropriate framework for testing if the selection response is different from zero. In particular, REP variation must be controlled, and a "pure error" term based on reps should be used for testing. SAS code for implementing this in PROC MIXED is given, with REPs declared as random. REP*GENERATION creates the correct error term for testing the linear regression over generations. CLASSGEN is used to create dummy variables that address all other differences among generation means, ensuring this variation does not affect statistical tests (it did affect the regression testing above). Results suggest weak evidence for a non-zero selection response (P=.11). The slope of –8.87 has been affected by the presence of CLASSGEN in the model, and this has also made the standard error unusable.

❸ By dropping CLASSGEN from the model, the selection response and standard error now match the correct regression results. Note that CLASSGEN did not greatly affect the test (P=.11), but this in general may not be true.

Output 3.4 Modeling results for genetic gain in *Tribolium* body weight.

```
                        The REG Procedure
                      Dependent Variable: bw
                       Parameter Estimates
                    Parameter       Standard
Variable      DF    Estimate         Error      t Value    Pr > |t|
generation     1    -10.12417        1.05611     -9.59      <.0001

                        The REG Procedure
                     Dependent Variable: diff
                       Parameter Estimates
                    Parameter       Standard
Variable      DF    Estimate         Error      t Value    Pr > |t|
generation     1    -12.47250        1.46788     -8.50      <.0001

                        The REG Procedure
                      Dependent Variable: bw
                       Parameter Estimates
                    Parameter       Standard
Variable      DF    Estimate         Error      t Value    Pr > |t|
generation     1    -6.87684         1.18723     -5.79      <.0001
control        1    -1.38282         0.37674     -3.67      0.0023

                        The MIXED Procedure
                     Solution for Fixed Effects
                                      Standard
Effect       classgen    Estimate      Error      DF   t Value   Pr > |t|
generation               -8.8719       916359      1    -0.00    1.0000

                    Type 1 Tests of Fixed Effects
                           Num      Den
                 Effect     DF       DF     F Value    Pr > F
                 control     1        6     130.36     <.0001
                 generation  1        1      33.61     0.1087
                 classgen    6        6       1.23     0.4023

                        The MIXED Procedure
                     Solution for Fixed Effects
                                Standard
Effect         Estimate          Error       DF    t Value   Pr > |t|
generation     -6.8768           1.1984       1     -5.74     0.1098

                    Type 1 Tests of Fixed Effects
                           Num      Den
                 Effect     DF       DF     F Value    Pr > F
                 control     1       13     127.92     <.0001
                 generation  1        1      32.93     0.1098
```

3.4 Inbreeding and Relationship

Inbreeding is the mating of individuals that are related by having common ancestry. Inbreeding coefficients are represented by the symbol F as defined by Sewall Wright (1922). The inbreeding coefficient represents the probability of alleles being identical by descent. F represents "fixation" of an allele, where one allele for a gene has a frequency of 100%, or complete loss of genetic variation. For an individual to be inbred, its parents must be related. Without using molecular techniques it is not possible to actually measure homozygosity changes; one can only estimate probabilities. It is important to note that individuals may have the same inbreeding coefficient, but may not be homozygous at the same loci.

Inbreeding generally has adverse effects on lowly heritable traits or those associated with fitness, known as inbreeding depression. Such traits include reproductive and survivability traits. Inbreeding effects are observed in both plants and animals, though animals generally show more inbreeding depression. The general formula for calculating inbreeding is

$$F_X = \sum \left[(1/2)^n (1 + F_A) \right],$$

where F_X is the inbreeding coefficient of individual X, Σ means that summation occurs across all common ancestors, n is the number of individuals in the path connecting the sire and dam of X through the common ancestor A (including sire and dam), and F_A is the inbreeding coefficient of the common ancestor.

The inbreeding coefficient is measured relative to a particular breed or generation at a specified time. It is common to trace a pedigree back six generations or more. Hence, F represents the increase in homozygosity as a result of mating related individuals since the reference date six generations ago. However, if a pedigree can be traced back only three generations, then F represents the increase in homozygosity as a result of mating related individuals since the reference date three generations ago. It is important to note that F represents only the relationship of an individual back to some point where the parentage is known.

There are several forms of inbreeding that will result in variation in the accumulation of inbreeding in a population. Selfing, or cloning, will result in the most intense form of inbreeding. Selfing is common among some plant species; however, it is not naturally possible with animals. Full-sib or parent–offspring types of matings are the most intense form of inbreeding possible with animals. Half-sib, grandparent–grandoffspring, uncle–niece, and aunt–nephew types are equal in inbreeding and would result in less inbred individuals. Lastly, cousins could be mated together and result in even lower inbreeding coefficients than those previously described.

Relationships between any pair of individuals within a pedigree can be computed easily. Inbreeding of an individual is equal to the relationship of its parents. When inbreeding calculations are made, relationships must be available. The general form of the relationship equation, which is very similar to that seen in the calculation of inbreeding, is as follows:

$$R_{XY} = \frac{\sum \left[(1/2^n)(1 + F_A) \right]}{\sqrt{(1 + F_X)(1 + F_Y)}},$$

where R_{XY} is the coancestry coefficient between individual X and individual Y, Σ means that summation occurs across all common ancestors, n is the number of individuals in the path connecting X and Y (inclusive), F_A is the inbreeding coefficient of the path's common ancestor, F_X is the inbreeding coefficient of individual X, and F_Y is the inbreeding coefficient of individual Y.

If Wright's relationship coefficient is needed, it is simply twice the coancestry R given here.

SAS PROC INBREED allows users to calculate the inbreeding and relationship coefficients from a defined pedigree. Inbreeding coefficients can be calculated for very large pedigree files. PROC INBREED can conduct an inbreeding analysis assuming that individuals belong either to the same generation or to non-overlapping generations. This example shows input of a simple pedigree, with codes identifying the individual and both parents, if known.

```
options ls=78;
data one;
  input  indiv mom dad;
datalines;
 5 1 .
 6 1 .
 8 5 6
 9 8 .
10 8 .
11 9 10
;
proc inbreed matrix ;
run;
```

A period is used to indicate missing data, as with any SAS data. Here individual 8 is from a mating of half sibs, which then produces half sibs that are mated to give individual 11. Note that individuals should be ordered from oldest to youngest, so that an individual has defined parents before it is used as a parent. Otherwise, undefined parents are automatically assigned as unknown and unrelated. PROC INBREED assumes the first three unused variables in the data set are codes for individual and parents, unless specified otherwise with the VAR statement. Results are shown in Output 3.5; note the inbreeding coefficients on the diagonal of the matrix. Individual 8 has "1" as a common ancestor, with three individuals in the chain from mother to father of "8," giving $F=1/8$. Individual 11 also has three individuals in the chain running through the common ancestor "8," but since "8" is inbred, the inbreeding of "11" is $F=(1/8)*(1+1/8) = 9/64$.

Off-diagonal elements in Output 3.5 are relationships between individuals represented by the row and column labels. The relationship values of 0.25 for full sibs and 0.125 for half sibs can be recognized. The relationship of individuals 9 and 10 is 0.140625, illustrating the fact that the inbreeding of an individual, in this case "11," equals the coancestry relationship of its parents.

Output 3.5 Genetic relationships from a sequence of half-sib matings.

```
                      The INBREED Procedure

                       Inbreeding Coefficients
indiv  mom  dad        1          5          6          8          9         10
1                                 0.2500     0.2500     0.2500     0.1250     0.1250
5      1               0.2500     .          0.1250     0.3125     0.1563     0.1563
6      1               0.2500     0.1250     .          0.3125     0.1563     0.1563
8      5    6          0.2500     0.3125     0.3125     0.1250     0.2813     0.2813
9      8               0.1250     0.1563     0.1563     0.2813     .          0.1406
10     8               0.1250     0.1563     0.1563     0.2813     0.1406     .
11     9    10         0.1250     0.1563     0.1563     0.2813     0.3203     0.3203

                       Inbreeding Coefficients
               indiv  mom  dad        11
               1                      0.1250
               5      1               0.1563
               6      1               0.1563
               8      5    6          0.2813
               9      8               0.3203
               10     8               0.3203
               11     9    10         0.1406

               Number of Individuals     7
```

Figure 3.1 A complex pedigree.

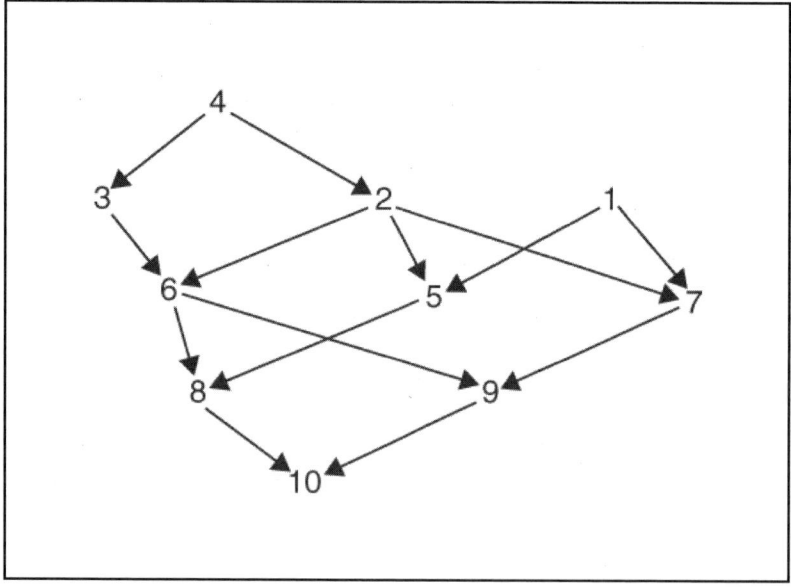

As a second example, a slightly more complex pedigree, shown in Figure 3.1, is analyzed by the following SAS program.

```
data one;
 input   gen sex$ indiv mom dad cov;
datalines;
 0 F 3  4  .    .05
 0 M 2  4  .    .05
 0 F 1  .  .    .05
 1 F 5  1  2    .
 1 M 6  3  2    .
 1 F 7  1  2    .
 2 F 8  5  6    .
 2 M 9  7  6    .
 3 F 10 8  9    .
;
proc inbreed ;
 var indiv dad mom cov;
 matings 3/4 , 8/9 , 2/3;
run;
proc inbreed average init=.05 matrix;
 class gen;
 gender sex;
 var indiv dad mom;
run;
```

This program includes information on generation (GEN), gender (SEX), and initial relationships (COV) existing in the base generation. The first PROC INBREED uses a VAR statement to identify the variables containing pedigree information. VAR is required here because pedigree codes are not the first variables in the data set. Additionally, the COV variable is listed, containing covariances (equals two times coancestry relationship) between parents for that individual. This is particularly useful when parents are unknown. For larger pedigrees, calculation of all relationships may not be of interest, so the MATRIX option is not used. Instead, the MATING statement is used to choose specific pairs for relationship calculation. Results from the first PROC INBREED are in Output 3.6. The relationship of parent–offspring for "3" and "4" is increased from the usual 0.25 due to the COV variable specifying that the parents of "3" are related. For the relationship of "8" and "9," the reader is challenged to identify the seven paths through the three common ancestors.

Output 3.6 Selected relationships for the Figure 3.1 pedigree.

```
                       The INBREED Procedure

               Inbreeding Coefficients of Matings
                      dad      mom      Coefficient
                       3        4          0.2687
                       8        9          0.2991
                       2        3          0.1468

                  Number of Individuals    10
```

The second PROC INBREED in the program above illustrates using generation and gender codes to obtain summaries. Results for the last two generations are shown in Output 3.7. If parents in one generation cannot be used in subsequent generations, relationships among all individuals in the pedigree are not useful. Instead, PROC INBREED reports relationship matrices within generation. Summaries of relationships and inbreeding are printed for the gender combinations by specifying the AVERAGE option.

Output 3.7 Inbreeding and relationship summaries for a complex pedigree.

```
                         The INBREED Procedure

                              gen = 2
                        Inbreeding Coefficients
          indiv     dad     mom             8                 9
            8        6       5           0.1772            0.2991
            9        6       7           0.2991            0.1772

       Averages of Inbreeding Coefficient Matrix in Generation 2
                                 Inbreeding          Coancestry
             Male X Male            0.1772             0.0000
             Male X Female             .               0.2991
             Female X Female        0.1772             0.0000
             Over Sex               0.1772             0.2991

                      Number of Males         1
                      Number of Females       1
                      Number of Individuals   2

                              gen = 3
                        Inbreeding Coefficients
          indiv     dad     mom            10
            10       9       8           0.2991

       Averages of Inbreeding Coefficient Matrix in Generation 3
                                 Inbreeding
             Male X Male            0.0000
             Male X Female             .
             Female X Female        0.2991
             Over Sex               0.2991

                      Number of Males         0
                      Number of Females       1
                      Number of Individuals   1
```

Other uses for information provided by PROC INBREED may be of interest. The inbreeding coefficients can be used in other SAS procedures to account for differences in degree of inbreeding among individuals. This is particularly important when analyzing data that might be influenced by the degree of inbreeding an individual has, such as traits with low heritability. The inbreeding coefficient or the deviation from population mean inbreeding might be used as a covariate in the analysis using PROC GLM or PROC MIXED.

PROC INBREED can be used as a selection tool to examine the inbreeding that exists in a current breeding population. Additionally, PROC INBREED can be used to determine the inbreeding coefficient of any particular mating. The user could create dummy individuals, and PROC INBREED would determine inbreeding coefficients of these matings with the MATINGS statement as above. Matings that minimize the accumulation of inbreeding in the population can be selected for implementation.

Another important consideration may be to determine effective population size. The reason effective population size is discussed in this chapter is because of its relationship to the increase or buildup of inbreeding in a population. This shows the extreme effect on the increase of inbreeding that occurs in a population that will occur when working with populations that are small. With the use of artificial insemination, embryo transfer, and other reproductive technologies, the number of breeding animals needed becomes greatly reduced and affects population size. Additionally, when effective population size becomes small, there is a greater chance that genes may be lost because of random genetic drift.

Effective population size can be reduced substantially when related animals are extensively used to produce the next generation of individuals, effectively reducing the average inbreeding coefficient. However, it does not substantially affect the rate at which inbreeding accumulates (Falconer and Mackay, 1996). Effective population size, usually represented by N_e, denotes the number of individuals that would give rise to the calculated sampling variance or rate of inbreeding if the animals bred in the manner of the idealized population (Falconer and Mackay, 1996). In many mammalian breeding populations, males are allowed to mate with more than one female. This gives rise to the case where family size differences exist between males and females. When this is the case, the general form of calculating effective population size is denoted (Hill, 1979) by

$$N_e = \frac{8N}{V_{km} + V_{kf} + 4} \ ,$$

where N_e is the effective population size, N is the number of breeding individuals, V_{km} is the variance of male family size, and V_{kf} is the variance of female family size.

If the variance of the family size does not differ between males and females in the population, this equation reduces substantially. See Falconer and Mackay (1996) for deviations from this equation. The effect of effective population size on the accumulation of inbreeding in a given population is given by the approximation

$$\Delta F = \frac{1}{8N_m} + \frac{1}{8N_f} \ ,$$

where ΔF is the change in average inbreeding in a population, N_m is the number of breeding males, and N_f is the number of breeding females. In many laboratory and animal breeding experiments, it is desirable to minimize inbreeding. This can be done by appropriate choice of individuals to become parents of the next generation. This will reduce the variation in family size (V_k) in the formula to calculate effective population size. It should be noted that avoiding close matings in any one generation will reduce the accumulation of inbreeding in that generation, but it does not reduce the overall rate of inbreeding accumulation.

3.5 Heterosis, or Hybrid Vigor

The use of hybrid seed corn was popularized by former United States Secretary of Agriculture Henry Wallace. Wallace founded what is now Pioneer Hybrid International, one of the largest seed suppliers in the world. Today, many animal and plant breeders take advantage of the heterosis first described in the seed corn industry. In fact, most animals used for commercial production are the result of breed or line crosses designed to take advantage of heterosis.

Why is it important to maximize heterosis? Because it is a free source of improved performance and profits. Producers need only to develop a planned mating system in which breeds or lines of plants are chosen appropriately to capture the heterosis from crossing lines. It is important to remember that the expression of heterosis occurs only with continual crossing of pure lines or breeds. Heterosis can be maximized only when highly inbred lines or divergent breeds are crossed.

Offspring produced by crossing of inbred lines will increase the productivity of traits that were shown to suffer from inbreeding depression in the inbred lines. Heterosis, or hybrid vigor, can be described as the increased performance of crossbred offspring over the average performance of pure parents. This phenomenon is the result of increased heterozygosity of the offspring that results from the crossing of inbred strains. The frequency of unfavorable homozygous genotypes is reduced when crossing occurs that makes the animals or plants more vigorous and adaptable to a wider range of environments. This adaptability and increased vigor results in increased performance.

A strict definition of heterosis is the difference of offspring performance from average performance of parents. There are really three types of heterosis. The first type is individual heterosis, which is described as the performance advantage of a crossbred offspring over purebred parents. The second type is maternal heterosis, which is described as the advantage of a crossbred mother over a purebred mother. The last type is paternal heterosis, described as the advantage of a crossbred father over a purebred father. This type is not as important as maternal heterosis, particularly in commercial animal production. Individual heterosis can be calculated by

$$H = \frac{\frac{(A \times B)+(B \times A)}{2} - \frac{(A \times A)+(B \times B)}{2}}{\frac{(A \times A)+(B \times B)}{2}} * 100 ,$$

where $A \times B$ and $B \times A$ represent the performance for a given trait from an individual or the mean of a group of individuals produced from the reciprocal cross of pure lines A and B, and $A \times A$ and $B \times B$ represent the performance for a given trait from an individual or the mean of a group of individuals produced from the matings of pure lines.

Maternal heterosis is the advantage of having a crossbred dam compared to a purebred dam and is usually if not exclusively observed in animal species. Maternal heterosis is exhibited one generation after the cross is made to produce the crossbred female. This is the result of better performance of traits like milking ability, number of individuals born in litter-bearing species, etc. Table 3.1 shows how maternal heterosis can be captured as compared to matings where no maternal heterosis exists. Notice that the offspring in every case captures 100 percent of the individual heterosis. Paternal heterosis can be estimated in a similar manner. However, paternal heterosis is not as important in the commercial livestock industries as a whole as it once was, because of the widespread use of artificial insemination.

Table 3.1 Example matings illustrating occurrence of maternal heterosis.

Maternal Line	Sire Line	Offspring	Amount of Maternal Heterosis (%)
B×B	A	A×(B×B)	0
C×C	A	A×(C×C)	0
B×C	A	A×(B×C)	100
C×B	A	A×(C×B)	100

The analysis of this type of data can easily be done using PROC GLM or PROC MIXED and the ESTIMATE statement. As the formulas above suggest, heterosis calculations are simply a series of comparisons among means. As an example, an experiment studying five corn lines and their first generation crosses was conducted by Dr. Dennis West at the University of Tennessee. The study did not include reciprocal crosses, meaning if the cross Male 1 by Female 2 was made, then the cross Female 1 by Male 2 was not. Parts of the program are shown here, with only estimates involving the first three lines shown to save space:

```
data one;
  input plot entry rep    year loc$ par1 par2 lodge height earht standpcnt buyield
kgyield;
datalines;
108 1      1     1998  KnoxTN 1  2   15.6  1.98  1.12   94        121.2   7604
207 2      1     1998  KnoxTN 1  3   16.4  2.37  1.31   99        132.4   8303
106 3      1     1998  KnoxTN 1  4   6     2.22  1.31   99        141.5   8871
109 4      1     1998  KnoxTN 1  5   7.5   2.16  1.28   99        139.1   8722
...more datalines...
;
proc mixed data=one;
  class par1 par2 year loc rep;
  model kgyield = par1*par2;    ❶
  random loc year(loc) rep*year(loc);   ❷
  estimate 'pure line mean' intercept 5 par1*par2 1 0 0 0 0   1 0 0 0   1 0 0  1
0  1/divisor=5 ;    ❸
  estimate 'avg heterosis' par1*par2  -4 2 2 2 2   -4 2 2 2   -4 2 2  -4 2   -
4/divisor=20;
  ** avg of 4 heterosis values per line ;
  estimate 'heterosis 1' par1*par2  -4 2 2 2 2  -1 0 0 0   -1 0 0  -1 0    -
1/divisor=8;
  estimate 'heterosis 2' par1*par2  -1 2 0 0 0  -4 2 2 2   -1 0 0  -1 0    -
1/divisor=8;
  estimate 'heterosis 3' par1*par2  -1 0 2 0 0  -1 2 0 0   -4 2 2  -1 0    -
1/divisor=8;
  ** deviation of line heterosis from avg heterosis;
  estimate 'dev heterosis 1' par1*par2  -12 6 6 6 6    3 -4 -4 -4    3 -4 -4  3 -4
3/divisor=40;
  estimate 'dev heterosis 2' par1*par2   3 6 -4 -4 -4   -12 6 6 6    3 -4 -4  3 -4
3/divisor=40;
  estimate 'dev heterosis 3' par1*par2   3 -4 6 -4 -4    3 6 -4 -4   -12 6 6  3 -4
3/divisor=40;
***** cross heterosis;
  estimate 'hij 1-2'  par1*par2 -1 2 0 0 0   -1 /divisor=2;
  estimate 'hij 1-3'  par1*par2 -1 0 2 0 0    0 0 0 0  -1/divisor=2;
  estimate 'hij 2-3'  par1*par2  0 0 0 0 0   -1 2 0 0  -1/divisor=2;
***** specific heterosis;
  estimate 'Sij 1-2' par1*par2  0    12   -4    -4    -4   0   -4   -4   -4  4  0  0
4  0  4  /divisor=20;
  estimate 'Sij 1-3' par1*par2  0   -4    12    -4    -4   4   -4    0    0  -4 -4
4  0  4  /divisor=20;
  estimate 'Sij 2-3' par1*par2  4   -4    -4    0     0   12   -4   -4    0  -4 -4
4  0  4  /divisor=20;

  lsmeans par1*par2;
run;
```

❶ After creating the working SAS data set ONE, a mixed model analysis is used to analyze the data. In order to work with each of the cross means, the model has only the interaction term of the two parent lines. Main effects of parents could be included in the model, but these would make the ESTIMATE statements that follow much more complex.

❷ As dictated by the experimental design, any random effects must be addressed. Here YEAR, LOCATION, and REP blocking terms are used to remove those sources of variation. The presence of random effects makes PROC MIXED the best choice for statistical analysis.

❸ ESTIMATE statements are used to produce comparisons of interest. After a label in quotes, coefficients are assigned to each cross mean (PAR1*PAR2) as needed to produce the desired information. Note that when a mean is being estimated, a coefficient for the intercept is needed in addition to the cross means. The DIVISOR option requests all coefficients be divided by the given number, allowing awkward fractions to be easily entered. As with any ESTIMATE or CONTRAST, order of the coefficients is critical, as the correct coefficient must be matched with the corresponding cross. An easy way to verify cross order is to request least squares means, the order there being identical. Besides estimated values of heterosis, variability explained by comparisons may be of interest. These can be investigated using CONTRAST statements, using a similar set of coefficients.

Hallauer and Miranda (1988) is a good general reference for diallel experiments, but Gardner and Eberhart (1966) should be consulted for statistical details. For more information on variance component estimation, see Chapter 2, "Estimation of Genetic Variances and Covariances by Restricted Maximum Likelihood Using PROC MIXED." In general terms, cross heterosis is as defined above: the deviation of individual cross mean from parental line means. All cross heterosis values for a line can be averaged to give line heterosis, the benefit of using that line in crosses. Then line heterosis values can be averaged to give the overall average heterosis. If the two parental line and average heterosis values are subtracted from cross heterosis, what remains is specific heterosis, heterosis that is specific to this cross above that expected from the two lines involved.

Output 3.8 contains the results from the diallel program, with estimates of the various types of heterosis. Again to save space only the first three line values are reported. On average, crosses gave 1749 kg/ha more yield than parents, and crosses involving line 2 averaged 2021 kg/ha more yield. However, the cross of line 2 with line 1 gave a 2317 kg/ha increase, with only 671 kg/ha of that being specific to the line 1 and 2 cross.

Diallel experiments such as this example are usually designed to include either (1) crosses, (2) crosses and parents, or (3) crosses, parents, and reciprocals. Increased information is available as more relatives are included. Naturally ESTIMATE statements will differ across these designs, and even within a design different ESTIMATE coefficients are needed depending on the number of lines involved. A crude beginning of a PROC IML program is included in the example file that will automatically generate ESTIMATE coefficients for "crosses and parents" experiments. It would be a welcome contribution for a reader to develop an easy-to-use macro for all experimental situations.

This type of analysis can also be used for a variety of crossing systems in animals. Commonly used crossing systems include terminal crosses, where the offspring are destined for market and not retained for further breeding purposes. There are several types of rotational systems that are commonly used, particularly in the commercial livestock industry. These systems include two-breed, three-breed, and four-breed crosses. These rotational systems do not obtain maximum heterosis, but they do have the advantage that replacement females are raised within the system.

Output 3.8 Partial results from the diallel analysis of heterosis.

```
                        The Mixed Procedure
                     Class Level Information
          Class    Levels    Values
          par1       5       1 2 3 4 5
          par2       5       1 2 3 4 5
          year       2       1998 1999
          loc        5       ColumMO KnoxTN LexingKY
                             MilanTN QuickKY
          rep        3       1 2 3

              Covariance Parameter  Estimates
                 Cov Parm             Estimate
                 loc                       0
                 year(loc)            922161
                 year*rep(loc)        207492
                 Residual            1109238

              Type 3 Tests of Fixed Effects
                        Num      Den
          Effect         DF       DF    F Value    Pr > F
          par1*par2      14      238      16.12   <.0001

                          Estimates
                          Standard
Label              Estimate    Error     DF    t Value    Pr > |t|
pure line mean      4695.53   421.36    238     11.14     <.0001
avg heterosis       1749.13   135.97    238     12.86     <.0001
heterosis 1         1715.08   186.18    238      9.21     <.0001
heterosis 2         2020.97   186.18    238     10.85     <.0001
heterosis 3         1508.86   186.18    238      8.10     <.0001
dev heterosis 1      -34.0514 127.19    238     -0.27     0.7891
dev heterosis 2      271.84   127.19    238      2.14     0.0336
dev heterosis 3     -240.27   127.19    238     -1.89     0.0601
hij 1-2             2317.53   304.03    238      7.62     <.0001
hij 1-3             1270.47   304.03    238      4.18     <.0001
hij 2-3             1408.78   304.03    238      4.63     <.0001
Sij 1-2             -671.31   210.64    238     -3.19     0.0016
Sij 1-3            -1308.68   210.64    238     -6.21     <.0001
Sij 2-3            -1415.09   210.64    238     -6.72     <.0001
```

The amount of hybrid vigor obtained in a rotational crossbreeding system at equilibrium using purebred or pure-line sires can be predicted by (Bourdon, 2000):

$$\%HybridVigor = \left(\frac{2^n - 2}{2^n - 1}\right) \times 100 ,$$

where *n* is the number of pure breeds or lines involved in the rotational system. When crossbred sires are used in the rotational system, the equation to predict equilibrium hybrid vigor differs (Bourdon, 2000) and is

$$\%HybridVigor = \left(\frac{m(2^n - 1) - 1}{m(2^n - 1)}\right) \times 100 ,$$

where *n* is the number of sire types involved in the system and *m* is the number of breeds present in each sire type. This formula assumes that no breed is present in more than one sire type.

It should be noted that the phenomenon of heterosis is generally lost if two F_1 individuals derived from the crossing of two pure line parents are crossed. The offspring produced from the mating of two F_1 individuals will often exhibit decreased performance values for the traits that exhibited the heterotic effect in the F_1 individuals. In other words, the superior performance observed in crossbred individuals is not transmitted upon mating. This is because the gene combinations are not transmitted to progeny; only individual genes are transmitted to progeny. The gene combinations are rearranged or lost when crossbred animals are mated together, because of random segregation of alleles during meiosis. Additionally, the crossing of different species can often result in reduced reproductive performance as exhibited by the sterility of offspring produced by crossing a horse and an ass.

3.6 References

Bourdon, R. M. 2000. *Understanding Animal Breeding*. 2d ed. Upper Saddle River, NJ: Prentice-Hall.

Falconer, D. S., and T. F. C. Mackay. 1996. *Introduction to Quantitative Genetics*. 4th ed. New York: John Wiley & Sons.

Gardner, C. O., and S. A. Eberhart. 1966. Analysis and interpretation of the variety cross diallel and related populations. *Biometrics* 22:439-452.

Hallauer, A. R., and J. B. Miranda Filho. 1988. *Quantitative Genetics in Maize Breeding*. 2d ed. Ames: Iowa State University Press.

Hartl, D. L., and A. G. Clark. 1989. *Principles of Population Genetics*. 2d ed. Sunderland, MA: Sinauer Associates.

Hill, W. G. 1979. A note on effective population size with overlapping generations. *Genetics* 92:317-322.

Lynch, M., and B. Walsh. 1998. *Genetics and Analysis of Quantitative Traits*. Sunderland, MA: Sinauer Associates.

Muir, W. M. 1986. Estimation of response to selection and utilization of control populations for additional information and accuracy. *Biometrics* 42:381-391.

Van Vleck, L. D., E. J. Pollak, and E. A. Branford Oltenacu. 1987. *Genetics for the Animal Sciences*. New York: W. H. Freeman.

Wright, S. 1922. Coefficients of inbreeding and relationship. *Amer. Naturalist* 56:330-339.

Chapter 4

Genetic Selection

Arnold M. Saxton

4.1 Introduction 55
4.2 Single-Trait Selection 56
    4.2.1 Individual Selection 56
    4.2.2 Selection on Relatives 57
    4.2.3 Indirect Selection 58
4.3 Independent Culling 59
4.4 Selection Index 62
4.5 Selection on BLUP 65
4.6 References 66

4.1 Introduction

A major application of quantitative genetics is the improvement through genetic selection of agricultural and model species. Typical examples are yield of corn, growth rate of beef cattle, milk production of dairy cattle, protein content of soybeans, and bristle number in fruit flies. In agriculture the goal is to create populations that are more productive or to produce a higher-quality product. Typical experience has been for trait means to increase by about 1% per year. These genetic gains are a critical component of meeting world food production needs. For model species, the goal is to study genetic mechanisms. By observing how traits respond to selection, genetic inferences can be made about number of genes involved and the importance of various forces on genetic variation (see realized heritability in Chapter 3). Selection also produces lines that differ genetically, allowing genes that contribute to this difference to be identified (Chapters 8–11).

Mean values of traits in a population are changed by choosing more desirable individuals as parents of the next generation. If the trait has a genetic basis, then frequency of "good" alleles will increase in the population, changing the mean. Genetic selection is also the mechanism for evolutionary changes seen in natural populations. A distinction is often made between the natural selection for evolution and the artificial selection conducted by man. But the mechanism is the same. Individuals that differ genetically in some way are chosen, by man or nature, to produce more offspring, and this changes gene frequencies and trait means.

This chapter focuses on artificial selection, starting with situations where just one trait is of interest, then covering multiple-trait selection methods of independent culling and selection index. Best linear unbiased prediction is widely accepted as the best predictor of genetic merit of individuals, so selecting individuals with high BLUP values is a common selection method. As with all chapters, the programs presented here are available at the companion Web site (support.sas.com/companionsites).

Geneticists have many choices in designing a genetic improvement program, as just outlined. In addition to the basic choices of single, multiple, or BLUP selection, there are many possible sources of information concerning which parents to choose. Phenotypic performance of the individual or of relatives, or multiple records on an individual could be used. Thus the primary focus of computer programs in this chapter is prediction of genetic gain. Given genetic parameters from Chapters 2 and 3, genetic gain can be predicted for any selection program. This allows an objective decision to be made as to which selection program should be implemented.

4.2 Single-Trait Selection

Falconer and Mackay (1996) present the rationale for the genetic gain formula

$$\Delta G = i * h^2 * \sigma_P / L,$$

where the expected change in the mean of a trait per unit of time is a function of the selection intensity (i), the heritability (h^2), the phenotypic standard deviation (σ_P), and the length of a generation interval (L).

Selection intensity is the mean of a truncated standard normal distribution, reflecting the mean of individuals selected as parents. This assumes truncation selection, with individuals below a chosen value not contributing any offspring to the next generation. For example, if the largest 10% of individuals in a population are selected, the mean value of selected individuals will be 1.755 standard deviations, and i=1.755. It is often more convenient to express selection intensity through the percent selected. The following examples show how either can be used.

Generation interval is the average time between the birth of an animal and the birth of its replacement(s). Generation interval may differ between males and females in breeding populations, in which case an average can be used. By reducing the generation interval, the response to selection per year can be maximized. Species with longer maturation times naturally have a longer generation interval, making minimizing generation interval more important.

4.2.1 Individual Selection

If selection decisions on individuals are based on observing those same individuals, then individual or mass selection is being done. The %SELECT1 macro implements this by giving values for heritability (H2), phenotypic standard deviation (PHENOSD), and either selection intensity (I) or percent selected (PSELECT). Optionally the generation interval (GENLEN) can be given, or the default value of 1 will be used. If selection intensity is given, then the standard formula can be used to calculate genetic gain.

However, the macro also computes percent selected, using an iterative process. If percent selected is given, then selection intensity is solved for directly, using SAS functions.

```
%select1(pselect=.10,h2=.2,phenosd=5);
%select1(i=.5,h2=.2,phenosd=5);
%select1(i=1,h2=.3,phenosd=50,genlen=1);
```

As the results in Output 4.1 demonstrate, higher selection intensity, corresponding to lower percent selected, and of course higher heritability or phenotypic standard deviation will produce greater genetic gain. Missing variables are involved with other ways to select for a single trait, which are discussed next.

Output 4.1 Examples of predicted genetic gain from single-trait selection.

Obs	i	h2	pselect	phenosd	genlen	Genetic Gain	family
1	1.75498	0.2	0.1	5	1	1.75498	.

Obs	nsib	sibr	sibt	corr	corrh2	INDIRECT Genetic Gain
1	.	.	.	.	.	.

Obs	i	h2	pselect	phenosd	genlen	Genetic Gain	family
1	0.5	0.2	0.69774	5	1	0.5	.

Obs	nsib	sibr	sibt	corr	corrh2	INDIRECT Genetic Gain
1	.	.	.	.	.	.

Obs	i	h2	pselect	phenosd	genlen	Genetic Gain	family
1	1	0.3	0.38109	50	1	15	.

Obs	nsib	sibr	sibt	corr	corrh2	INDIRECT Genetic Gain
1	.	.	.	.	.	.

4.2.2 Selection on Relatives

In some situations, selection based on the phenotypes of relatives is advantageous, or required. For example, when destructive sampling is necessary, the individual measured cannot be selected (unless gametes were saved). When heritabilities are low, selecting on family means may produce increased rates of genetic gain. Falconer and Mackay (1996) present several alternatives, and these have been incorporated into %SELECT1.

To estimate genetic gain from selection on relatives, %SELECT1 must be given a value for the FAMILY= option. Possible values are SIB for selection based on the phenotype of relatives, FAMILY for selection between families, WITHIN for selection within families, and COMB for a combination of between- and within-family selection. Formulas for genetic gain require additional options. SIBR is the relationship of the relatives, such as .25 for half sibs and .5 for full sibs. SIBT is the phenotypic correlation among relatives, and NSIB is the number of relatives measured to obtain the phenotypic mean for truncation selection.

This code produces Output 4.2:

```
%select1 (pselect=.20, h2=.3, phenosd=1, family=within, sibr=.25, sibt=.4,
nsib=8);
```

Output 4.2 Example of family selection.

Obs	i	h2	pselect	phenosd	genlen	Genetic Gain	family
1	1.39981	0.3	0.2	1	1	0.41994	WITHIN

Obs	nsib	sibr	sibt	within Genetic Gain	corr	corrh2	INDIRECTGenetic Gain
1	8	0.25	0.4	0.38035	.	.	.

Output echoes input values, and includes genetic gain for the chosen FAMILY method of selection, in addition to the GENETICGAIN variable, which estimates genetic gain for *individual selection* (selection based on the individual's own phenotype). A comparison of genetic gains predicted for different methods of selection will help in choosing the best selection method.

4.2.3 Indirect Selection

Indirect selection is a third variation on single-trait selection, where gain in one trait is obtained through selection on a different trait. This relies on the genetic correlation between the traits (Falconer and Mackay, 1996). To request these calculations in %SELECT1, specify CORR, the genetic correlation, and CORRH2, the heritability of the trait providing phenotypic values. In the code example that follows, both family and indirect selection options are requested. %SELECT1 simply produces genetic gains for each, but does not combine the two (Output 4.3). Thus %SELECT1 currently does not support indirect selection based on relatives.

```
%select1 (pselect=.20, h2=.3, phenosd=1, family=within, sibr=.25, sibt=.4,
nsib=8, corr=.3, corrh2=.1);
```

Output 4.3 Example of indirect selection.

Obs	i	h2	pselect	phenosd	genlen	Genetic Gain	family
1	1.39981	0.3	0.2	1	1	0.41994	WITHIN

Obs	nsib	sibr	sibt	within Genetic Gain	corr	corrh2	INDIRECTGenetic Gain
1	8	0.25	0.4	0.38035	0.3	0.1	0.072736

As mentioned earlier, it is often of interest to compare various methods of selection, either from a practical viewpoint to find the method that maximizes gain, or from a theoretical viewpoint to see what effect the various parameters have on selection methods. The following code shows how a loop can be programmed to examine values of SIBT ranging from .05 to 1.00. Other parameters can be examined using similar code. The SIBT values are given to %SELECT1, and the SELECT1ZZ data sets it produces are accumulated in DATA STORE, which is then plotted to produce Output 4.4. This clearly shows that as SIBT increases, within-family selection eventually produces genetic gains greater than those produced by individual selection.

```
%macro compare;
data store; run;
ods listing exclude all;
%do ii=1 %to 20;
%let sibt=%sysevalf(0+.05*&ii);
%select1 (pselect=.20, h2=.3, phenosd=1, family=within, sibr=.25, sibt=&sibt,
nsib=8 );
data store; set store select1zz;
run;
%end;
ods listing;
proc gplot;
  goptions ftext=swiss;
  symbol1 v=square i=join c=black;
symbol2 v=dot i=join c=black;
  legend1 frame;
  plot (geneticgain withingeneticgain)*sibt/overlay legend=legend1;
run; quit;
%mend;
%compare;
```

Output 4.4 Comparison of within-family and individual selection.

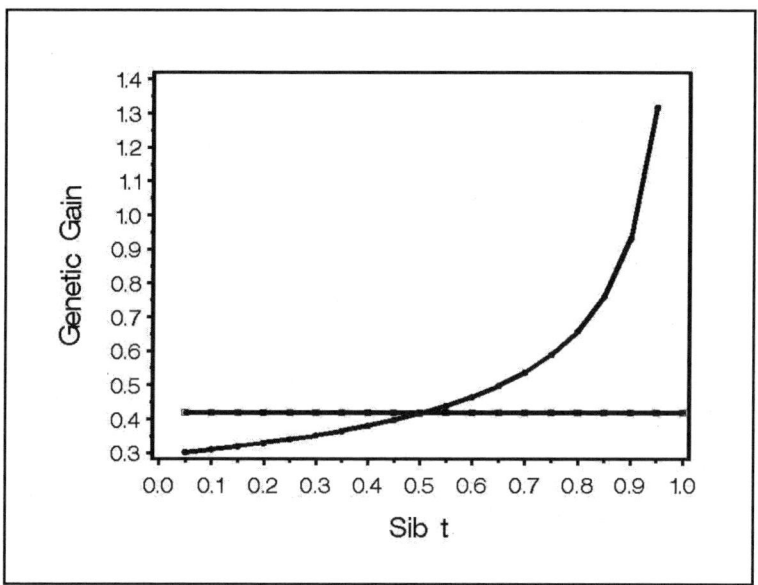

4.3 Independent Culling

It is more likely that genetic gain in several traits, rather than a single trait, will be of interest. In agriculture, economic value of a population often comes from the contribution of several traits. For dairy cows, increases in milk production and longevity are important, and fat content of milk must not decrease. For corn, high yield is important, but the plants must not fall over, and protein or oil content may be of interest.

The simplest multiple-trait selection method is independent culling. Individuals are chosen as parents if they exceed chosen cutoff values for all traits. Thus independent culling is a logical extension to multiple traits of truncation selection on a single trait. Young and Weiler (1960) presented the first complete description of the method, but development and applications have continued (e.g., Muir and Xu, 1991; Ducrocq and Colleau, 1989).

In practice, it is not uncommon for the cutoff values, or "culling levels," to be chosen arbitrarily. Thus dairy cows with at least 15,000 lbs. of milk, a productive life span of at least four years, and milk fat content above 2.7% might be chosen as parents. Culling levels are chosen to place more selection pressure on traits of more interest. Of course these culling levels would need to be adjusted to give the necessary selection intensity, retaining sufficient parents to maintain population size in the next generation. However, it is unlikely that arbitrary culling levels will produce the maximum gain, since genetic parameters are not used.

With multiple traits, it is convenient to express gain from selection in dollars, converting the gain in each trait to the economic value by multiplying by "economic weights" (Falconer and Mackay, 1996). This provides a measure of optimality, as the best culling levels are those that produce maximum dollar increase. In addition to economic weights, maximum genetic gain from independent culling requires knowledge of phenotypic and genetic variances and covariances, as estimated in Chapters 2 and 3, for example.

A general description for computing optimal culling levels is as follows. First take arbitrary culling levels, and a multivariate normal distribution of the traits. As seen in Figure 4.1, the multivariate normal ellipse for two traits is divided into four quadrants by the two culling levels. Trait means of individuals in the selected quadrant can then be calculated by integrating the distribution. These trait means are used to calculate genetic gain, based on the genetic parameters. Maximization of gain is accomplished by an iterative search of culling levels, shifting the culling levels up and down until a maximum gain quadrant is found. Of course each choice of culling levels must also produce the specified selection intensity.

Figure 4.1 Illustration of how two independent culling levels select part of a population.

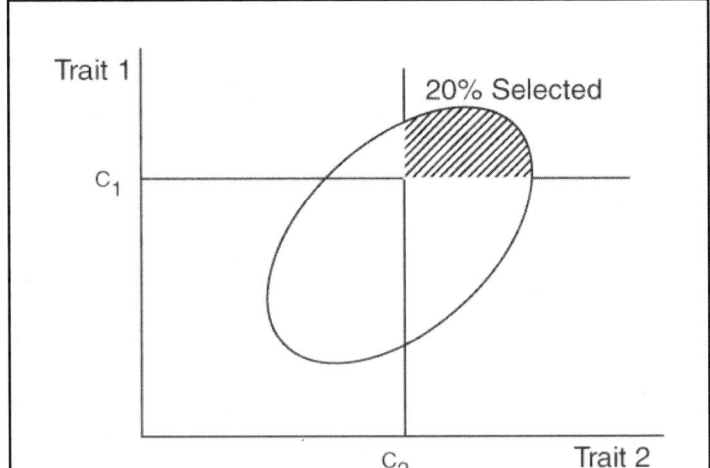

INDCULL.SAS contains a SAS macro for finding optimal culling levels. It is a direct translation of a FORTRAN program of the same name (Saxton, 1989). The following code shows an example of its use:

```
%indcull(2,trait1 trait2, 1.21 2.2 6.25, .605 .8609 .8609 2.2,1 1,.05);
```

Six arguments are required, in required order separated by commas:

1. number of traits selected on, and optionally the number of genetic traits;
2. trait labels (not separated by commas);
3. phenotypic (co)variance matrix, listed by rows of the lower triangle;

4. genetic (co)variance matrix, listed by complete rows;
5. economic weights;
6. fraction selected.

The example shows the simplest situation, where phenotypes of two traits are selected to genetically improve the same two traits. Output 4.5 gives the results from this program. Selection index gains (see next section) are printed for comparison, with a gain of 2.802 predicted in the aggregate genotype. A grid search calculates gains for independent culling with culling levels ranging across the possible values (only one pair is shown here). The best of these serves as a starting point for a non-linear iterative search for maximum gain. The best independent culling selection program found here produces a gain of 0.23, much less than selection index. This is achieved by culling trait 1 at 4 standard deviations below the mean (almost no selection pressure; note the 100% selected) and culling trait 2 at 1.64 standard deviations above the mean. Gains in the individual traits are reported in standardized and unstandardized units, and the latter when multiplied by economic weights sum to the total genetic gain.

Output 4.5 Independent culling on two traits.

```
                        PHENOTYPIC INFORMATION Covariances
                           trait1        1.21            2.2
                           trait2         2.2           6.25

                        GENETIC INFORMATION Covariances    P\\G
                                       trait1         trait2
                           trait1        0.605         0.8609
                           trait2        0.8609         2.2

                           Economic Weights            1
                                                       1
                                                                        SI
              5 percent selected, or selection intensity = 2.06271281

                                                  SIB        SIGAIN
                        Selection index results   0.8917888  2.80216672
                                                  0.17583434

                              ***************
                              *   Results   *
                              ***************
        According to input values, genetic gain is to be calculated with these weights
on the std. means
                                    -2.4818687 2.09797778

                               ****************
                               * Grid Results *
                               ****************
    Trait    Truncation    % Select   Std Mean   Std Gain   Unstd Gain     Total Gain
     trait1         -4   99.9968329  1.65017025  0.46813542  0.51494896
     trait2   1.64485363          5  2.06271281 -0.1131717  -0.2829292
                                                                          0.23201977

                    **********************************************
                    *         Maximum Gain.         2 iterations *
                    **********************************************

    Trait    Truncation    % Select   Std Mean   Std Gain   Unstd Gain     Total Gain
     trait1  -4.0667273  99.9976161  1.65017025  0.46813542  0.51494896
     trait2   1.64485363          5  2.06271281 -0.1131717  -0.2829292
                                                                          0.23201977
                                 Normal Termination
```

More complex selection programs are not currently implemented in the %INDCULL macro, but are briefly described. A common objective is to improve some traits, while forcing other traits to have a specified gain. For example, improvement of two traits while holding a third trait to zero gain would be of interest if indirect response in the third trait would have been negative. Another situation, also involving indirect selection response, is to select on some traits in order to improve others. This was discussed above for a single trait, and the same principle applies to multiple traits. Genetic variation may result in better progress from indirect selection, or practically speaking, a trait that is easier to measure might be selected on instead of the trait of direct economic interest. These complex selection programs are easier to program for selection indices, and are available in the SAS program discussed next.

4.4 Selection Index

The selection index implements multiple trait selection by calculating index weights that are applied to phenotypic measures on an individual. The index,

$$I = b_1 * p_1 + b_2 * p_2 + b_3 * p_3 + \ldots = \mathbf{b'p},$$

is known to produce the maximum genetic gain in the aggregate genotype,

$$H = a_1 * g_1 + a_2 * g_2 + a_3 * g_3 + \ldots = \mathbf{a'g},$$

when the index weights are calculated by

$$\mathbf{b} = \mathbf{P^{-1}Ga}.$$

Matrix algebra is useful for these calculations, with the inverse of the phenotypic (co)variance matrix multiplied by the genetic (co)variance matrix and the economic weights.

Since selection index produces the most genetic gain, why is independent culling still of interest? Note that for selection on the index to be possible, all phenotypes must be measured on each individual. This implies that individuals must be tagged so they can be identified, and individuals must be retained as long as needed to measure all phenotypes. If one trait is longevity, then all individuals must be kept for their entire natural life span, a potentially very expensive requirement. Independent culling, on the other hand, allows selection to be conducted as soon as any one trait is measured, because culling levels are used on each trait. This is multi-stage selection, and it is possible for independent culling to give more genetic gain per unit cost by taking advantage of this feature.

The calculations for index selection have been implemented in PROC IML, using its matrix algebra capabilities. SINDEX.SAS contains the %SINDEX macro version of these calculations. As can be seen by the following example, the minimum required information is the phenotypic (co)variance matrix, specified as a lower triangular matrix; the additive genetic covariance matrix between traits in the phenotype and traits in the aggregate genotype, specified as the full matrix; the proportion selected; and economic weights.

This example is from Cunningham et al. (1970):

```
%sindex(p=.25    0  .25     0 0 36,
   g=.0875  -.0047  .0887  -.0047 .0250 0  .0887  0 9,
   ew= 2 22 1,
   save=.10         );
```

Output 4.6 contains the printed results from running the program. Input values are echoed, the proportion saved is converted to selection intensity, and the selection index weights (**b**) are printed, along with genetic gain estimates for individual traits and aggregate genotype (SIGAIN). Also printed is the standard deviation of the index,

$$\sqrt{b'Pb},$$

which is the square root of aggregate genetic gain, and the relative contributions of individual traits to the aggregate gain, as described in Cunningham et al. (1970). For the third trait, the relative contribution of 40.7 indicates that if this trait were removed from the selection index, genetic gain in the aggregate genotype would be reduced by 40.7%.

Output 4.6 Selection index results.

```
            PHENOTYPIC INFORMATION Covariances
                        PP
        PTrait1      0.25          0           0
        PTrait2         0       0.25           0
        PTrait3         0          0          36


         GENETIC INFORMATION Covariances    P\\G
                        GG
                 GTrait1    GTrait2     GTrait3
        PTrait1   0.0875    -0.0047      0.0887
        PTrait2  -0.0047     0.025            0
        PTrait3   0.0887         0            9

                                                       SI
   10   percent selected, or selection intensity = 1.75498332


            Selection index and economic weights
             SIB                      EW
        PTrait1     0.6412            2
        PTrait2     2.1624           22
        PTrait3 0.25492778            1


        Individual trait gains and relative contribution
            INDGAINS              RELCONTR
        GTrait1  0.06330962     1.43334553
        GTrait2  0.04714144    17.7625188
        GTrait3  2.17136169    40.6568344

            Predicted total gain and StdDev(Index)
                 SIGAIN       SDINDEX
               3.33509265    1.90035575
```

To test this, the SELECTON option in the %SINDEX macro can be used to select on traits 1 and 2 only, as seen here. Simply list the traits by number that the selection will be based on.

```
%sindex(p=.25    0 .25     0 0 36,
    g=.0875  -.0047  .0887  -.0047 .0250 0  .0887  0 9,
    ew= 2 22 1,
    save=.10,
    selecton=1 2
    );
```

The aggregate genotype still has three traits, so genetic progress in the third trait is from correlated response to selection on the other two traits. This program calculates a genetic gain of 1.979 (not shown), a reduction of 40.7% from the 3.335 gain in Output 4.6.

Other options in the %SINDEX macro are illustrated by this example, from Yamada et al. (1975):

```
%sindex(
    p    = -.80
            -.10 -.30
             .80 -.40 -.10
             .10  .40  .40 -.05
            -.30  .20  .20 -.40  .20
              0    0    0    0  .10   0 ,
    g    =  -.60
            -.60 -.40
             .85 -.50  .30
              0   .30  .40   0
            -.50  .20  .30 -.50  .10
              0    0    0   .05 .10   0 ,
    h2=     .20  .20  .50  .30  .40  .30  .05 ,
    psd=    10    2    4   10   18   12   30,
    plabels=EP   FC   EW   EF   BW   SM   AV,
    glabels=EP   FC   EW   EF   BW   SM   AV,
    ew= 0 -1 .1 -.1 -.1 0 0,
    desired= 8 -3 0 max max max max,
    save=.10 ,
    selecton= 3 4 5 );
```

Here heritabilities are given (H2), which signals the macro to read in **G** as a lower triangular correlation matrix. Note this implies that the traits in **P** must be the same as those in **G**. Also given are phenotypic standard deviations (PSD), which indicates that **P** will be expressed as correlations, without the main diagonal. Phenotypic variances (PVAR) can be supplied instead of PSD, if desired. The %SINDEX macro currently requires that if H2 is specified, then PSD or PVAR must also be. In other words, if correlation input is used, both **P** and **G** must be in correlation form. Labels for the phenotypic and genetic traits can be specified, with PLABELS and GLABELS.

The last option, DESIRED, requests that genetic gains for traits in the aggregate genotype follow a specified pattern. Some traits may have no restrictions, and this is indicated with the keyword MAX. In the example, gains for the first three traits are requested to have the ratio of 8 to –3 to 0, while gains in the remaining traits are maximized, subject to the restriction. Output 4.7 shows output from this program. First unrestricted selection results are given, showing a large response in EW, and a total gain in the index of 1.539. With restriction, the primary goal is to have zero gain in EW, while maintaining a desirable ratio of gains in EP and FC. Results show this can be achieved with small effect on total gain, reduced only to 1.218.

Output 4.7 Example of desired gains selection index.

```
                                                      SI
         10   percent selected, or selection intensity = 1.75498332

                 Selection index and economic weights
                    SIB                     EW
                EW   0.10139679              0
                EF  -0.0017079              -1
                BW  -0.0545133               0.1
                                            -0.1
                                            -0.1
                                             0
                                             0

           Individual trait gains and relative contribution
                   INDGAINS              RELCONTR
              EP   -1.5709589            8.86877114
              FC   -0.5169724            0.01784718
              EW    0.19730709          98.173548
              EF    0.81955023
              BW  -11.237147
              SM    0.36751117
              AV   -0.8184348

               Predicted total gain and StdDev(Index)
                    SIGAIN      SDINDEX
                  1.57846271    0.8994175

        ***Tallis restricted index using phenotypic traits   EW EF BW
                 to produce desired genetic gains ***

                        Selection index weights
                              SIB
                      EW   0.38898814
                      EF   0.52602612
                      BW  -0.4314255

                         Individual trait gains
            INDGAINS                  DGINPUT              EW
              EP    1.53942203          8                   0
              FC   -0.5772833           -3                  -1
              EW    2.7356E-16          0                   0.1
              EF    3.38454191         max                  -0.1
              BW   -9.7950879          max                  -0.1
              SM   -2.0257197          max                   0
              AV   -0.4480331          max                   0

               Predicted total gain and StdDev(Index)
                    TOTALGAIN      SDINDEX
                   1.21833786     9.12021932
```

4.5 Selection on BLUP

In many farm species, choice of parents is made through a genetic evaluation, which produces estimates of genetic merit for individuals. Genetic merit has various names, like expected progeny difference (EPD), transmitting ability, breeding value, and the statistical term best linear unbiased prediction (BLUP). Genetic merit is an estimate of the genetic value that an individual transmits to the next generation. Thus there is no need for the mathematics of previous sections. BLUPs are calculated for each trait being selected, and these are combined using economic weights to give the aggregate genotype. Selection is then based directly on the predicted genotype rather than some function of the phenotype.

The file SELECTBLUP.SAS contains an illustration of how this might be accomplished using SAS. A copy of the file should be obtained for the following brief description to be meaningful. Basic elements of the program are as follows:

- Read in the pedigree and use PROC INBREED to produce covariances among individuals.
- Read in the phenotypic data. Note that phenotypic data are not available for all individuals in the pedigree, as is often the case.
- Use PROC IML to reduce the pedigree covariance matrix to contain only those individuals with phenotypic data. Comments indicate where the user may need to adapt the code for different problems. Of particular importance is the scaling of the covariance matrix by the genetic-to-error variance ratio.
- The %GETBLUPS macro runs an animal model to produce BLUPs. The pedigree covariances are read into PROC MIXED, as they represent the "known" covariance matrix among animals. This macro is run for each trait being selected, with BLUPs accumulated in the AGGENO data set.
- Finally, economic weights are applied in AGGENO to calculate the aggregate genotype (AG). These values are sorted and printed for use in selecting parents.

For the example included in SELECTBLUP.SAS, Output 4.8 is produced. Two or three individuals appear quite superior in both the positive and negative selection directions.

Output 4.8 Results from running SELECTBLUP.SAS.

Obs	Effect	ind	blup_milk	blup_age	blup_fat	ag
1	ind	124	0.000143	0.2801	0	0.28143
2	ind	102	0.000101	0.2130	0	0.21395
3	ind	114	-0.00012	0.2084	0	0.20731
4	ind	123	0.000124	0.1257	0	0.12682
5	ind	103	-0.00013	0.05860	0	0.05746
6	ind	104	-0.00016	0.05860	0	0.05716
7	ind	113	-0.00018	0.05395	0	0.05232
8	ind	121	0.000197	0.01259	0	0.01436
9	ind	101	-0.00013	-0.09584	0	-0.09702
10	ind	122	0.000192	-0.1419	0	-0.14013
11	ind	112	-0.00007	-0.3267	0	-0.32736
12	ind	111	-0.00005	-0.4812	0	-0.48160

4.6 References

Cunningham, E. P., R. A. Moen, and T. Gjedrem. 1970. Restriction of selection indexes. *Biometrics* 26(1):67-74.

Ducrocq, V., and J. J. Colleau. 1989. Optimum truncation points for independent culling level selection on a multivariate normal distribution, with an application to dairy cattle selection. *Genetics, Selection, Evolution* 21:185-198.

Falconer, D. S., and T. F. C. Mackay. 1996. *Introduction to Quantitative Genetics*. 4th ed. Essex, UK: Longman.

Muir, W. M., and S. Xu. 1991. An approximate method for optimum independent culling level selection for *n* stages of selection with explicit solutions. *Theoretical and Applied Genetics* 82:457-465.

Saxton, A. M. 1989. INDCULL Version 3.0: Independent culling for two or more traits. *Journal of Heredity* 80:166-167.

Yamada, Y., K. Yokouchi, and A. Nishida. 1975. Selection index when genetic gains of individual traits are of primary concern. *Japanese Journal of Genetics* 50:33-41.

Young, S. S. Y., and H. Weiler. 1960. Selection for two correlated traits by independent culling levels. *Journal of Genetics* 57:329-338.

Chapter 5

Genotype-by-Environment Interaction

Manjit S. Kang, Mónica G. Balzarini, and Jose L. L. Guerra

5.1 Introduction 69
5.2 Modeling Genotype-by-Environment Interaction 72
    5.2.1 ANOVA Model with Fixed GEI 72
    5.2.2 Linear-Bilinear Models with Fixed GEI 76
    5.2.3 Linear Mixed Model Approach to GEI Analysis 82
    5.2.4 Generalized Linear Models to Explore GEI 86
5.3 Smoothing Spline Genotype Analysis (SSGA) 90
5.4 References 94

5.1 Introduction

Genotype-by-environment interaction (GEI) refers to the differential responses of different genotypes across a range of environments (Kang, 2004). This is a universal issue relating to all living organisms, from bacteria to plants to humans (Kang, 1998), and it is important in agricultural, genetic, evolutionary, and statistical research. Gene expression is dependent upon environmental factors and may be modified, enhanced, silenced, and/or timed by the regulatory mechanisms of the cell in response to internal and external factors (Rédei, 1998). A range of phenotypes may result from a genotype when it is exposed to different environments. This phenomenon is called *norms of reaction,* or phenotypic plasticity. Norms of reaction represent the expression of phenotypic variability in individuals of a single genotype. The lack of phenotypic plasticity is called canalization.

Quantitative traits, especially those which are controlled by several genes, are in general highly influenced by environmental factors and display a continuous variation. Most agronomically and economically important traits, such as grain yield and meat production, are quantitative or multigenic in nature. Because of the universal presence of GEI of quantitative traits, genotype evaluations in at least advanced plant breeding stages are carried out in multiple environments in planned experiments called multi-environment trials (MET). The major objectives of MET are (1) to compare genotype performances in two inference bases—i.e., broad inference, the general performance of a genotype (across environments), and environment-specific or narrow inference, the performance of a genotype within a specific environment; and (2) to estimate the GEI component to gauge heritability and its impact on selection, to select test sites and mega-environments, to identify genotypes specifically adapted to target environments, and to establish breeding objectives (Yan and Kang, 2003). Genotype-by-environment interaction has long been an important and challenging issue among plant breeders, geneticists, and agronomists engaged in crop performance evaluation (Kang, 1990; Kang and Gauch, 1996; Cooper and Hammer, 1996; Kang, 2002; Annicchiarico, 2002). Typical MET in plant breeding involve an evaluation of a large number of genotypes for several economically important traits in replicated field tests that are conducted across several years (seasons) at several locations in a region and/or diverse management systems (Annicchiarico, 2002). At each location, a randomized complete-block design is commonly used. Animal breeders also are concerned about GEI in the context of breed evaluation (Montaldo, 2001). Knowledge of genetic effects of breeds and their crosses in various climatic-forage systems can be used to identify optimal breed combinations and crossbreeding systems for existing markets (Franke et al., 2001).

The traditional modeling approach for MET is based on a general linear model of analysis of variance (ANOVA) involving genotype (G), environment (E), and GEI effects for a response variable (trait values). Traditionally, the model error terms are assumed to be normally distributed with constant variances. When GEI is significant, use of appropriate breeding/management strategies is called for, as the usefulness of overall genotype means is reduced (Kang, 1990; 1998; 2002). Hence, the concept of stability is extremely important to identify consistent performing and high-yielding genotypes. Because the most stable genotype may not be the highest yielding, integration of performance and stability to select superior genotypes is important (Kang, 1993).

A closer examination of GEI can reveal whether or not genotype rank changes occurred in different environments. Thus, GEI can be grouped into two broad categories: crossover and non-crossover interactions (Baker, 1990; Cornelius et al., 1996). The differential responses of genotypes to diverse environments when genotype ranks change from one environment to another are referred to as crossover or qualitative interaction. Since the presence of crossover interaction has strong implications for breeding for specific adaptation, it is important to assess its frequency (Singh et al., 1999). The non-crossover interactions represent quantitative changes in the magnitude of genotype performance, but rank order of genotype across environments remains unchanged—i.e., genotypes that are superior in one environment maintain their superiority in other environments.

In a classical GEI analysis, a basic assumption is that each environment yields a vector of g measurements for the trait of interest (g is the number of genotypes tested in that environment). Each vector element usually is the mean value (across replications in replicated trials) of a given genotype in the environment. These measurements can be decomposed into G and GEI effects. Most of the classical analytical procedures used to quantify a genotype's contribution to the overall GEI are based on a fixed effects model approach, and they are applicable only to complete and highly balanced data sets. The frequent deletion and substitution of entries (genotypes) in a series of trials, however, result in incomplete data. The fixed model analysis runs into problems when the data are incomplete or unbalanced. Thus, the fixed ANOVA approach to MET does not use all the available information, and narrow inference in a specific environment is possible only for genotypes that have been tested in the target environment. Estimation procedures under the mixed ANOVA model circumvent this problem (Piepho, 1998; Balzarini, 2002).

Environments in MET might be reasonably assumed as random effects. The environments are generally selected to represent a relatively large target population of environments. Various aspects or elements of environments may represent stochastic variation between or among environments. On the contrary, genotype effects during advanced breeding stages might be treated as fixed since only a highly selected set of genotypes is evaluated in replicated tests. When at least one of the two main factor effects is regarded as random, GEI terms are random entities and a mixed model approach to GEI analysis becomes appropriate.

It is natural to assume that a set of observations from one environment tends to be correlated. Latent variables associated with each environment may cause dependencies among genotype responses in a given environment. Moreover, genotype performance across environments may generate a structured pattern of interdependence among GEI terms. Therefore, a covariance matrix for the genotypic means within an environment should receive particular attention. If GEI is significant, we can invoke an appropriate GEI analysis, which, in the linear mixed model framework, is based on the best linear unbiased predictor (BLUP) of GEI random effects (Balzarini, 2002). By further modeling the variance-covariance structure of GEI random effects, well-known stability measures can be expressed as parameters of a general mixed model (Piepho, 1998). This unified approach to GEI analysis represents a substantial advantage over the traditional fixed model.

Debate continues, however, on whether GEI is real or a function of deviations from an assumed model—namely, that the variance is independent of the mean (normal distribution–based models). If the variance can be said to be some function of the mean, there are likely to be consequences for the inferences that can be drawn from the normal approach to GEI analysis. One consequence of this is that a nonsignificant interaction may be misrepresented as significant or that a significant interaction may not be detected via the traditional ANOVA methodology. The detection or non-detection of GEI can be no more than a reflection of the departure from homoscedasticity or other structural flaw of the model (Snee, 1982). Trying to avoid measuring deviations from the underlying assumptions instead of interactions, Lefkovitch (1990) suggested the use of the gamma distribution, pointing out that the commonly used continuous responses share a common feature in that a negative record is impossible. Moreover, in cases where the responses are true discrete variables, such as counts or scores, the consequences of assuming normality and homoscedasticity can be costly. For example, in the case of counts (Poisson distribution), as the mean of the responses decreases or approaches zero, the distribution becomes positively skewed because the variance is the mean. Hence, significant results are likely caused by small variation when the responses approach zero. The opposite is true as the counts increase—the variance will increase and nonsignificance is likely attributable to a large variance. For the above-mentioned cases, the generalized linear models can be fitted using PROC GENMOD under a fixed model framework and using PROC NLINMIXED if GEI terms are regarded as random.

Another generalized method that can be used to conduct GEI analyses is smoothing spline analysis. A *smoother* is essentially a tool to analyze trends in the response variable; the essence of smoothing is to produce a trend that is less variable than the response itself (Hastie and Tibshirani, 1990). GEI analyses can be performed through smoothing to properly rank a group of genotypes when GEI is real and significant. With smoothing splines, there is the problem of making the GEI nonexistent because we can remove the *E* by fitting a high-degree polynomial to the data. Hence, by fitting the data with a 50-degree (or higher) polynomial curve, we make all *E* effects the same. It is, however, not advisable to work with such high-degree polynomials, because this creates an over-parameterized model. An optimum level of smoothing should be carefully determined. In this methodology, the *G* effect will be fitted not as a spline, but as a regular class parameter. For that reason, the procedure can be called semiparametric smoothing spline genotype analysis. We will show some examples of how to use this procedure to explore GEI.

Under the mixed model, we formulate the general model for correlated data using explicit models for the covariance structure of the GEI random effects, each of them related to stability analyses commonly used in testing of genotypes. We use PROC MIXED to deal with linear mixed models for normally distributed responses. The normal case is presented here as a special case of a generalized linear mixed model. For theoretical aspects, the reader is referred to classical literature on linear mixed models (Searle et al., 1992; Khuri et al., 1998) and generalized linear models (McCullagh and Nelder, 1983). For practical issues about mixed model fitting with SAS, see Littell et al. (1996). We will devote most of this chapter to SAS commands for GEI analyses.

5.2 Modeling Genotype-by-Environment Interaction

Assuming several years, locations, and replications within locations are used for testing genotype performance, the basic model for MET is

$$y_{ijlk} = \mu + Y_j + L_l + YL_{jl} + R_{k(jl)} + G_i + GY_{ij} + GL_{il} + GYL_{ijl} + \varepsilon_{ijlk},$$

where y_{ijlk} is the observation of the ith genotype (G_i) in the kth replicate ($R_{k(jl)}$) in the lth location (L_l) for the jth year (Y_j); μ is the overall mean, and the other terms, except ε_{ijlk}, which represents a random error term, are interaction effects between main effects. All interactions that include the genotype effect are of particular interest in exploring GEI.

This model can be used to analyze a series of trials conducted over several years or seasons or in several management systems. Commonly, MET data are highly unbalanced relative to genotype-by-year and location-by-year combinations. So, statistical analyses are usually conducted by year, allowing GEI to be explored from genotype-by-location (two-way) tables, which are frequently balanced. Another common practice is to re-parameterize the model by including only the terms related to environment, replication-within-environment, genotype, and GEI effects, with environments defined as combinations of years and locations.

5.2.1 ANOVA Model with Fixed GEI

The ANOVA model with environment and genotype main effects allowing replications within environment and a fixed GEI term is regarded as a first step in providing GEI analysis of a complete set of yield trials. The principal virtue of this approach is its computational simplicity.

The use of PROC MIXED for analyzing a set of peanut variety trials is illustrated by a data set containing peanut yields (see the data set PEANUT; data courtesy of National Institute of Agricultural Technology, INTA, Argentina). The data set includes ten genotypes: six with a long crop cycle (numbered 1–6) and four with a short crop cycle (numbered 7–10), evaluated across 15 environments, with four replications per genotype within each environment. In Table 5.1, the genotype names and yield mean at each environment are shown.

Table 5.1 Plot yield means for 10 peanut genotypes evaluated in 15 environments.

Env	Florman 1	Tegua 2	mf484 3	mf485 4	mf487 5	mf489 6	manf393 7	mf447 8	mf478 9	mf480 10
1	0.80	0.96	1.16	1.12	0.87	1.11	1.24	0.95	1.37	1.41
2	2.17	2.04	1.08	0.58	1.52	0.86	1.57	1.29	2.15	3.27
3	2.43	2.58	2.64	2.24	2.30	2.20	2.47	2.34	2.19	2.19
4	2.71	2.26	2.14	1.88	1.72	2.18	1.77	1.61	2.15	2.04
5	1.13	1.14	1.71	0.85	1.24	1.21	1.55	1.86	1.98	1.61
6	3.08	3.22	3.05	2.90	2.94	2.57	2.90	2.59	2.36	2.43
7	2.81	2.88	2.91	2.53	2.73	2.90	2.96	3.41	3.20	2.96
8	1.74	1.73	2.86	2.13	1.60	2.29	2.16	1.44	2.20	0.95
9	2.16	2.44	2.73	3.00	3.18	3.25	3.30	3.01	3.37	2.53
10	4.29	4.21	4.45	4.46	4.24	4.03	3.55	3.84	3.53	3.22
11	1.82	1.71	2.53	1.87	1.71	2.27	2.16	1.88	2.09	1.91
12	5.33	4.93	5.57	5.43	4.99	4.67	4.69	4.16	4.70	3.57
13	1.18	1.32	2.45	1.78	1.54	2.00	2.24	1.63	1.54	1.15
14	4.39	4.40	4.28	3.77	4.17	4.75	4.13	3.79	4.33	3.72
15	3.41	3.45	2.81	3.15	3.84	3.54	2.22	2.46	3.09	2.61

As a first step in our analysis, we present the following SAS code, which relates to the traditional ANOVA model for a complete genotype-by-environment data table. It generates means for each genotype across environments, which allow for broad inferences when GEI is not significant or, at least, when there is not crossover GEI.

```
options nodate ls=75;
proc import datafile="c:\peanut.xls" out=peanut replace;   ❶
run;

proc mixed data=peanut covtest;
class   rep gen env;
model yield=  env gen gen*env ;   ❷
random rep(env);
lsmeans gen /pdiff;
ods output diffs=p lsmeans=m;
ods listing exclude diffs lsmeans ;
run;
%include 'c:\PDMIX800.sas';   ❸
%pdmix800(p,m,alpha=.05, sort=yes);
run;
```

❶ Data in a spreadsheet are imported using the IMPORT procedure to create a SAS data set.

❷ A yield analysis under the fixed GEI model framework is done with PROC MIXED. In yield trials the genotype assignment on replications is usually done under a randomized complete block design. The random term REP(ENV) models variability among reps within an environment.

❸ To compare yield means, we use the SAS macro %PDMIX800, which lists means along with group letters according to their statistical differences. This macro formats pairwise differences from SAS PROC MIXED, created by the PDIFF option in the LSMEANS statement. The differences are used to create groups of similar means, represented by letters A, B, etc. Program documentation is included as comments at the top of the file (Saxton, 1998).

Test results for all fixed effects, including GEI, are significant ($P < .0001$), indicating that mean comparisons regarding GEI effects as random should be more appropriate. Nevertheless, we show results of mean separation produced by the %PDMIX800 macro in Output 5.1.

Output 5.1 Statistical differences among genotype means for the PEANUT data set under the fixed GEI model approach.

```
Obs    gen    Estimate    Error      Group
 1      3     2.8248      0.05925    A
 2      9     2.6840      0.05925    AB
 3      6     2.6544      0.05925    BC
 4      1     2.6312      0.05925    BC
 5      2     2.6185      0.05925    BC
 6      7     2.5946      0.05925    BC
 7      5     2.5729      0.05925    BC
 8      4     2.5134      0.05925    CD
 9      8     2.4176      0.05925    D
10     10     2.3708      0.05925    D
```

Under the fixed model approach, the model term GEN*ENV (GEI)—i.e., the non-additivity of the ith genotype and the jth environment—is estimated as the residual of the additive model, $\bar{y}_{ij} - \bar{y}_{i.} - \bar{y}_{.j} + \bar{y}_{..}$. The error term, $\bar{\varepsilon}_{ij}$, associated with the ith genotype in the jth environment is the mean of the k errors of measurements of this genotype in that environment. $\bar{\varepsilon}_{ij}$ is assumed to be an independent normal entity with constant variance. As a consequence, standard errors for the genotype means are equal. The MIXED procedure uses restricted maximum likelihood to estimate the residual variance. Given that sets of observations are independent and normally distributed, results are equivalent to least squares estimation (classical ANOVA). The independence among error terms is crucial, as correlation between observations requires special methodology (Cullis et al., 1998; Smith et al., 2002a,b). The assumption of independence and homogeneity of variances for the error terms is important; otherwise, estimation of the interaction terms can be misleading if residual variances in each environment are not the same. For example, when trials are conducted with different precision at different locations, we can modify the variance-covariance matrix of the error terms to obtain heterogeneous within-environment residual variances as follows:

```
proc mixed data=peanut covtest;
class   rep gen env;
model yield=  env gen gen*env ;
random rep(env);
repeated/group=env;      ❶
lsmeans gen /pdiff;
ods output diffs=p lsmeans=m;
ods listing exclude diffs lsmeans ;
run;
```

❶ The REPEATED statement is used here to allow heterogeneity of residual variances among test environments.

The output obtained from this approach shows that standard errors of means are slightly modified with respect to the homogeneous residual variance (homoscedastic) model. The heterogeneous residual variance model seems to fit better than the homogeneous model according to AIC fitting criteria, since AIC is 680.9 for the heterogeneous residual variance model and 722.0 for the homoscedastic model, and as stated in the output, models with smaller AIC values are better. The BIC value (a more conservative fitting criterion, since it takes into account the increase in the number of parameters to be estimated with the proposed model) does not favor the heterogeneous residual variance model in the same magnitude as AIC (726 vs. 714; homogeneous vs. heterogeneous model).

The fixed GEI model approach of MET (split-plot model with E and G factors) is useful to obtain the additive contributions of G, E, and GEI effects on the trait. But, after knowing that GEI is significant, ranking of genotypes using means across environments is not recommended. The fixed split-plot model involves $(g–1) \times (e–1)$ independent parameters relating to the GEI (where g and e are the number of levels for G and E), which do not provide much information about GEI patterns.

First attempts to explore GEI have used regression models to break down GEI into a series of regression terms and deviations from them, each one related to the specific interaction between a given genotype and the environments. These univariate approaches fit GEI as a linear regression function of additive environmental effects—i.e., gen*env$_{ij}$=β*env$_j$+d_{ij}, where d_{ij} is a regression deviation and (1+β) is the coefficient of the linear regression of responses of the ith genotype on the environmental mean. By fitting the additive model, i.e., the model without the interaction term, we can treat residuals as rough estimates of the GEI terms and regress them on functions of environmental means such as ($\bar{y}_{.j} - \bar{y}_{..}$). The following SAS code enables one to obtain β$_i$ parameters (parameter estimate) for the given set of genotypes in the PEANUT data set:

```
options nodate ls=75;
proc import datafile="c:\peanut.xls" out=peanut replace;
proc means data=peanut noprint mean;
  var yield;
  output out=yppp mean=yppp;
proc sort data=peanut; by env;
proc means noprint mean; by env;
  var yield;
  output out=yipp mean=yipp;
proc sort data=peanut; by gen;
proc means noprint mean; by gen;
  var yield;
  output out=ypjp mean=ypjp;
data enveff;
  merge yipp yppp;
  by _TYPE_;
  ei=yipp-yppp;
proc glm data= peanut noprint;
  class env gen rep;
  model yield=env rep(env) gen/p;
  output out=pred r=zij;
proc sort data=pred;by env;
proc sort data=enveff;by env;
data Ze;
  merge pred enveff;
  by env;
proc sort out=GEI;by gen;
proc reg data=GEI outest=estimate;by gen;   ❶
  model zij=ei/noint;
data estimate;
set estimate;
betai=ei;keep gen betai;
proc print data=estimate;
run;
```

❶ The REG procedure was used to estimate a linear regression fit of GEI estimates on environmental effects for each genotype. The slopes of such regression are commonly used as stability statistics.

The output of fitted regressions shows variability of slopes among genotypes. Except for genotypes 6 (mf489), 3 (mf484), and 8 (mf447), all genotypes show linear changes with changing environmental effects ($P<0.05$). Genotypes 7 (manf393), 9 (mf478), and 10 (mf480), all short-crop-cycle varieties, yield less under high-yield environments (negative regression coefficient). Under the regression-based procedures, GEI is expressed as heterogeneity of slopes, and the optimal stability is represented by high mean response, moderate to high response to favorable environments, and low deviations from regression.

Many statistics have been proposed for stability analysis under the univariate regression approach to explain GEI; most commonly used are those proposed by Finlay and Wilkinson (1963) and Eberhart and Russell (1966). Hussein et al. (2000) provided a comprehensive SAS program for computing more than 15 stability-related statistics for complete or balanced data sets obtained from the fixed model framework. To simultaneously evaluate for production and stability, genotype means need to be integrated with stability statistics, such as the regression coefficients.

5.2.2 Linear-Bilinear Models with Fixed GEI

Another approach to GEI analysis is the quantification of the portion of GEI accounted for by the heterogeneity of genotype regressions on environmental means in the context of linear-bilinear models (Gollob, 1968). Such models are based on principal component analysis (PCA) to analyze GEI in more than one dimension. The linear-bilinear models (LBM) represent a multivariate version of regression-based procedures. The response of the ith genotype in the jth environment is modeled by a systematic part that involves G and E additive main effects (linear components) and one or more multiplicative terms to explain the pattern in interaction terms (bilinear components). The random part of the model involves both the residual interaction term (part of the GEI not explained by the multiplicative model) and the error term. The expected value for a cell mean under this model is

$$E(y_{ij}) = \mu + G_i + E_j + \sum_k \lambda_k u_{ki} v_{kj}$$

The sum of K multiplicative terms, $\sum_k \lambda_k u_{ki} v_{kj}$, models GEI. Specifically, each interaction term, GEI_{ij}, is modeled as $\sum_k \lambda_k u_{ki} v_{kj} + \rho_{ij}$, with ρ_{ij} representing the portion of the ijth interaction term not explained by the multiplicative model. The kth bilinear term of GEI_{ij} has three components: the interaction parameter for the jth environment or the environment score, denoted by v_{kj}; the ith genotype score for the same principal component, denoted by u_{ki}; and the singular value associated with the kth axis, denoted by λ_k. The singular value represents a measure of correlation between the environment and the genotype scores. The score u_{ki} can be interpreted as genotypic sensitivities to latent environmental factors (in the jth environment, represented by v_{kj}).

Under a fixed linear-bilinear model, the interaction parameters are estimated via a singular value decomposition (SVD) of the matrix, **Z**, containing the residuals from the additive model following the least squares estimation of the main effects parameters. The components of the SVD of **Z** are ordered with respect to the singular values (from larger to smaller). The ratio of the sum of the first M singular values to the sum of all singular values represents the proportion of total variability in **Z** explained by the first M principal components (PC1 to PCM). Usually the first two multiplicative terms are interpreted, since they more likely explain real interaction patterns. The remainder of variability is regarded as noise. The linear-bilinear models were designated as "additive main effects and multiplicative interaction," or AMMI, models by Zobel et al. (1988) and Gauch (1988). When principal component analysis is used, genotypes with a first principal component axis value close to zero can be regarded as generally adapted to the test environments.

The genotype and environment scores of linear-bilinear models are commonly visualized in biplots (Gabriel, 1971), which allow identification of major environmental and genotypic causal factors of GEI. For a linear-bilinear model, a biplot contains both genotype and environment scores in an M-dimensional space, often $M=2$. Under the fixed-model approach, the genotype and environment scores are the elements of the eigenvectors of the left side and right side, respectively, of the SVD of **Z**. Price and Shafii (1993) gave a SAS/IML program to obtain biplots. Macchiavelli and Beaver (1999) generalized this program to obtain sequential likelihood ratio tests to determine *M*, i.e., the number of axes that should be retained in explaining the interaction pattern. The following SAS code invokes a version of that macro, which we updated for SAS Version 8 or later to produce GEI biplots under the fixed model framework.

```
goptions cback=white;
options nodate ls=75;
proc import datafile="c:\peanut.xls" out=peanut replace; run;  ❶
%include 'c:\FixedBiplot.sas';  ❷
%FixedBiplot(peanut,15,10,4,yield,env,gen,rep,geno);
```

❶ Data in a spreadsheet are imported by the IMPORT procedure to create a SAS data set.

❷ The %INCLUDE statement invokes a macro called %FIXEDBIPLOT, which takes nine required values: DATA (the name of the data set to be used), ENV_N (the number of environments), GEN_N (the number of genotypes), REP_N (the number of replicates), VAR (the name of the response variable), ENV (the name of the environment variable; should be a numeric variable), GEN (the name of the genotype variable; should be a numeric variable), REP (the name of the replication variable; should be a numeric variable), and GENCOD (a character variable containing the genotype names that will be printed on the biplot). To run this code, users must first download the FIXEDBIPLOT.SAS file to C: or redefine the path that calls this macro.

In Output 5.2, generated by the %FIXEDBIPLOT macro, the first column gives λ_k values (VAL) for $k=1,\ldots,K$, with K equal to the min(GEN_N, ENV_N). The proportion of total residual variance from the additive model explained by each eigenvalue (PROPEIG), and the accumulated proportion of GEI explained by the first *M* multiplicative components are given on the *m*th row (Output 5.2). For the PEANUT data set, we can see that the first two multiplicative components explain 75.7% of total variation in the residual matrix, and we assume this is due to a pattern generated by GEI. With three multiplicative components, this AMMI model explains 85.5% of total variation in the matrix of residuals from the additive model.

Output 5.2 Proportions of residual variance explained by each multiplicative component.

```
           VAL     PROPEIG    PROPACUM
     43.316965   0.4902006   0.4902006
     23.608472   0.2671675   0.7573681
     8.6682112   0.0980946   0.8554628
     5.1413597   0.0581827   0.9136455
     2.8253850   0.0319737   0.9456192
     2.2091424   0.0250000   0.9706192
     1.4715713   0.0166532   0.9872724
     0.8825353   0.0099873   0.9972597
     0.2421512   0.0027403           1
     2.484E-26   2.811E-28           1
```

Next, likelihood ratio tests (LRTs) are produced in Output 5.3, which can be used to judge if one, two, or three PCs should be retained for explaining the underlying GEI pattern. The first row of the table corresponds to the LRT for the model without GEI; a *p*-value less than 0.05 indicates that GEI should be modeled. The second row shows the LRT for the model with one PC to explain GEI, and so on until a model with three PCs is shown.

Output 5.3 Goodness-of-fit likelihood ratio tests for adding PCA terms.

```
        Obs      pca       chisq        df       p_value
         1        0       447.236      126             0
         2        1       272.247      104             0
         3        2       148.670       82     .000009445
         4        3        93.259       62     .006271239
```

For the PEANUT data set, the LRTs for selecting the numbers of principal axes suggest that a third PC could still be useful for discovering associations that explain GEI. PC1, PC2, and PC3 explain 85% of the total residual variability that is attributable to GEI.

Beside this information in the output, the macro prints the matrix of residuals from the additive model that was subjected to SVD to obtain the PCs. In the graphics windows three useful plots are automatically displayed: the plot of the first multiplicative components (PC1) vs. the trait means, the second vs. the first multiplicative component plot (PC2 vs. PC1), and the third vs. the first multiplicative component plot (PC3 vs. PC1). Here we show the more commonly used biplot, showing genotype and environment scores on the first two multiplicative components (Output 5.4).

The biplot (Output 5.4) reveals that genotype 10 (mf480) performs quite differently than the other genotypes. Its contribution to GEI is high since in environment 2 it performs relatively better than in other environments. Genotypes 4 (mf485) and 3 (mf484) show positive correlations with environments 12, 10, and 8, and negative correlation with performance of genotype 10 (mf480). If we take into account the mean yield, it is possible to conclude that genotypes 4 (mf485) and 3 (mf484) show a relative high performance in high-yielding environments, while genotypes 10 (mf480) and 9 (mf478) are winner genotypes only in environments 2 and 5, both low-yielding environments. The second multiplicative term allows setting up of differences in the performance of genotypes 1 (Florman) and 2 (Tegua), both check cultivars, and genotype 5 (mf487) with respect to short-cycle genotypes. Thus, genotype 5 (mf487) showed the most similar pattern of performance across environment compared to the check cultivars. Differences between GEI contributions of genotype 3 (mf484) and genotype 6 (mf489) can be detected if one incorporates the third axis (PC3) but not on the first two PCs.

Output 5.4 Biplot for a fixed-effects AMMI model. Names represent genotype codes and numbers denote environment codes.

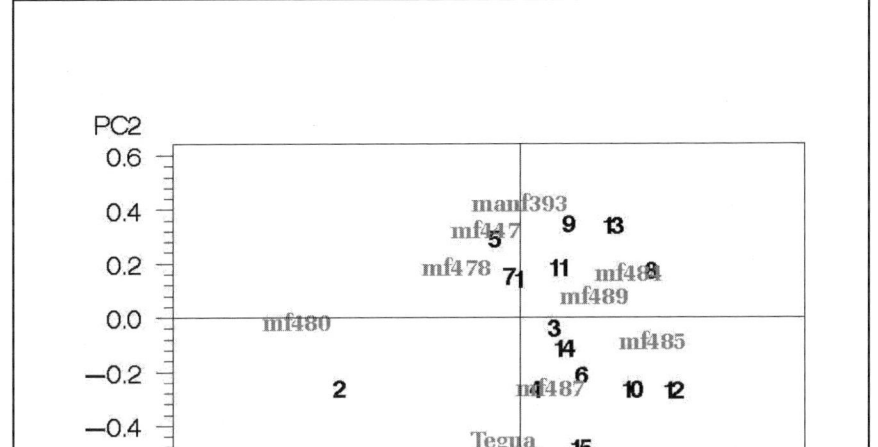

Usually, PC1 genotype and/or environment scores are plotted against the mean response to visually identify the more stable and high-yielding genotypes (or environments) as those with low scores (Output 5.5).

For this data set, like many others, variability among environments is larger than variability among genotypes, so using standardized environment values may improve the biplot. Here, the genotypes manf393, Florman, and Tegua do not seem to make an important contribution to GEI (PC1 close to zero), and they are the three check cultivars for this data set. The genotype mf480 is unstable and low yielding in general.

Output 5.5 PC1 scores from a fixed-effects AMMI model vs. mean response of genotypes (yield). Names represent genotype codes and numbers denote environment codes.

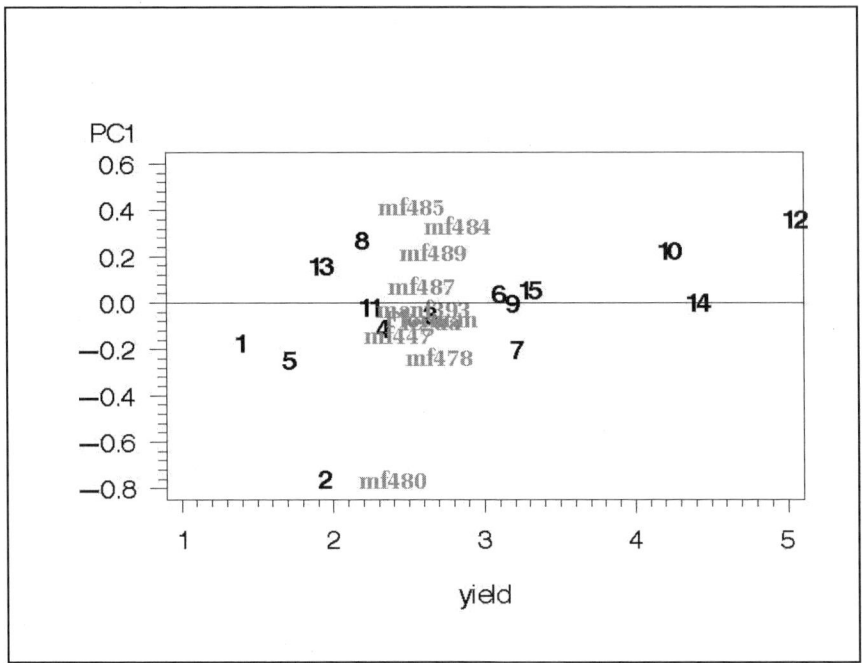

Other commonly used linear-bilinear models for GEI analyses are the genotype regression (GREG) model and the sites regression (SREG) model (Cornelius et al., 1996). For the GREG model, the expected value of the response variable is

$$E(y_{ij}) = \mu + G_i + \sum_k \lambda_k u_{ki} v_{kj},$$

whereas for the SREG models, the expected response is

$$E(y_{ij}) = \mu + E_j + \sum_k \lambda_k u_{ki} v_{kj}.$$

In both cases, we can build the matrix **Z**, whose elements are the residuals obtained by fitting a reduced MET model with GEI, but without either E or G effects. In the SREG models, the multiplicative terms contain the main effect of genotypes plus the GEI (Crossa and Cornelius, 1997). Yan et al. (2000) suggested that for cultivar evaluation, both G and GEI effects must be considered simultaneously to assess both broad and environment-specific adaptation of genotypes. Using an SREG model, they repartitioned the combined G and GEI effects, denoted as GGE, into crossover and non-crossover GEI. The biplot of the first two PCs from the SREG model (GGE biplot) can be used to identify winner genotypes in each subset of environments. As Yan et al. (2000) pointed out, when environment-centered responses of MET are subjected to SVD, the portion of variation that is relevant to cultivar evaluation is partitioned into two components, represented by the first two principal components, PC1 and PC2. Assuming a high correlation among the first PCA scores (PC1) and the genotype means, the PC1 scores are interpreted as representing genotype main effects. Arbitrarily, a positive value is assigned to the PC1 scores so that the genotype PC1 scores are positively correlated with the mean genotype response. Environmental PC1 scores with only positive values imply that the response is always higher for genotypes with larger PC1 scores. The differences among genotypes relative to PC1 will be greater in environments with larger PC1 scores. Thus, PC1 represents a non-crossover GEI or a rank-consistent genotype response. Unlike PC1, the

PC2 scores of genotypes and environments may take both positive and negative values. A genotype that has large positive PC2–based interactions with an environment must have a large negative interaction with some other environment, which is a major source of variation for any crossover GEI. Under this situation, genotypes with large PC1 scores tend to give higher mean yield, and environments with large PC1 scores and near-zero PC2 scores facilitate identification of such genotypes (Yan and Hunt, 2001).

The SREG model and the corresponding GGE biplot can be obtained by deleting, besides GEI, the *G* term from the model to obtain the elements of the **Z** matrix. Heterogeneity of within-environment error variances can still be handled by using the REPEATED statement with the GROUP=ENV option. The following SAS macro gives the environmental and genotype scores for the three linear-bilinear models: GREG, SREG, and SREGH (SREG with heterogeneous residual variances among environments):

```
goptions cback=white;
options nodate ls=75;
proc import datafile="c:\peanut.xls" out=peanut replace; run;  ❶
%include 'c:\GLBM.sas';
%GLBM(DATA=peanut, ENV_N=15 ,GEN_N=10, REP_N=4,
      VAR=yield, ENV=env ,GEN=gen, REP=rep,COMMENT=SREG);  ❷
%include 'c:\PC1PC2PLOT.sas';
%PC1PC2PLOT(DATA=peanut,VAR=yield, ENV=env ,GEN=gen,
         SCORE_e=c.vece, SCORE_g=c.vecg, GENCOD=geno);  ❸
```

❶ Data in a spreadsheet are imported by the IMPORT procedure to create a SAS data set.

❷ The %INCLUDE statement invokes a macro called %GLBM, which saves the environment and genotype scores in two permanent SAS data sets named C.VECA and C.VECG, respectively. They can be used as arguments to macro %PC1PC2PLOT to produce a biplot for the selected general linear-bilinear model. The macro %GLBM takes nine required values: DATA (the name of the data set to be used), ENV_N (the number of environments), GEN_N (the number of genotypes), REP_N (the number of replicates), VAR (the name of the response variable), ENV (the name of the environment variable; should be a numeric variable), GEN (the name of the genotype variable; should be a numeric variable), REP (the name of the replication variable; should be a numeric variable), and COMMENT (one of the model names GREG, SREG, and SREGH to run a genotype regression, a site regression, or a heteroscedastic site regression model). To run this code, users must first download the GLBM.SAS file into C: or redefine the path that calls this macro.

❸ After obtaining the genotype and environment scores for the desired linear-bilinear model from macro %GLBM, SAS users can call the macro %PC1PC2PLOT to yield the typical PC1 vs. PC2 biplot. This statement calls the %PC1PC2 macro to obtain a biplot of the first two PCs of the SREG model for the PEANUT data set. To run this code, users must first download the PC1PC2PLOT.SAS file to C: or redefine the path that calls this macro.

Output 5.6 shows the resulting GGE biplot and shows that genotype PC1 scores are positively correlated with the mean genotype response. Here the extreme PC1 scores correspond to genotypes 10 (mf480) and 3 (mf484), which are the highest-yield and lowest-yield cultivars. The differences among these genotypes, relative to PC1, will be greater in environments with larger absolute PC1 scores (environments 2 and 12). The genotypes 1 (Florman) and 2 (Tegua) that have larger PC2 scores showed a large negative interaction with environment 5, a low-yield environment where the short-cycle check performed better.

Output 5.6 GGE biplot for the PEANUT data set.

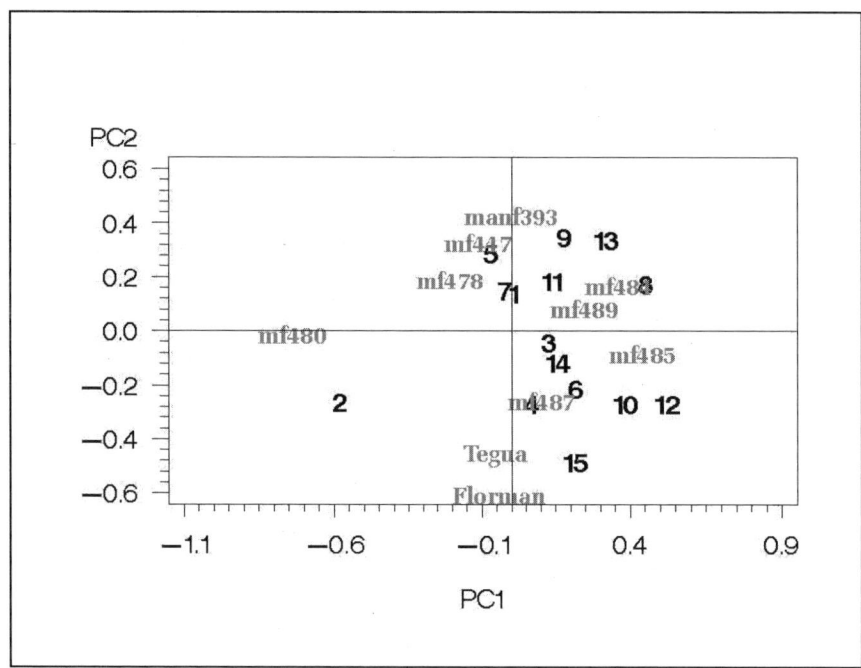

5.2.3 Linear Mixed Model Approach to GEI Analysis

A usual feature of all MET is the representation of a relatively large number of representative elements (Littell et al., 1996). In MET, environments might be reasonably assumed to be random effects. However, the genotype effects might be treated as fixed since only a few highly selected genotypes are usually involved in replicated tests. Therefore, a mixed model approach with environmental effects and GEI effects as random and with genotype effects as fixed is more tenable to fit MET data. The assumptions for the random effects, other than GEI, are that environmental effects, E_j, are independently, identically distributed, or iid, $N(0, \sigma_e^2)$ and replication effects, $R_{k(j)}$, are iid $N(0, \sigma_r^2)$. Error terms are usually assumed to be iid $N(0, \sigma_e^2)$, but a heterogeneous variance model for the error terms is permissible. Environment, replication, interaction, and error effects are independent of one another. The GEI terms are also regarded as normal random effects with zero means but with a variance-covariance matrix not necessarily implying independence and homogeneity of variances. In modeling variances and covariances of the random GEI terms, one might compare the mean performance in a more realistic manner and obtain stability statistics using an appropriate GEI analysis.

The trait means of any two genotypes in the jth environment, $\overline{y}_{ij}$ and $\overline{y}_{i'j}$, have covariance

$$\text{cov}(\overline{y}_{ij}, \overline{y}_{i'j}) = \sigma_E^2 + \text{cov}(\text{GEI}_{ij}, \text{GEI}_{i'j}), \text{ for } i \neq i'.$$

A set of potential mixed models for MET is made by different variance-covariance structures that can be imposed on the interaction term, $\text{cov}(\text{GEI}_{ij}, \text{GEI}_{i'j})$. The mixed model framework allows modeling of the variability of yields from environment to environment and the correlation of genotype values within an environment. The best linear unbiased predictors (BLUPs) of GEI random terms render environment-specific inferences more efficient (Balzarini et al., 2001).

It is important to note that the narrow inference under the fixed ANOVA model for MET relies on comparisons of genotypic means in specific environments. Unfortunately, this procedure does not use all the available information. It is possible to make inferences about the performance of only those genotypes that have been tested in a specific environment. By using BLUP of genotype performance in each environment instead of genotypic mean, we use information from the entire data set to obtain environment-specific inferences, allowing prediction of genotype performance even in environments where the genotype was not tested.

For MET with normal responses, PROC MIXED runs weighted least squares parameter estimation of fixed parameters and restricted maximum likelihood for dispersion parameters (variances and covariances) related to random terms. All random effects are regarded as normal variables with zero mean and specific covariance matrices. PROC MIXED allows different covariance structures for the random environment and GEI terms as well as for the residual variance. These estimation procedures facilitate handling of unbalanced and incomplete genotype-by-environment data sets.

Kang and Magari (1996) used likelihood-based estimation (restricted maximum likelihood, or REML) under a mixed model to estimate stability variances in unbalanced data sets for analyzing GEI. They used a mixed model with fixed genotypes, random environmental and GEI effects, and the assumptions of independent GEI terms with heterogeneous (by genotype) variances. The REML variance components, assignable to each genotype, estimated the same statistics as Shukla's stability variance (Shukla, 1972). The mixed model with heterogeneous GEI terms is *a priori* more tenable than the traditional mixed analysis of variance (independent and heterogeneous variance GEI terms within environment) because it allows different GEI variances for each genotype.

By further modeling the variance-covariance structure of environment and interaction random effects, several stability measures can be derived from the fitted mixed models (Piepho, 1998). The common regression approaches for studying genotype sensitivities to environmental changes with multiplicative models for GEI can be handled by using a factor-analytic variance-covariance structure (Jennrich and Schluchter, 1986) for the GEI terms of the same environment. The following SAS programs exemplify the fitting of several mixed MET models that can be used for GEI analysis:

```
/* Mixed ANOVA for fixed genotypes, random environmental and
replication(environment) effects. Assumptions for the GEI terms: Homogeneous
variance and independence */;

proc mixed;
class env gen rep;
model y=gen;
random env rep(env) gen*env;

/* Mixed Shukla model for fixed genotypes, random environmental and
replication(environment) effects. Assumptions for the GEI terms: Heterogeneous
by genotypes variances and independence */;

proc mixed;
class env gen rep;
model y=gen;
random int rep/subject=env;
random gen/subject=env type=UN(1);

/* Mixed AMMI(n) model for fixed genotypes, random environmental and
replication(environment) effects. Assumptions for the GEI terms: Heterogeneous
by genotypes variances and covariances between GE terms of two genotypes in the
same environment */;

proc mixed;
class env gen rep;
model y=gen;
```

```
random int rep/subject=env;
random gen/subject=env type=FA0(n);
```

/* Mixed Eberhart & Russell model for fixed genotypes, random
replication(environment) effects. Assumptions for the GEI terms: Heterogeneous
by genotypes variances and covariances between GEI terms of two genotypes in the
same environment */;

```
proc mixed;
class env gen rep;
model y=gen;
random rep/subject=env;
random gen/subject=env type=FA1(1);
```

/* Mixed HetR model for fixed genotypes, random environmental and
replication(environment) effects. Assumptions for the GEI terms: Homogeneous
variance and independence. Assumptions for the error terms: Heterogeneous by
environment variance */;

```
proc mixed;
class env gen rep;
model y=gen;
random env rep(env) gen*env;
repeated/group=env;
```

The following SAS statements can be used to facilitate GEI analysis using mixed models. They invoke a SAS macro called %RUNMIXED to automatically run a classical two-way mixed model, the stability variance mixed model, and Finlay-Wilkinson, Eberhart-Russell, and AMMI(2) versions of a mixed model for a typical MET model. At the end of the output, users can find a table with several fitting criteria to select the most appropriate covariance model for the data.

```
options nodate ls=75;
proc import datafile="c:\peanut.xls" out=peanut replace; run;   ❶
/* creating base files for appending results*/;
        data crit;
        length structr $25;
        length descr $25;
        structr=' ';descr=' ';value=.;
        data LSM;
        length structr $25;
        length msgroup $20;
        structr=' ';gen=.;
        Estimate=.;
        msgroup=' ';
/* running macro runmixed with 5 mixed models for MET */;
        %let class=GEN ENV REP;
        %include   'c:\RUNMIXED.sas';   ❷
        *2 way Analysis;
        %runmixed(data=peanut, GEN=gen,ENV=env, REP=rep, VAR=yield,
        model=GEN,
        z=random ENV GEN*ENV REP(ENV),
        comment=2W, outfit=CRIT, outlsm=lsm, alpha=0.05);

        * Shukla stability variance;
        %runmixed(data=peanut, GEN=gen,ENV=env, REP=rep, VAR=yield,
        model=GEN,
        z=random int REP/subject=ENV ;
        random GEN/subject=ENV type=un(1),
        comment=SV, outfit=CRIT, outlsm=lsm, alpha=0.05);

        *Finlay-Wilkinson;
        %runmixed(data=peanut, GEN=gen,ENV=env, REP=rep, VAR=yield,
        model=GEN,
```

```
z=random int REP/sub=ENV;
random GEN/sub=ENV type=FA1(1),
comment=FW, outfit=CRIT, outlsm=lsm, alpha=0.05);

*Eberhart-Russell;
%runmixed(data=peanut, GEN=gen,ENV=env, REP=rep, VAR=yield,
model=GEN,
z=random REP/sub=ENV;
random GEN/sub=ENV type=FA1(1),
comment=ER, outfit=CRIT, outlsm=lsm,alpha=0.05);

*AMMI(2);
%runmixed(data=peanut, GEN=gen,ENV=env, REP=rep, VAR=yield,
model=GEN,
z=random int REP/sub=ENV;
random GEN/sub=ENV type=FA0(2),
comment=FA2, outfit=CRIT, outlsm=lsm, alpha=0.05);

proc print data=lsm; proc print data=crit noobs;run;
```

❶ Data in a spreadsheet are imported by the IMPORT procedure to create a SAS data set.

❷ The %INCLUDE statement invokes a macro called %RUNMIXED, which takes 11 required values: DATA (the name of the data set to be used), GEN (the name of the genotype variable; should be a numeric variable), ENV (the name of the environment variable; should be a numeric variable), REP (the name of the replication variable; should be a numeric variable), VAR (the name of the response variable), and other arguments that should not be modified by the user, since they define the different models to be run. The last argument is the alpha value to use for mean comparisons. To run this code, users must first download the RUNMIXED.SAS file to C: or redefine the path that calls this macro. The last part of the output produced by the statements (Output 5.7) allows comparing the fitted models to choose the best one for the data.

Output 5.7 Comparing alternative mixed models with random GEI.

Struct	Descr	value
2W	-2 Res Log Likelihood	957.398
2W	AIC (smaller is better)	965.398
2W	AICC (smaller is better)	965.468
2W	BIC (smaller is better)	968.230
SV	-2 Res Log Likelihood	941.855
SV	AIC (smaller is better)	967.855
SV	AICC (smaller is better)	968.498
SV	BIC (smaller is better)	977.060
FW	-2 Res Log Likelihood	910.761
FW	AIC (smaller is better)	938.761
FW	AICC (smaller is better)	939.504
FW	BIC (smaller is better)	948.673
ER	-2 Res Log Likelihood	927.677
ER	AIC (smaller is better)	953.677
ER	AICC (smaller is better)	954.320
ER	BIC (smaller is better)	962.882
FA2	-2 Res Log Likelihood	870.435
FA2	AIC (smaller is better)	914.435
FA2	AICC (smaller is better)	916.252
FA2	BIC (smaller is better)	930.013

The model with a factor analytic covariance structure for the genotype means within an environment, based on two multiplicative components (FA2), is the best one among the fitted mixed models for the PEANUT data set. All fitting criteria have smaller values for this model. Other models fail in declaring significant genotype effects. For the FA2 model, the p-value associated with genotype effect is $p=0.0002$.

The mean comparisons based on this mixed model made pairwise comparisons of genotype performance based on both overall yield mean and stability of production across environments. Thus, broad inference can be made (Output 5.8).

Output 5.8 Comparing overall genotype performance in MET under a mixed AMMI model with two multiplicative components.

```
Struct      msgroup     gen     estimate
FA2         AB          3       2.82199
FA2         AC          9       2.68944
FA2         CD          6       2.65422
FA2         ABCE        1       2.63256
FA2         ABCE        2       2.61929
FA2         ABC         7       2.59157
FA2         CDE         5       2.57293
FA2         CDE         4       2.51407
FA2         E           8       2.41963
FA2         BDE         10      2.36988
```

Genotype 3 (mf484) appears as significantly different (higher yielding, taking into account variability across environment) from genotypes 4, 5, 6, 8, and 10. Other models fail to detect all these differences.

For environment-specific inferences, users can ask for the BLUPs of each genotype in each environment using the ESTIMATE statement (Littell et al., 1996). Biplots under the mixed model framework can still be obtained. The estimated factor loadings of the factor analytic structure for the AMMI(2) model are used as genotype scores and the empirical BLUPs of the random environmental variables from each multiplicative term are used as environmental scores to plot genotypes and environments in a two-dimensional space. A plot with the environmental scores and the genotype scores for the first two multiplicative terms will play a role analogous to the traditional biplot of the first two principal component scores for genotype and environment in the fixed ANOVA model approach. Despite the differing procedures to obtain biplots under both approaches, they show the same interaction pattern with balanced data sets (Balzarini and Macchiavelli, 2000). An important advantage of the mixed model framework is that biplots can still be obtained with incomplete data.

5.2.4 Generalized Linear Models to Explore GEI

The previous linear models specify that the response is a linear function of G, E, and GEI effects with error terms normally distributed. In the generalized linear models, the distribution of the response variable is assumed to be any member of the exponential family of distributions. For example, it could be binomial, Poisson, exponential, or gamma, in addition to normal. Also, the relationship between the expected response and the linear predictor is specified by a non-linear link function $g(\mu)$, where μ is the expected response; the g function must be monotonic and differentiable. For generalized linear models in which the response variable follows the normal distribution, the canonical link function is the identity. The identity link establishes that the expected mean of the response variable is identical to the linear predictor, rather than to a non-linear function of the linear predictor. The canonical link functions for probability distributions, which could model the random component of MET models for several trait types, are given in Table 5.2.

Table 5.2 Common trait probability distributions and canonical link functions in MET.

Probability distribution	Canonical link function
Normal	Identity
Binomial	Logit or probit
Poisson	Log
Gamma	Reciprocal
Negative binomial	Log

The parameters in a generalized linear model can be estimated via maximum likelihood-based methods. Analogous to the residual sum of squares in the linear models, the goodness of fit of a generalized linear model can be measured by a quantity, in this case called the scaled deviance:

$$D(y;\hat{\mu}) = 2[l(y;y) - l(\hat{\mu};y)],$$

where $l(y;y)$ is the maximum likelihood achievable for an exact fit in which the fitted values are equal to the observed values, and $l(\hat{\mu};y)$ is the log-likelihood function calculated at the estimated parameters. The deviance function is quite useful for comparing two models when one model has parameters that are a subset of the second model.

Consider two nested models where the second one has some covariates omitted, and denote the maximum likelihood estimates in the two models by $\hat{\mu}_1$ and $\hat{\mu}_2$, respectively. Then the deviance difference, $D(y;\hat{\mu}_1) - D(y;\hat{\mu}_2)$, is identical to the likelihood ratio statistic and has an approximate chi-square distribution with degrees of freedom equal to the difference between the numbers of parameters in the two models. Thus, the composition of the likelihood ratio test for hypothesis testing can be realized using the deviances as estimated with PROC GENMOD.

There are a number of distribution models for dealing with discrete responses. Two common approaches for count data are the Poisson model and its modifications to deal with overdispersion (e.g., the negative binomial model). The Poisson model uses a one-parameter model to describe the distribution of the dependent variable since the variance is expected to be equal to the mean. The negative binomial model adds an "overdispersion" parameter to estimate the possible deviation of the variance from the expected value. For the negative binomial model, the variance is expected to be larger than the mean. A potentially very useful model is the gamma distribution model. Even though the gamma is a continuous model, it represents a natural alternative with respect to the negative binomial model. In the gamma model, the expected value and the variance of the response variable are regarded as simultaneous outcomes, and each may be estimated as a function of the set of independent variables.

The fixed ANOVA models for MET can be adjusted using PROC GENMOD in accordance with the trait probability distribution. To properly rank genotypes in light of GEI, covariance parameters can be calculated even when data are non-normal. The traditional mixed approach to GEI analysis, which splits GEI into variance components that are assignable to each genotype, is implemented here under a generalized linear model.

The following SAS statements replicate the GEI analysis given in the paper by Kang and Gorman (1989) for normal data as a particular case of a generalized linear model. The data set in the paper involves codes for environment index (ENVINDEX), genotypes (GEN), and the values of the response variable (YIELD).

```
%include 'c:glimmix8.sas' / nosource;  ❶
%glimmix (data=ONE,  ❷
procopt=method=REML  maxfunc = 1000 maxiter = 300 lognote,
stmts=%str(
class GEN ENVINDEX;
model yield = ENVINDEX;
random GEN/SOLUTION ;
repeated/group= gen;
parms /eqcons=1 to 18/EQCONS=3;),
error=normal,
converge= 1e-10,
link=identity);
run;
```

❶ The %INCLUDE statement invokes a macro called %GLIMMIX, which takes a SAS data set and fits a generalized linear mixed model. To run this code, users must have the GLIMMIX8.SAS file on C: or redefine the path that calls this macro.

❷ The %GLIMMIX statement gives the required arguments to the macro. Data set ONE contains data published in Kang and Gorman (1989). The subsequent statements use the %GLIMMIX macro to decompose the variation associated with each genotype under the normal model, but they can be easily adapted to other probability distributions, properly changing the error and the link statements. From this analysis, the most stable genotypes are identified.

The same conclusions are reached by the set of estimates provided by %GLIMMIX (Table 5.3). The correlation between σ^2_i and %GLIMMIX is $r=0.98$ ($P=0.001$).

Table 5.3 A comparison of results from Kang and Gorman (1989) with those from %GLIMMIX.

σ^2_i	%GLIMMIX	σ^2_i	%GLIMMIX
2.47	0.57	2.27	0.54
1.63	0.46	2.56	0.62
0.57	0.18	1.56	0.39
2.61	0.64	3.48	0.82
1.86	0.41	5.16	1.24
3.58	0.88	2.38	0.58
3.58	0.98	3.45	0.90
2.72	0.72	3.53	0.95
4.25	0.96		

Proving Whether or Not GEI Is Real

The following analyses conducted with PROC GENMOD and PROC MIXED are useful to demonstrate how GEI can be verified as real and whether the proposed model is measuring its own inadequacy for the data rather than the intended hypothesis. Note that with some minor modifications the same analysis could be performed using %GLIMMIX. The data used here are from Huehn (1990).

The first step in this example is to determine the distribution that better approximates the Huehn data—i.e., to select a distribution that fits the data.

```
options nodate ls=75;
proc import datafile="c:\Huehn.xls" out=HD replace;   ❶
run;

proc genmod data=HD;   ❷
class genotype location;
    model yield=genotype location / dist=normal ;      * Model A ;

proc genmod data=HD;
class genotype location;
    model yield=genotype location / dist=poisson;     * Model B;

proc genmod data=HD;
class genotype location;
    model yield=genotype location / dist=gamma   ;    * Model C;
run;
```

❶ Data in a spreadsheet are imported by the IMPORT procedure to create a SAS data set.

❷ PROC GENMOD is run for generalized linear models based on a normal, Poisson, or gamma response. No structural GEI components have been added to the SAS code. We are investigating the structure of the dispersion parameters. Model A is the ordinary least squares model using a homogeneous variance independent of the mean. Model B is a heterogeneous variance model. For model C, the expected value and the variance of the dependent variable are regarded as simultaneous outcomes. Models A and C are continuous models, and model B is a discrete model.

A second step is to investigate the nature of the inference we can make using these different probability models including a GEI term in the model.

```
proc genmod;
class genotype location;
model yield =genotype location genotype*location / dist=normal type3 ;
output out=normal predicted=yhat;

proc genmod;
class genotype location;
model yield =genotype location genotype*location/ dist=poisson type3;
output out=poisson predicted=yhat;

proc genmod;
class genotype location;
model yield =genotype location genotype*location/ dist=gamma type3;
output out=gamma predicted=yhat;
run;
```

Finally, we compare the restricted maximum likelihood of two mixed models, with the location effects as random, one with GEI and the other without GEI; i.e., we simply correct for the heterogeneity of variance caused by locations.

```
proc mixed ;
class location genotype;
model yield = genotype;                      *[model E] ;
random location;
run;

proc mixed ;
class location genotype;
model yield = genotype;                      *[model F];
random location location*genotype;
run;
```

The restricted likelihood calculated for each model can be used to compose the likelihood ratio test for departures from the initial assumption of homogeneous variances. We have 460.1[model E] − 458.9 [model F] = 1.2. This statistical value is distributed in accordance with the chi-square distribution, and with one degree of freedom this result is not statistically significant at the $p = .05$ level. Thus, we conclude that the GEI is statistically nonsignificant, and demonstrate that the original normal model was indeed measuring departures from the assumption. Hence, finding genotypes that perform better than others in one environment does not necessarily mean local adaptation.

5.3 Smoothing Spline Genotype Analysis (SSGA)

The data set used here is available in Kang (1994, p. 131). The data set contains the variables LOCATION, GENOTYPE, and YIELD. We apply smoothing spline analysis. A smoother is essentially a tool to analyze trends in the response variable; the essence of smoothing is to produce a trend that is less variable than the response itself (Hastie and Tibshirani, 1990).

```
options nodate ls=75;
proc import datafile="c:\kang.xls" out=SSGA replace;  ❶
run;

proc gam data=SSGA;  ❷
class genotype;
model yield =  param(genotype) spline(location,df=100) ;
output out=estimate p all;
id loc gen;
run;
```

❶ Data in a spreadsheet are imported by the IMPORT procedure to create a SAS data set.

❷ PROC GAM is run with 100 degrees of freedom to look for the maximum level of smoothing the data can allow. We can see from the SAS output that the degrees of freedom for SPLINE(LOCATION) are 11.28. Next, we are going to try to reduce them and develop a likelihood ratio test to see if the reduced model is tenable.

Rerunning the SAS statements above, using 11 degrees of freedom in place of 100, generates the mentioned likelihood test. The final residual sum of squares changes from 161 (for 12.28 df) to 171 (for 11 df); the difference is greater than a chi-square value with one degree of freedom. Hence, the reduction in the smoothing level does significantly influence the fit of the model. We can think of degrees of freedom as a measure of how much smoothing is needed. As the degrees of freedom increase, bias is reduced and standard error bands of the predictors are increased. Hence, we can have some control over bias and truly obtain an unbiased estimator. If we add the following SAS code, we can visualize the model fit as shown in Output 5.9 and Output 5.10:

```
legend1 frame cframe=ligr cborder=black label=none
position=center;
axis1    label=(angle=90 rotate=0);
axis2    minor=none;
symbol1 color=red interpol=join value=none line=1;
proc gplot data=estimate;
plot p_yield*location = 1 /overlay legend=legend1
frame cframe=ligr vaxis=axis1 haxis=axis2;
```

Output 5.9 Linear fit of the predicted values by location.

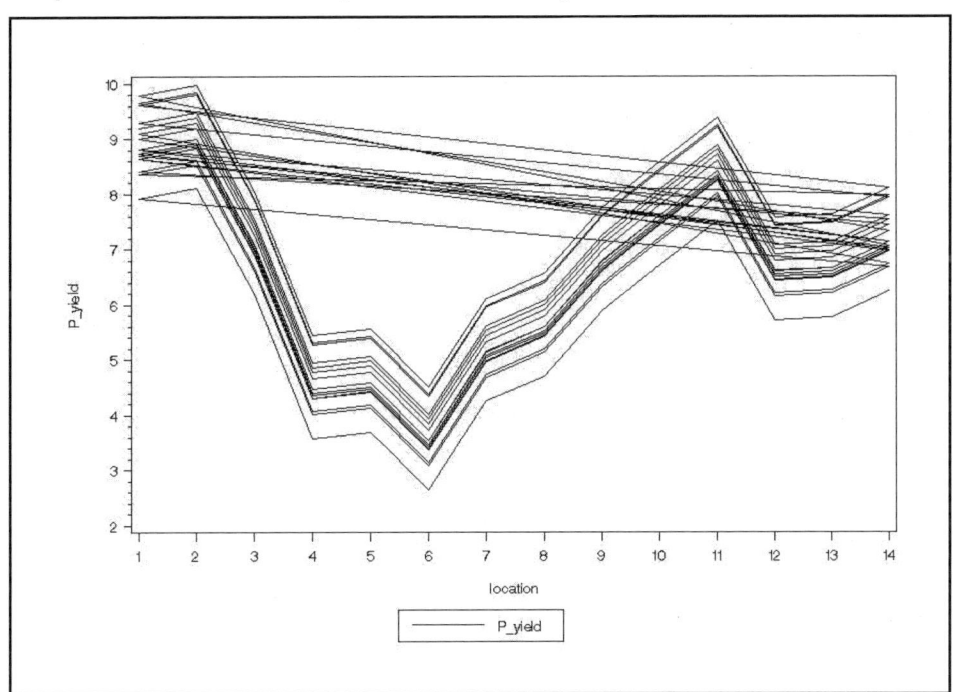

Output 5.10 Predicted yield by location (linear model).

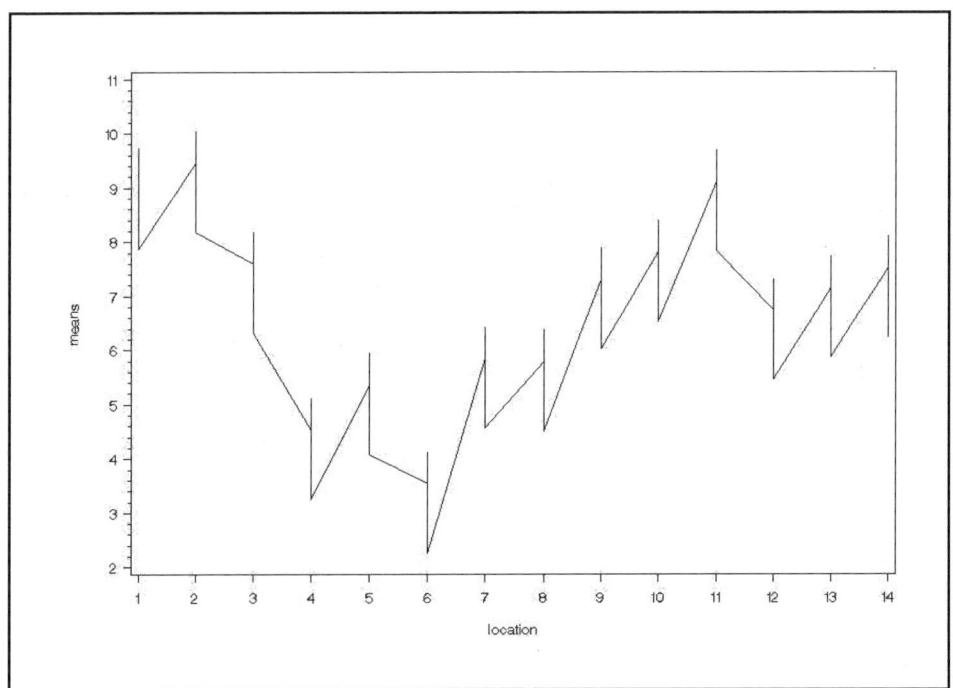

The linear or Gaussian model gives a clear idea of what is happening. But the nonparametric model relaxes the parametric assumptions, and its heterogeneous slope property allows the response variable to vary freely between locations without the need for treating location as a class variable. The fit of the nonparametric model does approximate the parametric linear model and it is able to describe more of the underlying variation. It is clear that the linearity is not the main trend here. The GEI is significant in the linear model because a heterogeneous slope does improve the fit of the model, but at the cost of our making improper inference if we are only detecting deviations from initial assumptions. Let us take a look at the raw data in three dimensions using the following SAS code to produce Outputs 5.11, 5.12, and 5.13.

```
proc g3d data=two;
   plot gen*location=yield ❶/ zmin=2 zmax=11;
proc g3d data=Lestimate; plot gen*location=yhat; ❷
proc g3d data=estimate; plot gen*location = p_hat; ❸
run;
```

❶ This plots the observed values (Output 5.11).

❷ This plots the predicted values from the linear model (Output 5.12).

❸ This plots the predicted values from the nonparametric model (Output 5.13).

Output 5.11 Three-dimensional display of raw data.

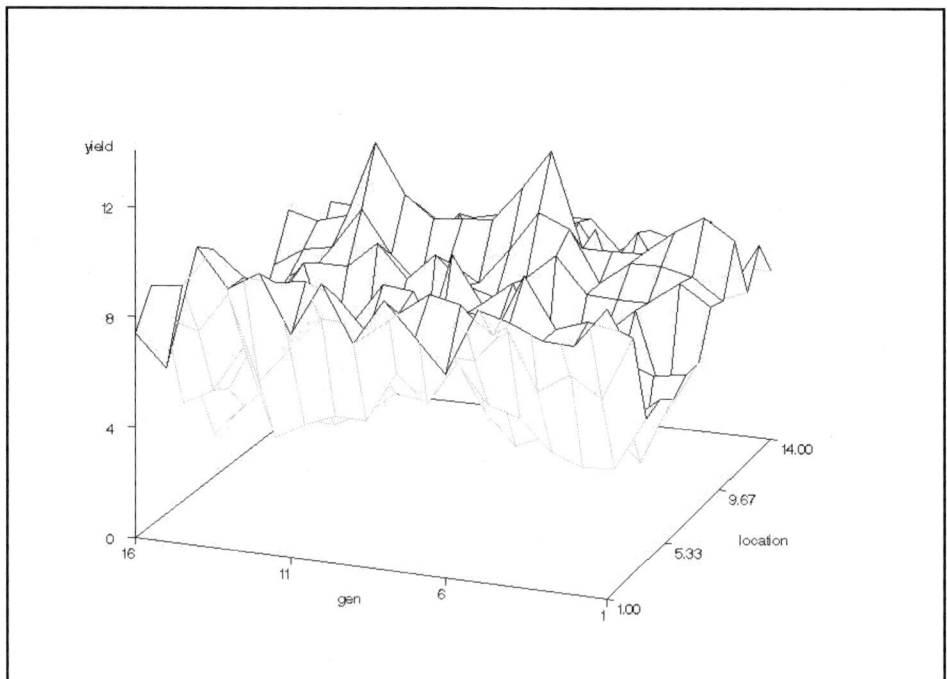

Output 5.12 shows how the linear (Gaussian) model adjusts the data. The adjustment here is the classical statistical adjustment. The comparison of the linear adjustment of the data with the nonparametric adjustment of the data (Output 5.13) reveals the heterogeneity of variance bias that the regression line introduces in the model. A 12-degrees-of-freedom nonparametric smoother spline as seen in Output 5.13 is less biased but more variable. Thus, if properly adjusted, it would be more appropriate to describe the data. With a proper level of smoothing (something less than the infinite smoothing of the regression line), we can be confident that our predictions will be less biased. In Table 5.4, we can see how the statistics computed by the nonparametric model compared with the Pi statistics (Lin and Binns, 1988) for the same data.

Output 5.12 Linear adjustment of the raw data.

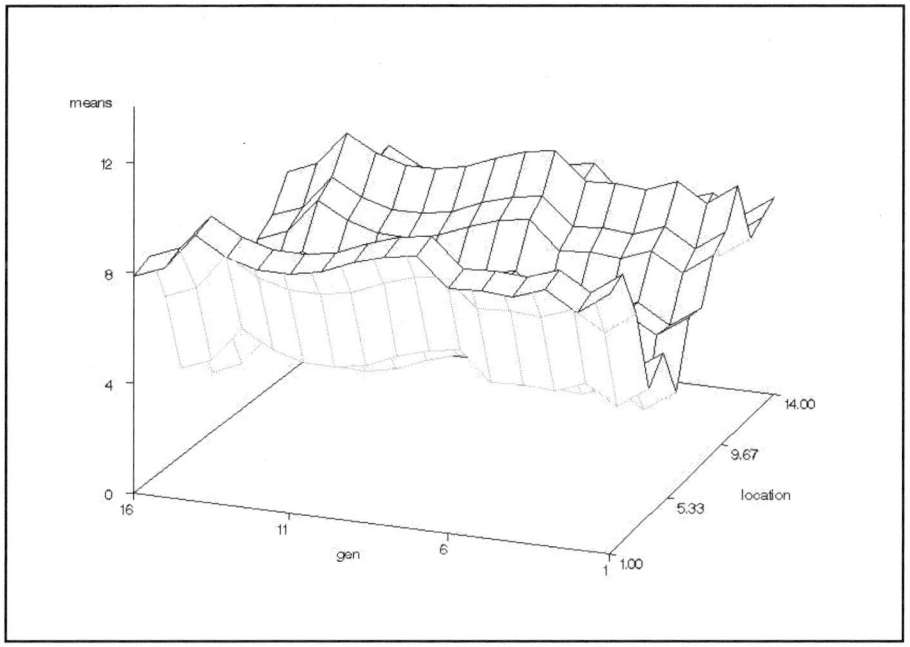

Output 5.13 Nonparametric adjustment of the raw data.

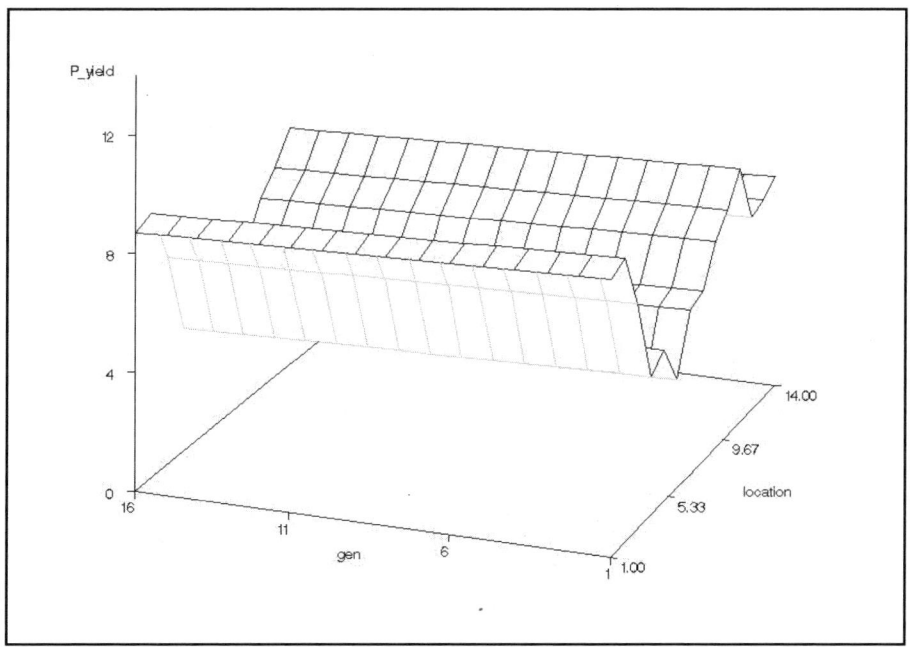

Table 5.4 SSGE parameters and the Pi statistics.

Parameter	Parameter Estimate	Pr > |t|	Pi
gen g1	−0.08286	0.8093	1.429
gen g10	−0.46857	0.1734	2.257
gen g11	−0.65143	0.0590	2.548
gen g12	−0.63500	0.0656	2.611
gen g13	−0.28714	0.4035	2.380
gen g14	0.35714	0.2990	0.938
gen g15	−0.93786	0.0068	3.244
gen g16	−1.36714	<.0001	4.383
gen g2	−0.87714	0.0113	3.161
gen g3	−0.17786	0.6046	1.625
gen g4	−0.58500	0.0896	2.465
gen g5	−0.48071	0.1626	2.303
gen g6	−0.54286	0.1151	2.953
gen g7	0.49214	0.1529	0.811
gen g8	0.33143	0.3350	0.840
gen g9	0.00000	.	1.639

The correlation between the parameter estimates and Pi equals −0.977.

As the parameter estimates of SSGE increase, Pi decreases. The smallest parameter estimate for SSGE (genotype g16) gives the largest Pi statistics. A reason behind the inverse relationship is that all parameters in SSGE are expressed as deviations from genotype g9. SSGE has the advantage of relaxing many of the assumptions of the traditional linear model. It is a semiparametric model, but a full bivariate nonparametric model can be used.

5.4 References

Annicchiarico, P. 2002. Genotype x environment interaction: Challenges and opportunities for breeding and cultivar recommendations. FAO Plant Production and Protection Paper 174. Rome: Food and Agriculture Organization of the United Nations.

Baker, R. J. 1990. Crossover genotype-environmental interaction in spring wheat. In *Genotype-by-Environment Interaction and Plant Breeding*, ed. M. S. Kang, 42-51. Baton Rouge: Louisiana State University Agricultural Center.

Balzarini, M. 2002. Applications of mixed models in plant breeding. In *Quantitative Genetics, Genomics, and Plant Breeding*, ed. M. S. Kang, 353-365. New York: CABI Publishing.

Balzarini, M., and R. Macchiavelli. 2000. Mixed AMMI models for investigating genotype-environment interaction. *Proceedings of the Symposium on Variance Components Analysis*, Gainesville, FL.

Balzarini, M., S. B. Milligan, and M. S. Kang. 2001. Best linear unbiased prediction: A mixed model approach in multi-environment trials. In *Crop Improvement: Challenges in the Twenty-first Century*, ed. M. S. Kang, 102-113. Binghamton, NY: Food Products Press.

Cooper, M., and G. L. Hammer, eds. 1996. *Plant Adaptation and Crop Improvement*. Wallingford, UK: CABI Publishing; Patancheru, India: ICRISAT; and Manila: IRRI.

Cornelius, P. L., J. Crossa, and M. S. Seyedsadr. 1996. Statistical test and estimates of multivariate models for GE interaction. In *Genotype-by-Environment Interaction*, ed. M. S. Kang and H. G. Gauch Jr., 199-234. Boca Raton, FL: CRC Press.

Crossa, J., and P. L. Cornelius. 1997. Sites regression and shifted multiplicative model clustering of cultivar trial sites under heterogeneity of error variance. *Crop Science* 37:405-415.

Cullis, B. R., B. J. Gogel, A. P. Verbyla, and R. Thompson. 1998. Spatial analysis of multi-environment early generation trials. *Biometrics* 54:1-18.

Eberhart, S. A., and W. A. Russell. 1966. Stability parameters for comparing varieties. *Crop Science* 6:36-40.

Finlay, K. W., and G. N. Wilkinson. 1963. The analysis of adaptation in a plant breeding programme. *Australian Journal of Agricultural Research* 14:742-754.

Franke, D. E., O. Habet, L. C. Tawah, A. R. Williams, and S. M. DeRouen. 2001. Direct and maternal genetic effects on birth and weaning traits in multibreed cattle and predicted performance of breed crosses. *J. Anim. Sci.* 79:1713-1722.

Gabriel, K. R. 1971. Biplot display of multivariate matrices with application to principal components analysis. *Biometrika* 58:453-467.

Gauch, H. G., Jr. 1988. Model selection and validation for yield trials with interaction. *Biometrics* 44:705-715.

Gollob, H. F. 1968. A statistical model which combines features of factor analytic and analysis of variance. *Psychometrika* 33:73-115.

Hastie, T. J., and R. J. Tibshirani. 1990. *Generalized Additive Models.* New York: Chapman & Hall.

Huehn, M. 1990. Nonparametric estimation and testing of genotype X environmental interaction by ranks. In *Genotype-by-Environment Interaction and Plant Breeding*, ed. M. S. Kang, 69-93. Baton Rouge: Louisiana State University Agricultural Center.

Hussein, M. A., A. Bjornstad, and A. H. Aastveit. 2000. SASG x ESTAB: A SAS program for computing genotype x environment stability statistics. *Agronomy Journal* 92:454-459.

Jennrich, R. L., and M. D. Schluchter. 1986. Unbalanced repeated measures models with structured covariance matrices. *Biometrics* 42:805-820.

Kang, M. S. 1990. Understanding and utilization of genotype-by-environment interaction in plant breeding. In *Genotype-by-Environment Interaction and Plant Breeding*, ed. M. S. Kang, 52-68. Baton Rouge: Louisiana State University Agricultural Center.

Kang, M. S. 1993. Simultaneous selection for yield and stability in crop performance trials: Consequences for growers. *Agronomy Journal* 85:754-757.

Kang, M. S. 1994. *Applied Quantitative Genetics.* Baton Rouge, LA: M. S. Kang Publishers.

Kang, M. S. 1998. Using genotype-by-environment interaction for crop cultivar development. *Advances in Agronomy* 62:199-252.

Kang, M. S. 2002. Preface to *Crop Improvement: Challenges in the Twenty-first Century*, ed. M. S. Kang. Binghamton, NY: Food Products Press.

Kang, M. S. 2004. Breeding: Genotype by environment interaction. In *Encyclopedia of Plant and Crop Science*, ed. R. M. Goodman. New York: Marcel Dekker.

Kang, M. S., and H. G. Gauch Jr., eds. 1996. *Genotype-by-Environment Interaction.* Boca Raton, FL: CRC Press.

Kang, M. S., and D. P. Gorman. 1989. Genotype x environment interaction in maize. *Agronomy Journal* 81:662-664.

Kang, M. S., and R. Magari. 1996. New developments in selecting for phenotypic stability in crop breeding. In *Genotype-by-Environment Interaction*, ed. M. S. Kang and H. G. Gauch Jr., 1-14. Boca Raton, FL: CRC Press.

Khuri, A. I., T. Mathew, and B. K. Sinha. 1998. *Statistical Tests for Mixed Linear Models.* Wiley Series in Probability and Statistics. New York: John Wiley & Sons.

Lefkovitch, L. P. 1990. Genotype-by-environment interaction, heterogeneous variance, and significance. In *Genotype-by-Environment Interaction and Plant Breeding*, ed. M. S. Kang, 20-27. Baton Rouge: Louisiana State University Agricultural Center.

Lin, C. S., and M. R. Binns. 1988. A superiority measure of cultivar performance for cultivar x location data. *Can. J. Plant Sci.* 68:193-198.

Littell, R. C., G. A. Milliken, W. W. Stroup, and R. Wolfinger. 1996. *SAS Systems for Mixed Models.* Cary, NC: SAS Institute Inc.

Macchiavelli, R., and J. Beaver. 1999. Analysis of genotype-by-environment interaction with AMMI models using SAS PROC MIXED. *Proceedings of the Kansas State University Conference on Applied Statistics in Agriculture*, Manhattan, KS.

McCullagh, P., and J. A. Nelder. 1983. *Generalized linear models.* London: Chapman & Hall.

Montaldo, H. H. 2001. Genotype by environment interaction in livestock breeding programs: A review. *Interciencia* 26(6):229-235.

Piepho, H. P. 1998. Empirical best linear unbiased prediction in cultivar trials using factor-analytic variance-covariance structures. *Theor. Appl. Genet.* 97:195-201.

Price, W. J., and B. Shafii. 1993. The use of biplots in diagnosing interaction patterns of two-way classification data. *Proceedings of the Eighteenth Annual SAS Users Group International Conference*, New York, NY.

Rédei, G. P. 1998. *Genetics Manual: Current Theory, Concepts, Terms.* River Edge, NJ: World Scientific.

Saxton, A. M. 1998. A macro for converting mean separation output to letter groupings in PROC MIXED. *Proceedings of the Twenty-third Annual SAS Users Group International Conference*, Nashville, TN. (Software available at http://animalscience.ag.utk.edu/faculty/saxton/software.htm.)

Searle, S. R., G. Casella, and C. H. McCulloch. 1992. *Variance Components.* New York: John Wiley & Sons.

Shukla, G. K. 1972. Some statistical aspects of partitioning genotype environmental components of variability. *Heredity* 29:237-245.

Singh, M., S. Ceccarelli, and S. Grando. 1999. Genotype x environment interaction of crossover type: Detecting its presence and estimating the crossover point. *Theor. Appl. Genet.* 99:988-995.

Smith, A., B. Cullis, D. Luckett, G. Hollamby, and R. Thompson. 2002a. Exploring variety-environment data using random effects AMMI models with adjustments for spatial field trend: Part 2: Applications. In *Quantitative Genetics, Genomics, and Plant Breeding*, ed. M. S. Kang, 337-352. New York: CABI Publishing.

Smith, A., B. Cullis, and R. Thompson. 2002b. Exploring variety-environment data using random effects AMMI models with adjustments for spatial field trend: Part 1: Theory. In *Quantitative Genetics, Genomics, and Plant Breeding*, ed. M. S. Kang, 323-336. New York: CABI Publishing.

Snee, R. D. 1982. Nonadditivity in a two-way classification: Is it interaction or nonhomogeneous variance? *J. Amer. Statist. Assoc.* 77:515-519.

Yan, W., and L. A. Hunt. 2001. Genetic and environment causes of genotype by environment interaction for winter wheat yield in Ontario. *Crop Science* 41:19-25.

Yan, W., L. A. Hunt, Q. Sheng, and Z. Szlavnics. 2000. Cultivar evaluation and mega-environment investigation based on the GGE biplot. *Crop Science* 40(3):597-605.

Yan, W., and M. S. Kang. 2003. GGE Biplot Analysis: A Graphical Tool for Breeders, Geneticists, and Agronomists. Boca Raton, FL: CRC Press.

Zobel, W. R., M. J. Wright, and H. G. Gauch. 1988. Statistical analysis of a yield trial. *Agronomy Journal* 80(3):388-393.

Chapter 6

Growth and Lactation Curves

Nicolò P. P. Macciotta, Aldo Cappio-Borlino, and Giuseppe Pulina

6.1 Introduction 97
6.2 Modeling Lactation and Growth Curves 98
6.3 Using PROC REG to Fit Lactation and Growth Data 101
6.4 Using PROC NLIN to Fit Non-linear Models 110
6.5 Repeated Measures Theory for Lactation and Growth Curves 123
6.6 PROC MIXED for Test Day Models 125
6.7 Prediction of Individual Test Day Data 138
6.8 The SPECTRA and ARIMA Procedures for Time Series Prediction 139
6.9 References 146

6.1 Introduction

The expression of complex traits often consists of several measurements taken sequentially on the same experimental unit even if they are usually synthesized by a single measure. Examples are cumulated milk yield for standardized lactation lengths (for example the 305-day yield in dairy cattle) and body weights at standard ages (birth, weaning, and mature age). However, in recent years there has been a relevant development of models able to analyze these traits as specific cases of "repeated measures"—i.e., as traits that change as a function of age or of some independent and continuous variable (Pletcher and Geyer, 1999). Such an approach is able to evaluate possible differences in the genetic control of the trait over time and to account for environmental factors that could act differently on each test day measure. This chapter is focused primarily on the two main technical aspects of the analysis of these infinite–dimensional traits (Kirkpatrick and Heckman, 1989): (1) the choice of suitable functions of time used to fit longitudinal data and (2) the use of linear mixed models for describing the (co)variance structure of repeated measures and for accounting for main environmental factors. These two points are developed through the most current genetic models, *random regression* and *covariance function* models. Random regression models are mixed models in which the sequential nature of data is accounted for by particular functions of time whose regression coefficients are allowed to vary from animal to animal (van der Werf and Schaeffer, 1997), whereas covariance functions are more focused on the correlation structure among repeated measurements. Finally, an approach based on the time series analysis that can represent a useful option for solving problems of future (or missing) data prediction, a problem that frequently occurs in repeated measures analysis, will be presented.

6.2 Modeling Lactation and Growth Curves

Daily milk yield at different stages of lactation in dairy species and body weight at different ages in beef animals are probably the most common examples of traits expressed along a trajectory of time. Models of lactation and growth curves have been traditionally used in the management of dairy and beef enterprises. They represent an essential subcomponent of random regression and covariance function models that, as previously stated, are currently the most innovative approaches for genetic analysis for both dairy and beef production traits.

The pattern of milk yield during lactation is the result of processes of synthesis and secretion of organic compounds and of active and passive blood filtration by specialized epithelial cells of the mammary gland. Starting from gestation there is a phase of rapid cellular activation, followed by a regression (cellular remodeling) at varying rates that finishes with the cessation of lactation, or dry-off. All these physiological mechanisms result in the typical pattern of the lactation curve (Figure 6.1), characterized by a phase of increasing production until the lactation peak, after which the production declines more or less rapidly toward the dry-off.

Figure 6.1 Typical lactation curve of dairy cattle (Ym = milk yield at lactation peak that occurs at time tm; tf = lactation length).

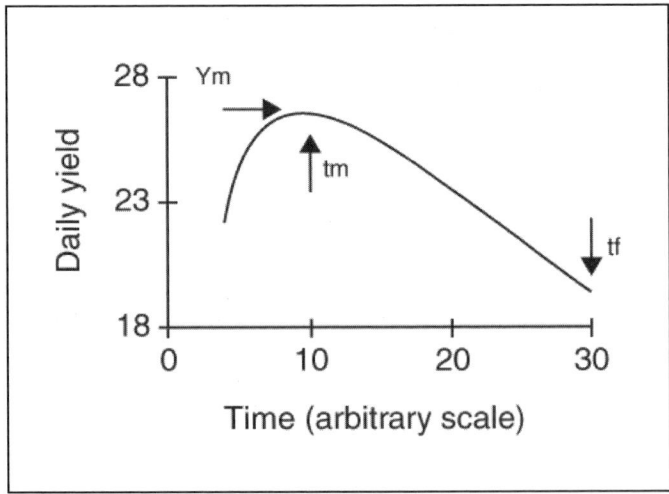

The dynamic changes that occur in the size, shape, and proportions of an animal as it grows arise from very complex mechanisms involving several organs and tissues. Three main functional phases can be recognized in the plot of the body weight of an animal as a function of time: a first, self-accelerating phase; a linear phase; and a final, self-decelerating phase (Lawrence and Fowler, 1997). These three phases are combined in the typical sigmoid shape common to many animal species, as shown in Figure 6.2.

Figure 6.2 The growth curve of a Sarda dairy breed lamb.

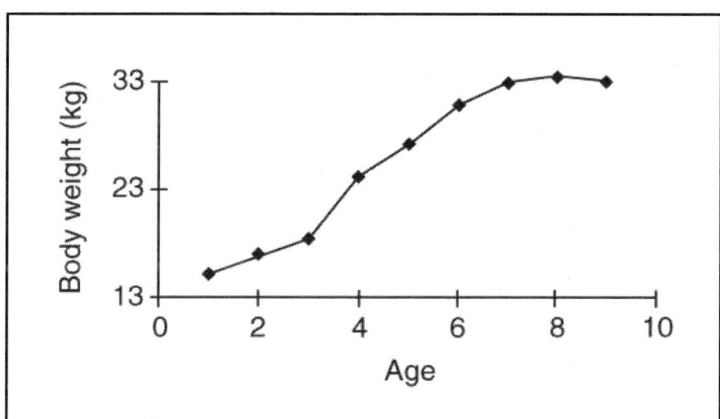

The self-accelerating phase can be related to a mechanism of cell proliferation that is unrestricted by the environment; cells double at regular intervals, and therefore the amount of growth at a certain period would be the square of that in the preceding period. However, as growth progresses, such acceleration is counterbalanced by the progressively greater complexity of the structures and the ability of the food supplied to meet body requirements. This antagonism can result in an extended linear phase of growth where the two forces are more or less in balance. Finally, in the self-decelerating phase the animal approaches its mature weight.

The mathematical modeling of milk production and growth usually concerns test day measures taken at different time intervals (approximately every 30 days) during lactation for dairy animals and birth, weaning, and yearling weight for beef animals. In the classic approach, a mathematical function of time is fitted to milk or growth traits:

$$y_t = f(t) + e,$$

where $f(t)$ is continuous and differentiable in the whole interval of time that corresponds to the lactation length or to the duration of the growth process, and e is the random residual. Mathematical functions are able to estimate the regular component underlying the phenomenon under study. Moreover, they are used to separate curves of homogeneous groups of animals and to make predictions of test day records. These methods also allow the evaluation of the main environmental factors that affect patterns of growth and milk production, which can be evaluated by variance analysis of the parameters that define the function actually used.

Several mathematical functions have been proposed for fitting milk test day data, differing in type (linear or non-linear) and in the number of parameters. Wood's incomplete gamma function (1967) is still one of the most frequently used models to fit milk test day (TD) data, mainly because a technical meaning can be assigned to its three parameters. The equation is defined by three parameters, related to main aspects of lactation curve shape:

$$y(t) = a\, t^b\, e^{-ct}, \tag{6.1}$$

where $y(t)$ is the average daily production at time t, and a, b, and c are parameters with positive values: a is an estimate of initial milk yield that offers a kind of benchmark of the total yield; b is a measurement of the rate of milk yield increase until the peak; and c is a measurement of the rate of decreasing milk yield after the peak. Combinations of parameter values allow for the estimation of essential features of the lactation curve:

- the lactation persistency p as $p = -(b+1)*\ln(c)$;
- the time at which the peak production is attained t_m, as $t_m = b/c$;
- the peak yield y_m, as $y_m = a(b/c)^b e^{-b}$.

The Wood function can be linearized by a simple natural logarithm transform:

$$\log(y(t)) = \log(a) + b\log(t) - ct. \tag{6.2}$$

Many other functions are able to fit data better than the Wood model, but they require a larger number of parameters whose technical meaning is difficult to define. An example is represented by the fitting of a third-order polynomial to the lactation pattern characterized by a false double peak, typical of animals of rangeland pasture systems. Although this function gives the best fit to experimental data, no relationships among its parameters and technical aspects of milk production pattern can be assessed.

The application of mathematical functions to describe growth patterns has been considerable. Growth models have been used to describe cattle growth for predictive purposes. Moreover, they allow for the calculation of some important features such as growth rate, maturation rate, and mature size that can be used in breeding and management programs in order to improve profits of the beef industry.

Several growth curve equations have been developed by integrating mechanistic models used to study biological phenomena such as enzymatic degradation of substrates or plant growth. These models rely on two keys that affect growth rate: (1) the quantity of growth machinery, usually proportional to the weight, and (2) the availability of substrates.

The more general growth mechanistic model has been proposed by Richards (1959), who developed the growth equation proposed by von Bertalanffy,

$$\frac{dW}{dt} = \frac{kW\left(W_f^n - W^n\right)}{nW^n},$$

where k, n, and W are constants in the equation. The general integral of the equation is

$$W_t = \frac{W_0 W_f}{\left[W_0^n + \left(W_f^n - W_0^n\right)e^{-kt}\right]^{\frac{1}{n}}} \tag{6.3}$$

where W_0 is the initial weight of the animal (for example, at birth), W_f is the asymptotic weight (at maturity), k is a parameter that characterizes the growth rate, and n is a shape parameter that gives a great flexibility to the Richards model: by changing its value, all the most frequently used growth curves, such as the logistic curve and the Gompertz functions, can be reconstructed.

The rationale of the logistic curve is that growth rate at time t depends on weight at the same time (W_t) and on the distance between W_t and the asymptotic mature weight W_f:

$$W_t = \frac{W_f}{1 + \left(\frac{W_f}{W_0} - 1\right) * e^{-kt}}, \qquad (6.4)$$

where W_0 is the initial weight, W_f is the asymptotic mature weight of the animal, k is a growth rate parameter, and t is the time. The logistic curve is one of the most versatile functions for fitting sigmoidal responses having a lower asymptote of zero and a finite upper asymptote. It is skew symmetric, with the weight at the inflection point fixed at $W_f/2$.

The symmetry of the logistic model about the inflection point can sometimes be an unrealistic constraint, but it can be overcome by using the Gompertz equation (Equation 6.5). This model assumes that growth rate at time t depends on weight at the same time (W_t), but with the effectiveness of the growth machinery that decays exponentially with time according to a first-order kinetic.

$$W_t = W_0 \exp\left[\mu\left(1 - e^{-Dt}\right)/D\right], \qquad (6.5)$$

where μ is the specific growth rate parameter whose decay is controlled by the additional parameter D. Such a decay can be due to the changes of the equilibrium among the phenomena of development, differentiation, and senescence.

The two models illustrated above consider both the accelerating and decelerating phases of growth. However, these two processes can be described separately, assuming that the growth in the first phase is a function of the growth already made, whereas in the self-decelerating phase it is a function of the growth yet to be made to reach maturity. In particular, the function of the second phase is a simple monomolecular growth equation (Brody, 1945):

$$W_t = W_f - (W_f - W_0)e^{-k(t-t_0)}, \qquad (6.6)$$

where k is a parameter that represents the ratio of maximum growth to mature weight, also called the maturing rate index. The Brody model of the self-decelerating phase of growth has been used to estimate genetic and environmental parameters for post-weaning weights, gains, and maturing rates.

6.3 Using PROC REG to Fit Lactation and Growth Data

The most common application of lactation curve fitting is the estimation of average lactation curves of homogeneous groups of cows. The following program creates a data set named PARITIES that contains average test day milk yields recorded monthly for three parity classes of cows and the milking days at which the test day milk yield occurs.

```
data parities;
input parity 1 milk 3-6 days 8-10;
datalines;
1 27.4 23
1 30.5 54
1 29.9 83
1 30.2 114
1 29.1 146
1 28.6 175
```

```
1 27.4 205
1 25.4 266
1 22.8 297
2 35.8 17
2 41.2 47
2 39.7 76
2 37.4 107
2 35.1 138
2 32.9 167
2 31.2 198
2 29.1 229
2 25.5 260
2 22.2 291
3 38.1 18
3 43.9 48
3 41.6 78
3 39.0 108
3 35.5 139
3 32.6 169
3 29.8 200
3 27.0 231
3 24.2 262
3 20.2 292
;
run;
```

The fitting of the Wood function in its original form (Equation 6.1) requires the use of a non-linear estimation procedure. However, as previously stated, it can be linearized via a logarithmic transformation into the form of Equation 6.2. The following statements create the logarithms of milk yield and milking days in order to use the linearized form of Wood equation:

```
data parities;  set parities;
logmilk=log(milk);
logdays=log(days);
run;
```

Finally, PROC REG is used to fit the data. The regression analysis is carried out separately for each parity class as indicated in the BY statement.

```
proc reg data=work.parities;
model logmilk= logdays days/dw;
by parity;
output out=predictions P=predicted R=residuals;
plot residual.*logdays=parity;
run;
```

The goodness of fit of the regression can be assessed by examining the adjusted R-square (0.96–0.99 in Output 6.1), which expresses the amount of variance explained by the regression corrected for the number of parameters used.

Output 6.1 Fitting of the Wood equation to dairy cattle test day data (PARITIES.SAS).

```
------------------------------------ parity=1 ------------------------------------
                              The REG Procedure
                           Dependent Variable: logmilk
                             Analysis of Variance
                                    Sum of           Mean
     Source                 DF     Squares         Square      F Value    Pr > F
     Model                   2     0.06919        0.03459        88.70    <.0001
     Error                   6     0.00234     0.00038999
     Corrected Total         8     0.07153

                 Root MSE              0.01975    R-Square     0.9673
                 Dependent Mean        3.32557    Adj R-Sq     0.9564
                 Coeff Var             0.59383

                             Parameter Estimates
                           Parameter     Standard
     Variable      DF       Estimate        Error    t Value    Pr > |t|
     Intercept      1        2.74397      0.08651      31.72      <.0001
     logdays        1        0.19842      0.02422       8.19      0.0002
     days           1       -0.00242   0.00021354     -11.35      <.0001

                    Durbin-Watson D                    2.415
                    Number of Observations                 9
                    1st Order Autocorrelation         -0.366
------------------------------------ parity=2 ------------------------------------
                             Analysis of Variance
                                    Sum of           Mean
     Source                 DF     Squares         Square      F Value    Pr > F
     Model                   2     0.34440        0.17220       239.75    <.0001
     Error                   7     0.00503     0.00071824
     Corrected Total         9     0.34943

                 Root MSE              0.02680    R-Square     0.9856
                 Dependent Mean        3.48010    Adj R-Sq     0.9815
                 Coeff Var             0.77009

                             Parameter Estimates
                           Parameter     Standard
     Variable      DF       Estimate        Error    t Value    Pr > |t|
     Intercept      1        3.00279      0.09424      31.86      <.0001
     logdays        1        0.22865      0.02727       8.39      <.0001
     days           1       -0.00401   0.00026261     -15.26      <.0001

                    Durbin-Watson D                    1.284
                    Number of Observations                10
                    1st Order Autocorrelation          0.242
------------------------------------ parity=3 ------------------------------------
                             Analysis of Variance
                                    Sum of           Mean
     Source                 DF     Squares         Square      F Value    Pr > F
     Model                   2     0.56837        0.28418       508.05    <.0001
     Error                   7     0.00392     0.00055936
     Corrected Total         9     0.57228

                 Root MSE              0.02365    R-Square     0.9932
                 Dependent Mean        3.47500    Adj R-Sq     0.9912
                 Coeff Var             0.68060

                             Parameter Estimates
                           Parameter     Standard
     Variable      DF       Estimate        Error    t Value    Pr > |t|
     Intercept      1        2.99083      0.08648      34.59      <.0001
     logdays        1        0.25746      0.02486      10.36      <.0001
     days           1       -0.00484   0.00023502     -20.59      <.0001

                    Durbin-Watson D                    1.805
                    Number of Observations                10
                    1st Order Autocorrelation         -0.052
```

A further indication of the goodness of fit, especially when time-dependent measurements are fitted, is represented by the absence of serial correlations among residuals. The visual inspection of the plot of residuals against time is the first step in assessing the possible existence of autocorrelation.

Output 6.2 Plot of residuals for first-parity cows.

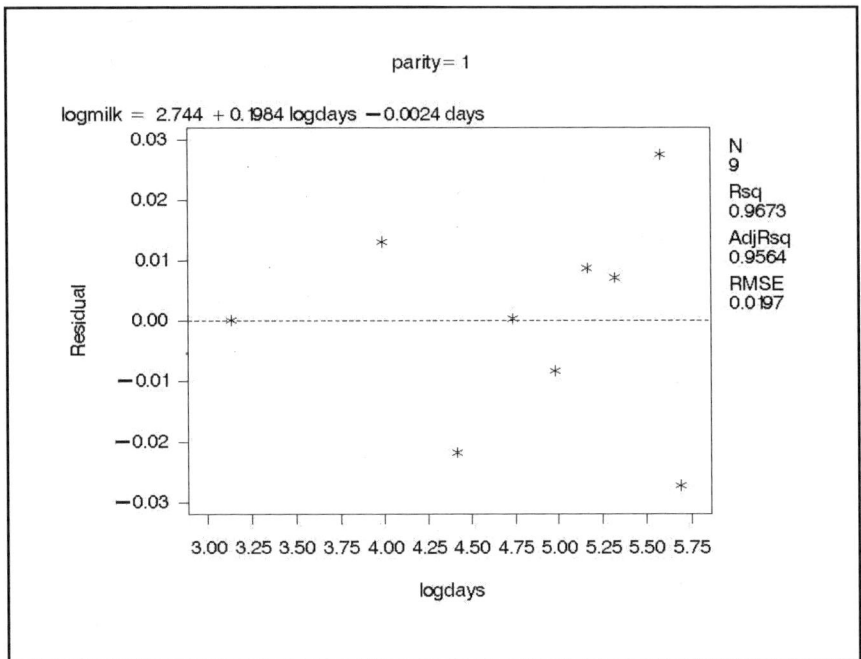

Output 6.3 Plot of residuals for second-parity cows.

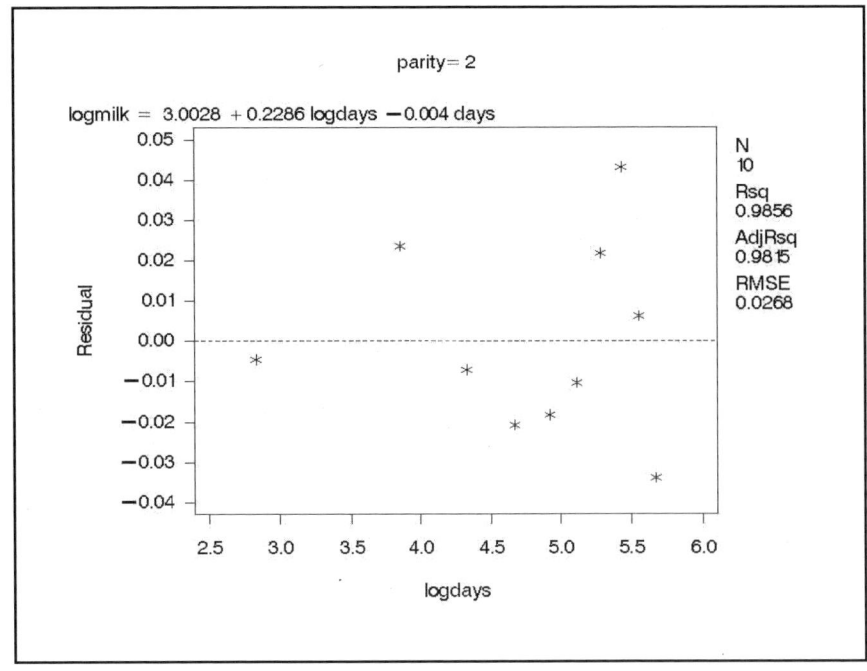

Output 6.4 Plot of residuals for third-parity cows.

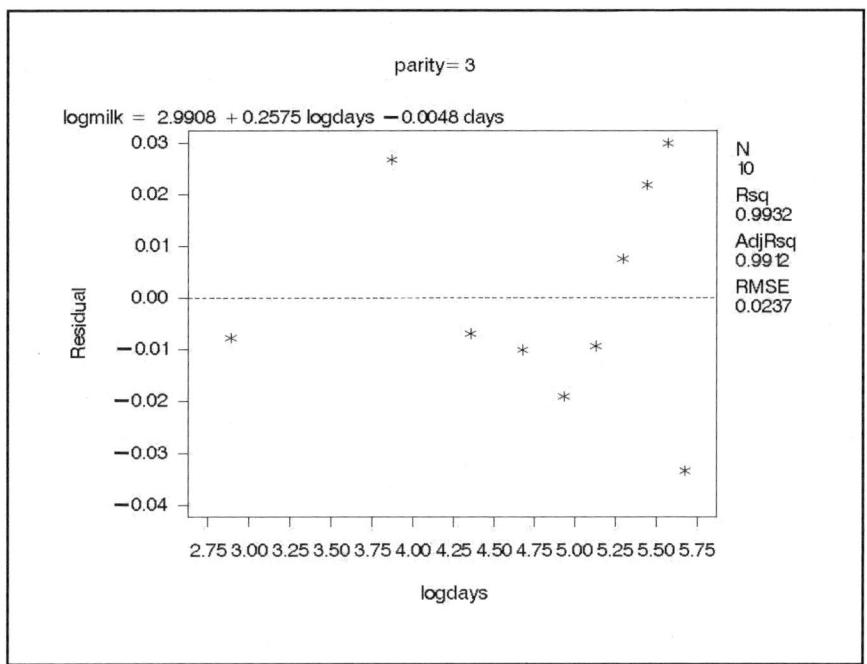

Patterns of residuals do not show definite trends in all three regressions. Such an impression is confirmed by the Durbin-Watson statistic (Output 6.1), produced by the DW option in the MODEL statement. When there is no serial correlation among residuals, the expected value of the DW statistic is approximately 2.0. As a practical rule, the value of DW <1.5 leads one to suspect a positive correlation, whereas values >2.5 show the existence of a negative autocorrelation. In the reported examples above, DW statistic values are all reasonably close to 2 for first- and third-parity cows.

From the results of the regression the usefulness of the technical meaning of the Wood function parameters can be seen. The *a* parameter (named INTERCEPT in Output 6.1) has the lowest value for first-parity cows and increases in higher parities, whereas *c* (named DAYS) shows an opposite trend. These results were expected because it is well known that production level increases until the third or fourth parity, whereas the persistency tends to decrease as the age of the animal increases.

Finally, this code produces a visual assessment of model fit by plotting predicted values against time, together with the residuals. Because the equation is in logarithmic form, a DATA step is first used to transform predicted values to the original unit of measure.

```
data predictions;
set predictions;
predmilk=exp(predicted);
resmilk=exp(residuals);
run;
symbol1 interpol=none color=black value=star;
symbol2 interpol=join color=black value=none;
proc gplot data=work.predictions;
plot milk*days predmilk*days/overlay;
by parity;
run;
quit;
```

Outputs 6.5, 6.6, and 6.7 show the good fit of the Wood function to average lactation curves even if we can notice the tendency, common to several three-parameter models (as, for example, the Wilmink function), to underestimate yields around the peak and to overestimate yields in the final part of lactation.

Output 6.5 Actual (*) and estimated (—) average lactation curve of first-parity cows.

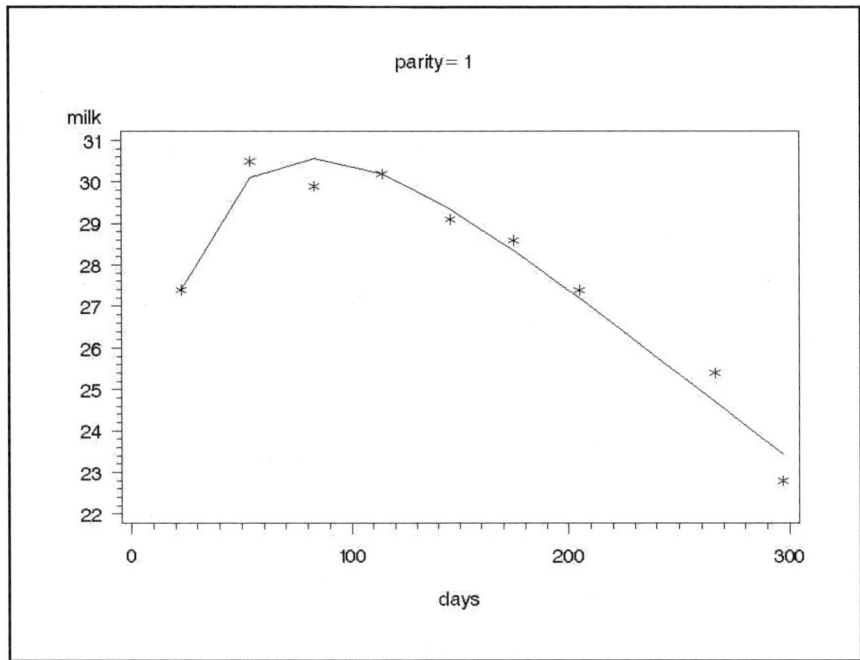

Output 6.6 Actual (*) and estimated (—) average lactation curve of second-parity cows.

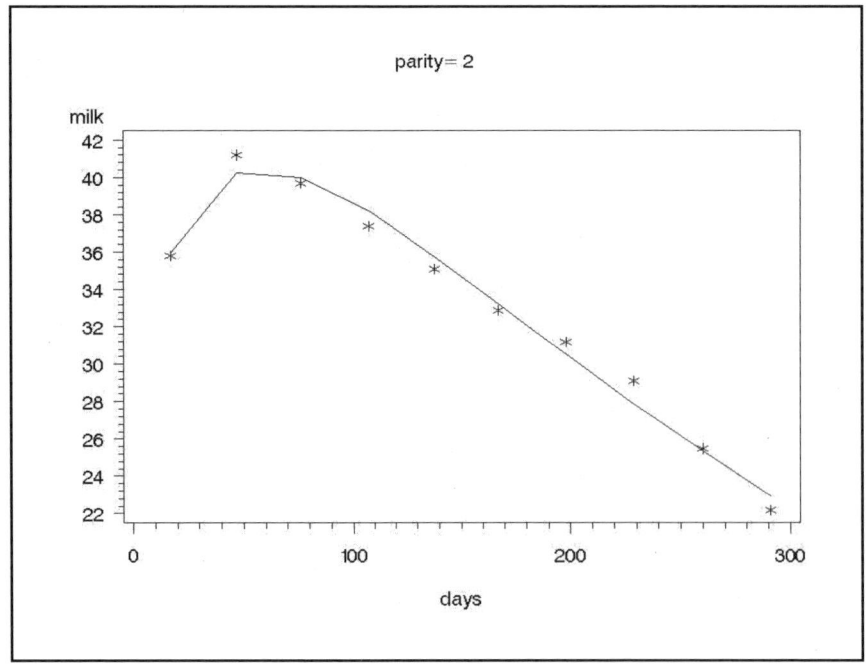

Output 6.7 Actual (*) and estimated (—) average lactation curve of third-parity cows.

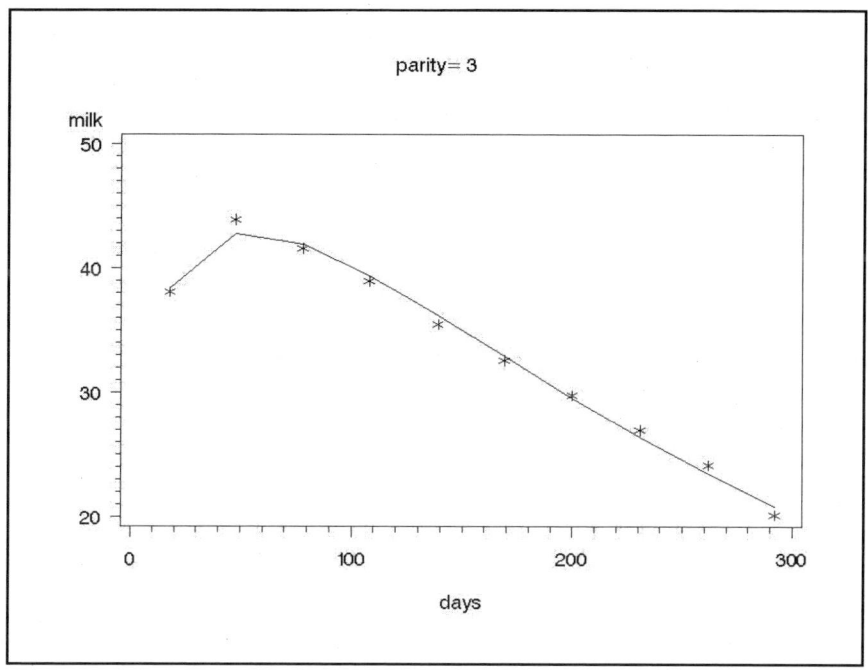

Even if the most common applications of lactation curve fitting concern average lactation curves of homogeneous groups of animals, fitting lactation curves to individual milk production data is useful for many practical purposes (genetic evaluations, individual feeding, health monitoring).

The following program (example COWS.SAS) fits the linearized form of the Wood equation to the individual lactation patterns of 142 Holstein cows of the data set COWS. The OUTEST statement yields a data set named ESTIMATE that contains several statistics of the model (Output 6.8), including parameter estimates and the adjusted R-square (as stated in the ADJRSQ option). The goodness of fit is assessed by examining the distribution of curves among different classes of adjusted R-square values. With this aim in view, a classification variable named FIT is created on the basis of adjusted R-square values.

```
data cows;   set cows;
logmilk=log(milk);
logdays=log(days);
run;
proc reg data=work.cows outest=estimate adjrsq;
model logmilk= logdays days/dw noprint;
by cow;
run;
data estimate;
set estimate;
drop _model_ _type_ _depvar_ logmilk;
run;
proc print data=estimate;
run;
data estimate;
set estimate;
if _adjrsq_<0.20 then fit=1;
if _adjrsq_>=0.20 and _adjrsq_<0.40 then fit=2;
if _adjrsq_>=0.40 and _adjrsq_<0.60 then fit=3;
if _adjrsq_>=0.60 and _adjrsq_<0.80 then fit=4;
if _adjrsq_>=0.80 then fit=5;
run;
proc freq data=estimate;
tables fit/list;
run;
```

Output 6.8 First lines of the ESTIMATE data set created with the OUTEST option.

```
Obs   cow   _RMSE_   Intercept   logdays     days    _IN_  _P_  _EDF_  _RSQ_    _ADJRSQ_
 1     1   0.06368   1.28729     0.56738  -0.004905   2    3     7    0.87930   0.84482
 2     2   0.06575   2.66563     0.31846  -0.004749   2    3     7    0.92243   0.90027
 3     3   0.04423   2.84557     0.21318  -0.002877   2    3     7    0.89033   0.85900
```

The frequency table (Output 6.9) shows the distribution of the curves among the different classes of fit; about 50% of fits have an adjusted R-square equal to or greater than 0.80.

Output 6.9 Frequency distributions of adjusted R-square classes.

```
                    The FREQ Procedure
                                    Cumulative    Cumulative
      fit    Frequency    Percent   Frequency     Percent

       1        12          8.45       12           8.45
       2         9          6.34       21          14.79
       3        14          9.86       35          24.65
       4        29         20.42       64          45.07
       5        78         54.93      142         100.00
```

Individual parameter estimates can be used to study the effects of environmental factors and to estimate genetic parameters of the different aspects of the lactation curve shape. Actually, genetic variation exists for parameters of the Wood curve indicating that the shape of the lactation curve can be modified in an economically desirable direction (Rekaya et al., 2000). A simple way to estimate the heritability (h^2) of Wood parameters is to use a linear mixed model that includes the effect of the sire of the cow, known as a sire model:

$$Y_{ijk} = P_i + S_j + e_{ijk},$$

where Y_{ijk} is the Wood parameter value (*a*, *b*, or *c*) of the *k*th cow, P_i is the fixed effect of the *i*th parity class (first, second, and third), S_j is the random effect of the *j*th sire, and e_{ijk} is the random residual. The additive genetic variance can then be estimated as four times the sire variance, i.e., $\sigma_A^2 = 4\sigma_S^2$ (Lynch and Walsh, 1998). Finally the heritability is calculated as the ratio $h^2 = \sigma_A^2/(\sigma_A^2 + \sigma_e^2)$.

The following program creates a CURVEPAR data set that contains for each cow estimates of the three Wood parameters plus the variables contained in the original COWS data set. A mixed model analysis is then carried out to test relationships between parameter *a* and parity of the cows and to estimate the sire variance.

```
data cows;
set cows;
drop days;
run;
data curvepar;
merge cows estimate;
by cow;
run;
proc sort data=curvepar nodupkey;
by cow;
run;
```

```
data curvepar;
set curvepar;
a=exp(intercept);
run;
proc mixed data=curvepar;
class parity;
model a=parity;
random sire;
lsmeans parity/ pdiff;
run;
```

Output 6.10 Result of sire model analysis for the Wood parameter *a* (COWS.SAS).

```
                        The Mixed Procedure
                         Model Information
          Data Set                     WORK.CURVEPAR
          Dependent Variable           a
          Covariance Structure         Variance Components
          Estimation Method            REML
          Residual Variance Method     Profile
          Fixed Effects SE Method      Model-Based
          Degrees of Freedom Method    Containment

                       Class Level Information
               Class      Levels    Values
               parity          3    1 2 3

                              Dimensions
                   Covariance Parameters         2
                   Columns in X                  4
                   Columns in Z                  1
                   Subjects                      1
                   Max Obs Per Subject         142
                   Observations Used           142
                   Observations Not Used         0
                   Total Observations          142

                           Iteration History
         Iteration    Evaluations    -2 Res Log Like     Criterion
                 0              1        1186.54284047
                 1              1        1185.57672283    0.00000000
                       Convergence criteria met.

                    Covariance Parameter Estimates
                       Cov Parm     Estimate
                       Sire           3.1159
                       Residual     270.44

                            Fit Statistics
               -2 Res Log Likelihood         1185.6
               AIC (smaller is better)       1189.6
               AICC (smaller is better)      1189.7
               BIC (smaller is better)       1185.6

                    Type 3 Tests of Fixed Effects
                             Num    Den
                  Effect      DF     DF    F Value    Pr > F
                  parity       2    138       0.62    0.5387

                         Least Squares Means
                                     Standard
      Effect    parity    Estimate      Error     DF    t Value    Pr > |t|
      parity    1          17.7566     3.3834    138       5.25      <.0001
      parity    2          21.3613     3.3265    138       6.42      <.0001
      parity    3          20.6030     3.3355    138       6.18      <.0001

                    Differences of Least Squares Means
                                          Standard
    Effect   parity   _parity   Estimate     Error    DF   t Value   Pr > |t|
    parity   1        2          -3.6047    3.3794   138     -1.07     0.2880
    parity   1        3          -2.8464    3.4298   138     -0.83     0.4080
    parity   2        3           0.7583    3.3416   138      0.23     0.8208
```

In Output 6.10, the additive genetic variance is $\sigma_A^2 = 4 \times 3.1159 = 12.44636$, and the heritability coefficient is 12.4636/(12.4636+270.44) = 0.044. This estimate is expected to be inaccurate, since the data set examined is quite small. However, it falls in the range of values reported in the literature (Rekaya et al., 2000). Parity did not significantly affect the parameter a, but it can be observed that first-parity cows had the lowest value, as expected.

6.4 Using PROC NLIN to Fit Non-linear Models

The original exponential form of the Wood function can be fitted to data by using the NLIN procedure. The non-linear estimation technique requires the input of preliminary values of parameters. Moreover, different algorithms can be chosen. The following SAS program (in PARITIES.SAS) fits the Wood equation to the data set PARITIES using the NLIN procedure and the Marquardt iterative method. Estimates obtained in the previous section are indicated as preliminary parameter values in the PARMS statement.

```
proc nlin data=work.parities method=marquardt;
model milk=a*days**(b)*exp(c*days);
parms a=15 b=0.19 c=-0.0012;
by parity;
output out=pred p=predict r=residuals;
run;
```

Results for first-parity cows are reported in Output 6.11.

Output 6.11 Results of non-linear fitting of the Wood equation to the average lactation curve of first-parity cows.

```
                        The NLIN Procedure
                         Iterative Phase
                       Dependent Variable milk
                         Method: Marquardt
                                                    Sum of
         Iter        a            b            c        Squares
          0       15.0000       0.1900      -0.00120     160.9
          1       16.3055       0.1811      -0.00215     2.6009
          2       15.7418       0.1943      -0.00238     1.6094
          3       15.7535       0.1943      -0.00238     1.6073
          4       15.7535       0.1943      -0.00238     1.6073
        NOTE: Convergence criterion met.

                         Estimation Summary
                Method                    Marquardt
                Iterations                       4
                R                          1.414E-7
                PPC(b)                     3.243E-8
                RPC(b)                     4.572E-6
                Object                     3.91E-10
                Objective                  1.607322
                Observations Read                 9
                Observations Used                 9
                Observations Missing              0
        NOTE: An intercept was not specified for this model.

                             Sum of      Mean                   Approx
        Source          DF   Squares    Square   F Value        Pr > F
        Regression       3   7066.0     2355.3   8792.24        <.0001
        Residual         6   1.6073     0.2679
        Uncorrected Total 9  7067.6
        Corrected Total  8   50.7356
```

(continued on next page)

Output 6.11 *(continued)*

```
                                   Approx
       Parameter    Estimate    Std Error    Approximate 95% Confidence Limits
          a          15.7535       1.3018        12.5681         18.9389
          b           0.1943       0.0231         0.1377          0.2509
          c          -0.00238      0.000209      -0.00289        -0.00187

                     Approximate Correlation Matrix
                            a               b               c
          a         1.0000000       -0.9891295       0.8786949
          b        -0.9891295        1.0000000      -0.9341059
          c         0.8786949       -0.9341059       1.0000000
```

Parameter estimates are quite similar to those obtained in the linearized form (remember that parameter *a*, "intercept" in the linearized form, is expressed as a logarithm). Values of the correlation matrix show the high degree of relationship that exists among the Wood curve parameters and that can result in a greater variance of parameter estimates. However, this is a common feature of almost all lactation curve models. A relevant difference in comparison with the linear regression is that confidence intervals are estimated for parameters. Since the R-square is no longer useful in non-linear regression (Ratkowsky, 1990), goodness of fit can be assessed by examining the magnitude of the residual variance and looking at a graph of the fitted curve superimposed on the actual data points. In this example, the residual variance is rather small in comparison with the total (0.2679 vs. 2355.3).

An interesting issue in lactation curve fitting, common to dairy cattle and sheep, is represented by the existence of two general shapes of the lactation curve. The first can be defined as regular and follows the typical pattern reported in Figure 6.1. The second shape, defined as atypical (Olori et al., 1999) and occurring rather frequently (from 30% to 50% of the cases in dairy sheep), lacks the lactation peak and therefore appears downhill-shaped (Shanks et al., 1981). Empirical mathematical models are usually able to describe the regular shape, whereas in the case of atypical curves they yield parameter estimates that fall beyond the range of biological significance. As an example, the fitting of the incomplete Wood's gamma function to atypical curves results in a negative estimate of parameter *b*. A possible solution is to impose a value greater than zero on the *b* parameter by using the BOUNDS statement in the NLIN procedure (example BOUNDS B>0), even if the parameter estimates completely lose their technical meaning.

An alternative can be represented by the use of mechanistic models. In contrast with empirical models that disentangle the main component of the phenomenon under study without any interest in the mechanisms that underlie the process itself, the mechanistic approach aims to increase the knowledge of a biological phenomenon. Thus, a mechanistic model attempts to translate in mathematical terms a hypothesis about the deep physiological and biochemical processes that regulate the phenomenon of interest. A useful mechanistic approach to lactation curve study based on a functional bicompartmental model of the mammary gland has been proposed by Ferguson and Boston (1993). It synthesizes the physiological theory according to which the quantity of milk produced at each time depends mainly on the number of active secretory cells that are in the mammary gland at the same time (Mepham, 1987). The first functional compartment represents a pool of inactivated milk secretory cells and the second a pool of activated secretory cells. The flow from the first to the second compartment corresponds to the cell activation process. The output of the model $q_2(t)$ is the number of secretory cells that are in the mammary gland at time (t), and it is considered proportional to the milk production at the same time. The final equations are

$$q_2(t) = Q_1 \, e^{-k_1 t} + Q_2 \, e^{-k_2 t} \tag{6.7}$$

$$q_2(t) = Q_2 \, e^{-k_2 t}, \tag{6.8}$$

where Q_1 is related to the number of cells that undergo activation at time t whereas Q_2 is related to the variation of the number of activated cells at the same time, k_1 and k_2 are positive parameters that measure cell activation rate and secretory cell inactivation rate, respectively, and t is the time. Equation 6.7 can be fitted to regular lactation curves, given that $Q_1>0$, $Q_2<0$ and $k_2>k_1$, whereas Equation 6.8 represents a decreasing curve that could be a good model for atypical lactation.

The following SAS program (in SHEEP.SAS) creates the SHEEP data set, which contains an example of the two shapes of lactation curves. Data are then plotted against time (in days). Ewe 1 shows a regular lactation pattern, whereas ewe 2 is characterized by an atypical shape, as shown in Output 6.12.

```
data sheep;
input ewe 1 milk 3-6 days 8-10;
datalines;
1 1.56 14
1 1.85 28
1 1.99 35
1 1.98 42
1 1.88 49
1 1.85 56
1 1.84 63
1 1.78 70
1 1.72 77
1 1.63 84
1 1.58 91
1 1.66 98
1 1.53 105
1 1.41 112
1 1.41 119
1 1.37 126
1 1.27 133
1 1.14 140
1 1.12 147
1 1.05 154
1 0.96 161
1 0.99 168
1 0.92 175
1 0.80 182
1 0.71 189
1 0.72 196
1 0.61 203
1 0.58 210
2 1.99 14
2 1.96 28
2 1.81 35
2 1.72 42
2 1.56 49
2 1.51 56
2 1.44 63
2 1.29 70
2 1.27 77
2 1.25 84
2 1.13 91
2 1.17 98
2 1.05 105
2 0.92 112
2 0.93 119
2 0.89 126
2 0.85 133
2 0.80 140
2 0.82 147
2 0.73 154
2 0.71 161
2 0.71 168
2 0.68 175
2 0.60 182
2 0.55 189
```

```
2 0.52 196
2 0.48 203
2 0.49 210
;

symbol1 interpol=join color=black value=star;
symbol2 interpol=join color=black value=none;
proc gplot data=work.sheep;
plot milk*days=ewe;
run;
quit;
```

Output 6.12 Examples of the two shapes of lactation curves in sheep (—*— is regular; — is atypical).

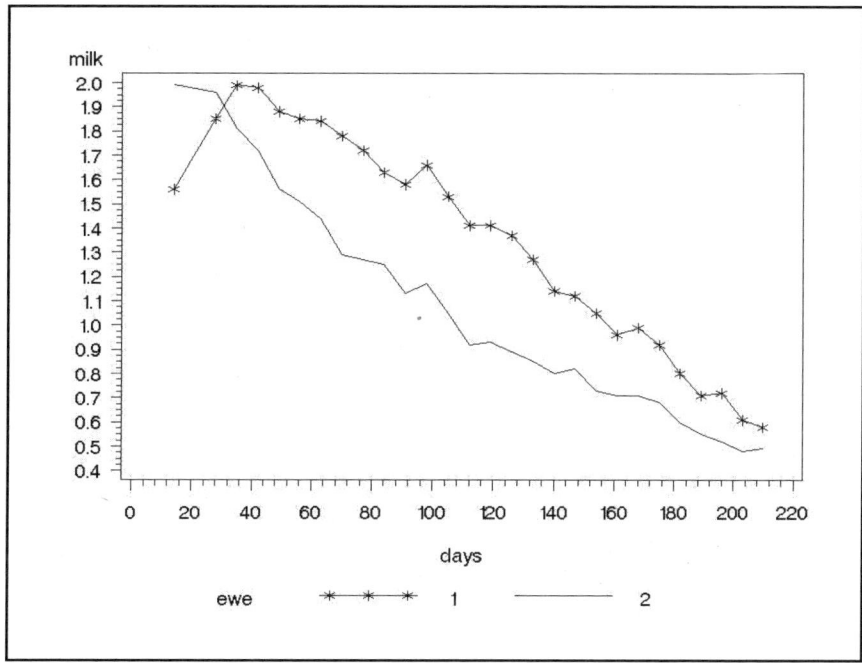

The following program (SHEEP.SAS) fits Equation 6.7 to the regular lactation curve and fits Equation 6.8 to the atypical pattern of ewe 2. Preliminary parameter values are taken from Cappio-Borlino et al. (1997). Fits and actual data are then plotted against time.

```
proc nlin data=sheep method=marquardt;
model milk=Q1*exp(-k1*days)+ Q2*exp(-k2*days);
parms Q1=7 Q2=-3.600 k1=0.000067 k2=0.00015;
where ewe=1;
output out=predictions1 p=predict r=residuals;
run;

symbol1 interpol=none color=black value=star;
symbol2 interpol=join color=black value=none;
proc gplot data=work.predictions1;
plot milk*days predict*days/overlay;
run;
quit;
run;

proc nlin data=sheep method=marquardt;
model milk=Q2*exp(-k2*days);
parms Q2=-4 k2=0.00005;
where ewe=2;
output out=predictions2 p=predict r=residuals;
run;
```

```
symbol1 interpol=none color=black value=star;
symbol2 interpol=join color=black value=none;
proc gplot data=work.predictions2;
plot milk*days predict*days/overlay;
run;
quit;
```

Output 6.13 Results of non-linear fitting of Equations 6.7 and 6.8 to a regular and an atypical lactation pattern, respectively.

```
                         The NLIN Procedure
                          Iterative Phase
                      Dependent Variable milk
                         Method: Marquardt
                                                            Sum of
   Iter         Q1         k1          Q2          k2      Squares
      0     7.0000   0.000067      -3.6000    0.000150      123.6
      1     5.5018   0.0539        -3.6000    0.1047       43.2957
      2     5.9294   0.0371        -4.7664    0.1042       32.3470
      3     6.1537   0.0276        -6.7946    0.0822       21.4589
      4     5.6517   0.0120        -6.2797    0.0322        1.4750
      5     5.7397   0.0106        -4.5951    0.0239        0.1340
      6     5.7279   0.0102        -4.5530    0.0217        0.0721
      7     5.9228   0.0103        -4.7459    0.0214        0.0711
      8     6.1004   0.0104        -4.9183    0.0210        0.0710
      9     6.2639   0.0105        -5.0777    0.0208        0.0709
     10     6.4148   0.0106        -5.2250    0.0205        0.0709
     11     6.5537   0.0106        -5.3608    0.0203        0.0709
     12     6.6805   0.0107        -5.4851    0.0202        0.0709
     13     6.7950   0.0108        -5.5975    0.0200        0.0708
     14     6.8967   0.0108        -5.6973    0.0199        0.0708
     15     6.9850   0.0109        -5.7841    0.0198        0.0708
     16     7.0596   0.0109        -5.8575    0.0197        0.0708
     17     7.1205   0.0109        -5.9174    0.0196        0.0708
     18     7.1681   0.0109        -5.9643    0.0196        0.0708
     19     7.2284   0.0110        -6.0237    0.0195        0.0708
     20     7.2549   0.0110        -6.0498    0.0195        0.0708
     21     7.2631   0.0110        -6.0578    0.0195        0.0708
     22     7.2656   0.0110        -6.0603    0.0195        0.0708
     23     7.2664   0.0110        -6.0611    0.0195        0.0708
   NOTE: Convergence criterion met.

                         Estimation Summary
             Method                       Marquardt
             Iterations                          23
             Subiterations                       21
             Average Subiterations         0.913043
             R                             7.443E-6
             PPC(Q2)                       0.000039
             RPC(Q2)                       0.000127
             Object                        7.57E-10
             Objective                     0.070825
             Observations Read                   28
             Observations Used                   28
             Observations Missing                 0
         NOTE: An intercept was not specified for this model.

                          Sum of        Mean                      Approx
   Source            DF  Squares       Square     F Value          Pr > F
   Regression         4  56.6265      14.1566     4797.15          <.0001
   Residual          24   0.0708      0.00295
   Uncorrected Total 28  56.6973
   Corrected Total   27   5.3699
```

(continued on next page)

Output 6.13 *(continued)*

```
++
                              Approx
     Parameter    Estimate    Std Error    Approximate 95% Confidence Limits
     Q1             7.2664      6.8047        -6.7776      21.3104
     k1             0.0110      0.00286        0.00507      0.0169
     Q2            -6.0611      6.7111       -19.9119       7.7898
     k2             0.0195      0.00708        0.00485      0.0341

                       Approximate Correlation Matrix
                      Q1             k1             Q2             k2
     Q1       1.0000000      0.9981316     -0.9999117     -0.9958982
     k1       0.9981316      1.0000000     -0.9985604     -0.9889011
     Q2      -0.9999117     -0.9985604      1.0000000      0.9948151
     k2      -0.9958982     -0.9889011      0.9948151      1.0000000


                            The NLIN Procedure
                              Iterative Phase
                          Dependent Variable milk
                             Method: Marquardt
                                                  Sum of
              Iter         Q2           k2        Squares
                0      -4.0000      0.000050       717.8
                1       1.9510     -0.00186       65.8706
                2       1.7777      0.00204        5.6951
                3       2.1017      0.00599        0.2486
                4       2.2763      0.00733        0.0485
                5       2.2906      0.00741        0.0480
                6       2.2907      0.00741        0.0480
     NOTE: Convergence criterion met.

                            Estimation Summary
                  Method                    Marquardt
                  Iterations                        6
                  R                         1.206E-6
                  PPC(k2)                   1.221E-7
                  RPC(k2)                   0.000038
                  Object                    2.751E-7
                  Objective                 0.048003
                  Observations Read               28
                  Observations Used               28
                  Observations Missing             0
          NOTE: An intercept was not specified for this model.

                              Sum of      Mean                      Approx
     Source              DF   Squares     Square     F Value        Pr > F
     Regression           2   37.3085    18.6542     10103.7        <.0001
     Residual            26    0.0480     0.00185
     Uncorrected Total   28   37.3565
     Corrected Total     27    5.5769

                              Approx
     Parameter   Estimate    Std Error    Approximate 95% Confidence Limits
     Q2            2.2907      0.0292        2.2307       2.3507
     k2            0.00741     0.000147      0.00710      0.00771

                       Approximate Correlation Matrix
                              Q2             k2
                  Q2    1.0000000      0.8339215
                  k2    0.8339215      1.0000000
```

The magnitude of the residual variance and the width of confidence intervals of parameter estimates in Output 6.13 indicate a general goodness of fit, also evidenced by the plots of fits superimposed to actual data in Outputs 6.14 and 6.15.

Output 6.14 Actual (*) and estimated (—) lactation curve with Equation 6.7.

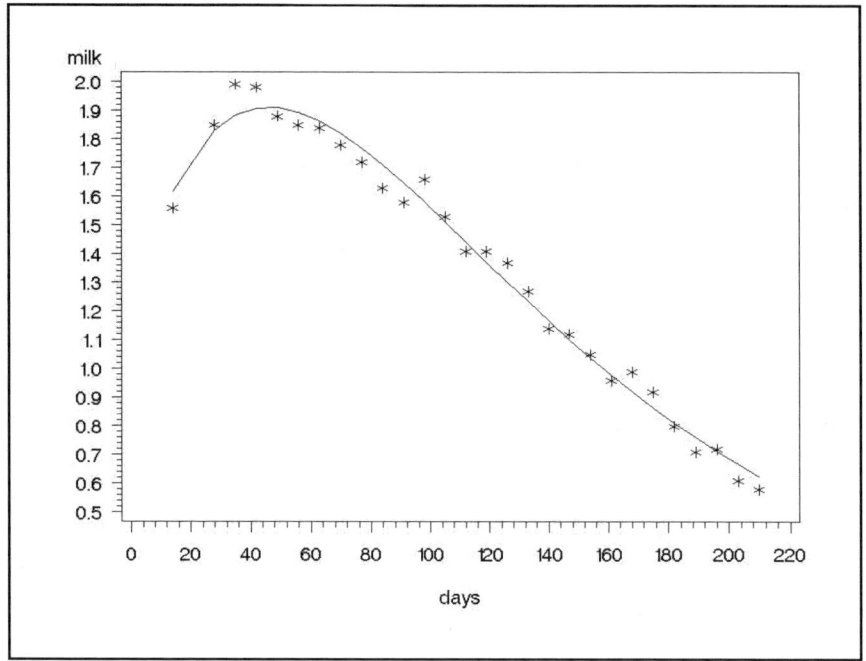

Output 6.15 Actual (*) and estimated (—) lactation curve with Equation 6.8.

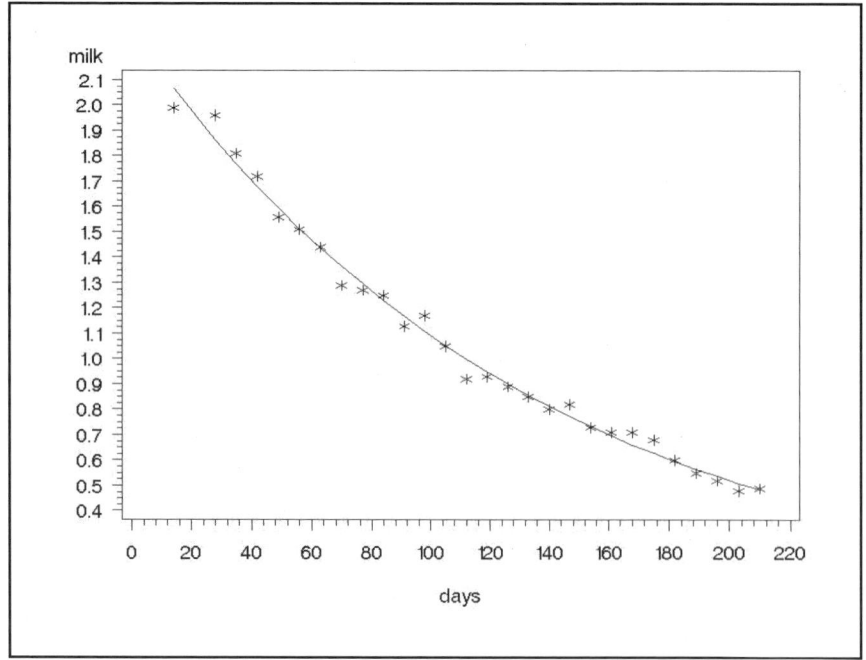

These results confirm the ability of the bicompartmental model of the mammary gland to explain the dimorphism of lactation curve shape in terms of the mechanism of activation and inactivation of secretory cells.

A non-linear estimation technique is generally necessary for growth curves, since these models often cannot be linearized. The following SAS program (GROWTH.SAS) creates the data set GROWTH with growth patterns for Chianina calves.

```
data growth;
input weight 1-3 age 5-7;
datalines;
227 185
251 224
282 251
325 282
362 308
414 343
454 369
510 402
550 430
596 461
634 491
666 521
696 550
;
run;
```

To fit the Brody equation (Equation 6.6) to the growth pattern of the GROWTH data set, the variable AGE (expressed in days) needs to be rescaled by defining the first age (185 d) as zero.

```
data growth;
set growth;
rescage=age-185;
run;
```

The following SAS program fits the Brody equation to the growth pattern reported above. The weight at birth (W_0) is 56 kg. As preliminary values, 700 is assumed for WF (since 696 is the final weight of the growth pattern) and 0.0024 for k (the ratio among the maximum observed growth rate, 1.7 kg/d, and the weight at maturity, 700 kg).

```
proc nlin data=growth;
model weight=Wf-(Wf-56)*exp(-k*rescage);
parms Wf=700 k=0.0024;
output out=brody predicted=p;
run;

symbol1 interpol=join value=none color=black;
symbol2 interpol=none value=star color=black;
proc gplot data=brody;
plot p*age weight*age/overlay;
run;
quit;
```

Output 6.16 Results of non-linear fitting of the Brody equation (Equation 6.6) to the GROWTH data set.

```
                        The NLIN Procedure
                         Iterative Phase
                     Dependent Variable weight
                       Method: Gauss-Newton
                                            Sum of
        Iter         Wf           k         Squares
          0        700.0       0.00240      481300
          1        709.7       0.00488      65598.4
          2        844.3       0.00401      40671.5
          3        863.3       0.00404      38841.3
          4        863.0       0.00404      38841.3
          5        863.0       0.00404      38841.3
      NOTE: Convergence criterion met.

                       Estimation Summary
          Method                  Gauss-Newton
          Iterations                       5
          R                          3.98E-6
          PPC(k)                    4.468E-6
          RPC(k)                    0.000048
          Object                    1.979E-9
          Objective                 38841.31
          Observations Read               13
          Observations Used               13
          Observations Missing             0
      NOTE: An intercept was not specified for this model.

                          Sum of        Mean                   Approx
   Source           DF    Squares       Square     F Value     Pr > F
   Regression        2    3017138      1508569      427.23     <.0001
   Residual         11    38841.3       3531.0
   Uncorrected Total 13   3055979
   Corrected Total  12    317126

                           Approx
    Parameter    Estimate  Std Error   Approximate 95% Confidence Limits
    Wf            863.0     158.0         515.3          1210.6
    k             0.00404   0.00137       0.00103        0.00705

                  Approximate Correlation Matrix
                             Wf              k
              Wf       1.0000000       -0.9805294
              k       -0.9805294        1.0000000
```

Output 6.17 Actual (*) and estimated (—) growth pattern with Equation 6.6.

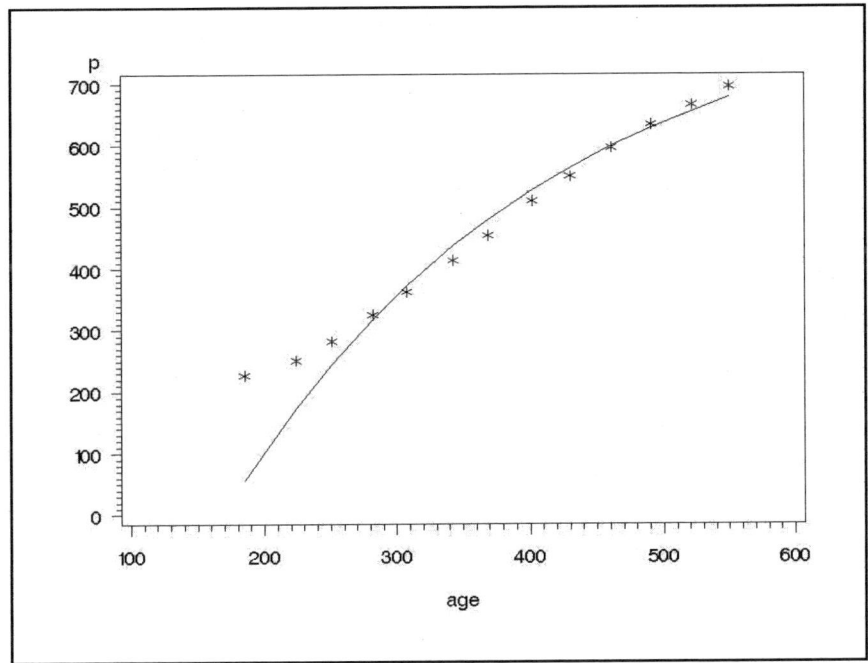

Output 6.16 suggests a poor fit. The plot of the data in Output 6.17 highlights the nature of the Brody equation. Actually this model is not able to adequately fit the complete growth pattern, because it is essentially devoted to describing the second decelerating (mature) phase. A better fit can be achieved by using sigmoidal functions such as the logistic model (GROWTH.SAS).

```
proc nlin data=growth;
model weight=a/(1+b*exp(-k*rescage));
parms a=700 b=3 k=0.013;
output out=curve predicted=p;
run;

symbol1 interpol=join value=none color=black;
symbol2 interpol=none value=star color=black;
proc gplot data=work.curve;
plot p*rescage weight*rescage/overlay;
run;
```

Output 6.18 Results of non-linear fitting of the logistic equation (Equation 6.4) to the growth pattern of the GROWTH data set.

```
                       The NLIN Procedure
                        Iterative Phase
                    Dependent Variable weight
                      Method: Gauss-Newton
                                                        Sum of
        Iter          a          b           k         Squares
         0          800.0      3.0000      0.0130      202037
         1          749.4      2.3363      0.00640      26182.3
         2          934.1      3.3422      0.00613       1490.3
         3          921.8      3.3675      0.00648        481.3
         4          924.4      3.3810      0.00648        478.8
         5          924.3      3.3807      0.00648        478.8
         6          924.3      3.3807      0.00648        478.8
    NOTE: Convergence criterion met.

                        Estimation Summary
             Method                   Gauss-Newton
             Iterations                          6
             R                           1.326E-6
             PPC(k)                      1.681E-7
             RPC(b)                      4.287E-6
             Object                      1.248E-9
             Objective                   478.7796
             Observations Read                 13
             Observations Used                 13
             Observations Missing               0

    NOTE: An intercept was not specified for this model.
                            Sum of        Mean                    Approx
    Source              DF  Squares       Square     F Value      Pr > F
    Regression           3  3055500       1018500     21272.8     <.0001
    Residual            10    478.8       47.8780
    Uncorrected Total   13  3055979
    Corrected Total     12   317126

                            Approx
    Parameter   Estimate   Std Error   Approximate 95% Confidence Limits
    a            924.3      35.6086         845.0            1003.7
    b            3.3807      0.1285         3.0943           3.6671
    k            0.00648     0.000298       0.00581          0.00714

                    Approximate Correlation Matrix
                           a              b              k
        a          1.0000000      0.8579413     -0.9512814
        b          0.8579413      1.0000000     -0.6769634
        k         -0.9512814     -0.6769634      1.0000000
```

The better fit of the logistic model is highlighted by both the magnitude of residual (Output 6.18) and the plot of fit superimposed on actual data (Output 6.19).

As previously stated in Section 6.2, the logistic model imposes the inflection of the curve at the midpoint between W_0 and W_f. However, the inflection of the curve could often occur at a lower weight.

Output 6.19 Actual (*) and estimated (—) growth pattern with Equation 6.6.

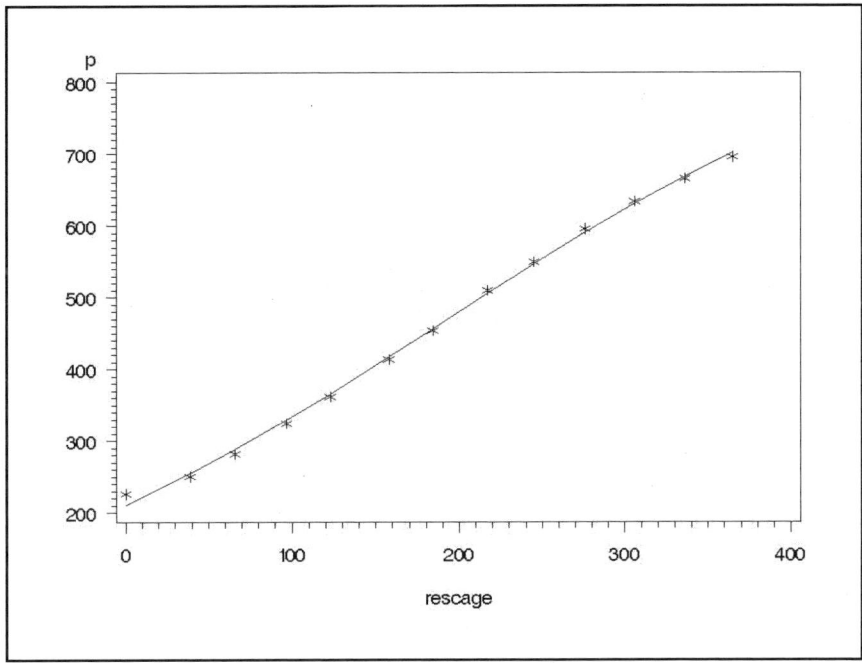

In these circumstances, an alternative option can be found in the Gompertz model. The following SAS program (GOMPERTZ.SAS) fits the Gompertz model to the growth patterns of Sarda dairy breed lambs. Weights are first rescaled, defining the initial weight to be zero. Excellent fit is noted by the low residual sums of squares in Output 6.20 and by the visual fit in Output 6.21.

```
data lamb;
input weight 1-5 age 7;
datalines;
4.44  0
4.86  1
6.44  2
8.14  3
9.88  4
11.40 5
12.78 6
13.14 7
14.86 8
;
run;
data lamb;
set lamb;
weight1=weight-4.44;
run;
proc nlin data=lamb;
model weight1=w0*exp(m*(1-exp(-D*age))/D);
parms w0=0 m=0.5 D=0.10;
output out=Gompertz predicted=p;
run;

symbol1 interpol=join value=none color=black;
symbol2 interpol=none value=star color=black;
proc gplot data=gompertz;
plot p*age weight1*age/overlay;
run;
quit;
run;
```

Output 6.20 Results of Gompertz equation (Equation 6.5) fit.

```
                    The NLIN Procedure
                     Iterative Phase
                Dependent Variable weight1
                    Method: Gauss-Newton
                                                    Sum of
       Iter          w0            m           D    Squares
        0             0        0.5000      0.1000    349.7
        1        0.7567        0.5000      0.1000   10.5947
        2        0.6338        0.6257      0.1597    7.5824
        3        0.5533        0.7284      0.1987    6.6394
        4        0.4972        0.8129      0.2263    5.9377
        5        0.4544        0.8861      0.2479    5.3339
        6        0.3853        1.0175      0.2836    5.1676
        7        0.3476        1.1051      0.3043    4.5980
        8        0.2944        1.2448      0.3349    4.3242
        9        0.2442        1.4069      0.3662    3.9645
       10        0.2080        1.5559      0.3915    3.0413
       11        0.1621        1.7910      0.4272    2.0920
       12        0.1551        1.8718      0.4362    0.6892
       13        0.1566        1.8666      0.4356    0.6857
       14        0.1566        1.8668      0.4357    0.6857
       15        0.1566        1.8668      0.4356    0.6857
    NOTE: Convergence criterion met.

                      Estimation Summary
          Method                       Gauss-Newton
          Iterations                             15
          Subiterations                          14
          Average Subiterations            0.933333
          R                                4.408E-6
          PPC(w0)                            5.3E-6
          RPC(w0)                          0.000065
          Object                           1.739E-9
          Objective                        0.685663
          Observations Read                       9
          Observations Used                       9
          Observations Missing                    0
    NOTE: An intercept was not specified for this model.

                         Sum of        Mean                      Approx
Source              DF   Squares       Square    F Value         Pr > F
Regression           3     349.0        116.3    1018.10         <.0001
Residual             6    0.6857       0.1143
Uncorrected Total    9     349.7
Corrected Total      8     114.8

                              Approx
Parameter      Estimate     Std Error   Approximate 95% Confidence Limits
w0               0.1566        0.0962      -0.0787            0.3918
m                1.8668        0.4742       0.7064            3.0271
D                0.4356        0.0559       0.2988            0.5725

                    Approximate Correlation Matrix
                        w0               m                D
          w0      1.0000000      -0.9871092       -0.9214068
          m      -0.9871092       1.0000000        0.9713281
          D      -0.9214068       0.9713281        1.0000000
```

Output 6.21 Actual (*) and estimated (—) growth pattern with Equation 6.5.

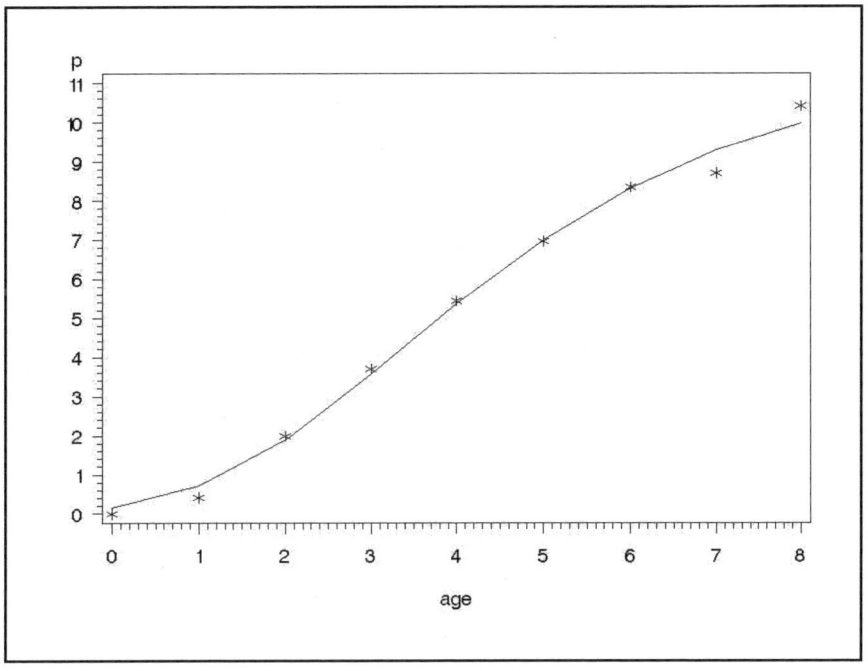

6.5 Repeated Measures Theory for Lactation and Growth Curves

The fit of mathematical functions to test day (TD) data allows for the reconstruction of the pattern of milk production or growth over time, but the analytical description loses its discriminant power when the regular component of the phenomenon is no longer separable from environmental effects and from residual random variation, such as when temporary environmental effects become predominant. This problem can be solved by using linear mixed models able to account for factors that could affect each TD differently. Test day models are currently used in phenotypic analysis to estimate average lactation curves of homogeneous groups of animals corrected for the effects of main environmental factors known to affect milk yield. In comparison with the method based on mathematical functions, this form of the TD model has the advantage that variable amounts of information from different lactations can be used, estimates of fixed effects can vary across stages, and these estimates can be adjusted for sampling data effects. However, the two approaches are not antithetical, because the lactation curve corrected by the effects of main environmental factors can be reconstructed by a TD model by including a DIM (days in milk) factor.

TD milk yields at different lactation stages or body weights recorded at different ages represent a case of *repeated measure design*, in which measurements are taken in sequence over time on the same experimental unit (the animal). In particular, they can be analyzed as a split-plot in time design: the animal is the main plot, and the different time intervals at which measurements are taken are the subplots. The essential aim of the experiment is to make inference about the experimental treatments on the mean response profile; in the specific case of milk or growth, it is usually aimed at investigating the effects of environmental factors such as parity, calving season, feeding, management system, etc. Repeated measure analysis allows the reconstruction of the "time-dependent" correlation pattern among responses on the same animal, or subject. Test day milk yields close in time are more closely correlated than those far apart in time. Actually, TDs are correlated because they share common contributions from the same animal, but there are also covariances between adjacent TDs due to environmental short-term effects such as dietary quality, intake, weather, minor injuries, and estrus (Ali and Schaeffer, 1987; Carvalheira et al., 1998; Wade et al., 1993), as well as genetic variation. These potential patterns of correlation and variation may combine to produce a complicated structure of covariance among TDs that, when ignored, may result in inefficient analysis or incorrect conclusions, especially as far as the test of the difference among treatment means is concerned.

The general linear mixed model methodology is able to address these issues by directly modeling the (co)variance structure among repeated measures. In matrix notation, the linear mixed model can be written in the well-known form

$$y = X\beta + Zu + \varepsilon,$$

where X and Z are known design matrices of fixed and random effects, respectively; β and u are unknown vectors of parameters of fixed and random effects, respectively; and ε is the random residual with $u \sim N(0, G)$ and $\varepsilon \sim N(0, R)$. The variance of y is therefore $V = ZGZ' + R$.

One way to fit the (co)variance pattern of repeated measurements is to model the G matrix by using the RANDOM statement in PROC MIXED, assuming a diagonal matrix $R = \sigma^2 I$. The other equivalent approach, developed in this section, is to eliminate the Z and G matrices and let the residual R matrix be block diagonal with identical $n \times n$ submatrices (R_i), each corresponding to a subject, where n is the number of measurements on each of s subjects.

$$R = \begin{bmatrix} R_1 & & & \\ & R_2 & & \\ & & \cdots & \\ & & & R_s \end{bmatrix}, \text{ where } R_1 = R_2 = \ldots R_s.$$

This approach can be achieved using the REPEATED statement in PROC MIXED. Actually, the basic difference between the two approaches is that in the first the inter-animal variance is modeled in the G matrix, whereas in the second it is absorbed in the general structure of R.

The SAS package provides several (co)variance structures that can be imposed on the R_i submatrices in order to model (co)variances within subjects.

The most complex structure in terms of number of parameters to be estimated is the unstructured (UN) structure. This structure makes no assumption of equal variances and covariances. As an example, the UN structure is reported for a repeated design with four time intervals:

$$R_i = \begin{bmatrix} \sigma_1^2 & \sigma_{12} & \sigma_{13} & \sigma_{14} \\ & \sigma_2^2 & \sigma_{23} & \sigma_{24} \\ & & \sigma_3^2 & \sigma_{34} \\ \text{symm} & & & \sigma_4^2 \end{bmatrix}.$$

The fitting of the UN structure can lead to severe computational problems due to the large number of parameters to be estimated: $t(t + 1)/2$, where t is the number of time intervals. Moreover, it does not exploit the existence of a trend of (co)variances over time, and this often results in erratic patterns of standard error estimates (Littell et al., 1998). However, the UN correlation pattern can represent a useful first step in the search of the most suitable (co)variance model.

The most frequently used (co)variance structure in split-plot in time designs is the compound symmetry (CS) structure. It assumes a constant variance (σ^2) for each measure and a fixed value of the covariance between measures for each time pairs. Thus, each block of **R** has the following form:

$$\mathbf{R}_i = \begin{bmatrix} \sigma^2+\sigma^2_1 & \sigma^2_1 & \sigma^2_1 & \sigma^2_1 \\ & \sigma^2+\sigma^2_1 & \sigma^2_1 & \sigma^2_1 \\ & & \sigma^2+\sigma^2_1 & \sigma^2_1 \\ \text{symm} & & & \sigma^2+\sigma^2_1 \end{bmatrix}.$$

The great advantage of the CS structure is that only two parameters, σ^2_1 and σ^2, have to be estimated. The correlation between two measures within an animal, regardless of the temporal distance between them, is calculated as the ratio $\sigma^2_1 / (\sigma^2_1 + \sigma^2)$.

However, it should be remembered that the actual (co)variance pattern of repeated measurements is not characterized by equal variances and covariances but by relationships among measures that tend to decrease with increasing time intervals (lag) across test days. A useful structure for this covariance pattern is the first-order autoregressive structure AR(1). Autoregressive processes have been proposed in several studies as suitable structures for modeling TD measures in dairy cattle, which give blocks of **R** with the pattern (Carvalheira et al., 1998; Wade et al., 1993).

$$\mathbf{R}_i = \begin{bmatrix} \sigma^2 & \sigma^2*\rho & \sigma^2*\rho^2 & \sigma^2*\rho^3 & \sigma^2*\rho^4 \\ & \sigma^2 & \sigma^2*\rho & \sigma^2*\rho^2 & \sigma^2*\rho^3 \\ & & \sigma^2 & \sigma^2*\rho & \sigma^2*\rho^2 \\ \text{symm} & & & \sigma^2 & \sigma^2*\rho \\ & & & & \sigma^2 \end{bmatrix}.$$

The AR(1) model assumes a constant variance for all measures, whereas the covariance decreases with increasing lags according to ρ^k, where ρ is the autoregressive parameter of the first order and k is the time lag. Actually, the AR structure smooths the original pattern of the correlations produced by the UN structure, with the advantage of fewer parameters to be estimated.

6.6 PROC MIXED for Test Day Models

The data set BUFFALOES (BUFFALO.SAS) reports test day data of milk production traits (milk yield, fat and protein percentages) of 493 lactations of Italian water buffaloes, used in the study of Catillo et al. (2002). Data are analyzed with a linear mixed model that includes date of the test, calving season, age at calving, DIM interval (lactation length has been divided into 10 DIM intervals of 30 days) as fixed effects, and cow and residual as random effects. Fixed DIM(AGE) and DIM(SEA) effects are included in order to estimate lactation curves of different age at calving and calving seasons. Two (co)variance structures, CS and AR(1), are fitted. The subjects of the **R** matrix are indicated by the SUB option in the REPEATED statement, and the type of (co)variance structure is indicated in the TYPE statement.

The fit of the UN structure is rather time-consuming. As an example, the **R** correlation matrix estimated with the UN structure is shown in Output 6.22.

Output 6.22 Unstructured R matrix for the first subject.

```
                    Estimated R Correlation Matrix for cow 1
Row    Col1      Col2      Col3      Col4      Col5      Col6      Col7      Col8
  1  1.0000    0.7036    0.6577    0.5734    0.4780    0.4317    0.4067    0.4033
  2  0.7036    1.0000    0.7217    0.5707    0.5120    0.4494    0.4294    0.4091
  3  0.6577    0.7217    1.0000    0.6561    0.5664    0.5386    0.4719    0.4817
  4  0.5734    0.5707    0.6561    1.0000    0.5859    0.5025    0.5020    0.4556
  5  0.4780    0.5120    0.5664    0.5859    1.0000    0.5517    0.4950    0.4312
  6  0.4317    0.4494    0.5386    0.5025    0.5517    1.0000    0.5464    0.4881
  7  0.4067    0.4294    0.4719    0.5020    0.4950    0.5464    1.0000    0.4878
  8  0.4033    0.4091    0.4817    0.4556    0.4312    0.4881    0.4878    1.0000
```

The **R** correlation matrix shows the typical pattern of repeated measurements, with a decrease in correlation values as the time interval (lag) between measurements increases. The decreasing trend is evident until lag 4 (column 5), after which the correlation tends to level out.

At first, the compound symmetry (CS) structure is fitted.

```
proc mixed data=buffaloes noclprint update info covtest;
class age testdate sea dim cow;
model milk=age sea dim testdate dim(sea) dim(age)/ddfm=kr;
repeated dim/ sub= cow type=CS r rcorr;
lsmeans age sea/diff adjust=tukey;
lsmeans dim(sea) dim(age);
ods output lsmeans=curves;
run;
```

The Kenward-Roger adjustment for degrees of freedom is indicated with the DDFM=KR option in the MODEL statement. This adjustment is recommended in mixed model analysis. Usually the SAS System standard errors are based on the assumption that estimated variances are the true variances. The Kenward-Roger correction inflates the standard errors of pairwise comparisons and changes the degrees of freedom in order to correct for the initial exclusion of variance uncertainty. Moreover, the Tukey-Kramer adjustment for multiple comparisons with the option ADJUST makes pairwise comparisons more conservative.

Least squares means for DIM(AGE) and DIM(SEA) are not reported in Output 6.23 but are plotted against DIM in the graphs to construct lactation curves.

Output 6.23 Results of mixed model analysis of BUFFALOES data.

```
                        The Mixed Procedure
                         Model Information
         Data Set                    WORK.BUFFALOES
         Dependent Variable          milk
         Covariance Structure        Compound Symmetry
         Subject Effect              cow
         Estimation Method           REML
         Residual Variance Method    Profile
         Fixed Effects SE Method     Prasad-Rao-Jeske-
                                     Kackar-Harville
         Degrees of Freedom Method   Kenward-Roger

                            Dimensions
                  Covariance Parameters      2
                  Columns in X             291
                  Columns in Z               0
                  Subjects                 493
                  Max Obs Per Subject       10
                  Observations Used       4187
                  Observations Not Used    578
                  Total Observations      4765

                         Iteration History
         Iteration    Evaluations    -2 Res Log Like       Criterion
             0             1         17953.05791910
             1             2         16509.61703549        0.00049100
             2             1         16507.09557206        0.00001622
             3             1         16507.01850284        0.00000002
             4             1         16507.01840539        0.00000000
                        Convergence criteria met.
                      Estimated R Matrix for cow 1
 Row    Col1     Col2     Col3     Col4     Col5     Col6     Col7     Col8
  1   4.9828   2.5259   2.5259   2.5259   2.5259   2.5259   2.5259   2.5259
  2   2.5259   4.9828   2.5259   2.5259   2.5259   2.5259   2.5259   2.5259
  3   2.5259   2.5259   4.9828   2.5259   2.5259   2.5259   2.5259   2.5259
  4   2.5259   2.5259   2.5259   4.9828   2.5259   2.5259   2.5259   2.5259
  5   2.5259   2.5259   2.5259   2.5259   4.9828   2.5259   2.5259   2.5259
  6   2.5259   2.5259   2.5259   2.5259   2.5259   4.9828   2.5259   2.5259
  7   2.5259   2.5259   2.5259   2.5259   2.5259   2.5259   4.9828   2.5259
  8   2.5259   2.5259   2.5259   2.5259   2.5259   2.5259   2.5259   4.9828

                 Estimated R Correlation Matrix for cow 1
 Row    Col1     Col2     Col3     Col4     Col5     Col6     Col7     Col8
  1   1.0000   0.5069   0.5069   0.5069   0.5069   0.5069   0.5069   0.5069
  2   0.5069   1.0000   0.5069   0.5069   0.5069   0.5069   0.5069   0.5069
  3   0.5069   0.5069   1.0000   0.5069   0.5069   0.5069   0.5069   0.5069
  4   0.5069   0.5069   0.5069   1.0000   0.5069   0.5069   0.5069   0.5069
  5   0.5069   0.5069   0.5069   0.5069   1.0000   0.5069   0.5069   0.5069
  6   0.5069   0.5069   0.5069   0.5069   0.5069   1.0000   0.5069   0.5069
  7   0.5069   0.5069   0.5069   0.5069   0.5069   0.5069   1.0000   0.5069
  8   0.5069   0.5069   0.5069   0.5069   0.5069   0.5069   0.5069   1.0000

                     Covariance Parameter Estimates
                                      Standard         Z
        Cov Parm    Subject  Estimate    Error      Value    Pr Z
        CS          cow       2.5259    0.1927      13.11   <.0001
        Residual              2.4569    0.05950     41.29   <.0001

                           Fit Statistics
                  -2 Res Log Likelihood         16507.0
                  AIC  (smaller is better)      16511.0
                  AICC (smaller is better)      16511.0
                  BIC  (smaller is better)      16519.4

                    Null Model Likelihood Ratio Test
                     DF    Chi-Square    Pr > ChiSq
                      1      1446.04        <.0001
```

(continued on next page)

Output 6.23 *(continued)*

```
                    Type 3 Tests of Fixed Effects
                          Num    Den
              Effect      DF     DF    F Value    Pr > F
              age          5     440    12.50    <.0001
              sea          3     470     5.29     0.0013
              dim          9    3465   427.13    <.0001
              testdate   169    3691     2.41    <.0001
              dim(sea)    27    3457     1.23     0.1943
              dim(age)    45    3437     5.01    <.0001
```

```
                           Least Squares Means
                                      Standard
Effect    age    sea    dim    Estimate    Error     DF    t Value    Pr > |t|
age        1                    7.0491    0.5579    440    12.63      <.0001
age        2                    7.1443    0.2124    478    33.63      <.0001
age        3                    8.3087    0.2340    460    35.51      <.0001
age        4                    8.7765    0.2507    457    35.01      <.0001
age        5                    8.8248    0.2729    456    32.33      <.0001
age        6                    8.4263    0.2179    455    38.67      <.0001
sea               1             8.5404    0.2423    465    35.25      <.0001
sea               2             8.1976    0.2547    465    32.19      <.0001
sea               3             7.6601    0.2175    480    35.22      <.0001
sea               4             7.9551    0.2235    475    35.59      <.0001
```

```
                        Differences of Least Squares Means
                                              Standard
Effect  age  sea  dim  _age  _sea  _dim  Estimate   Error     DF   t Value   Pr > |t|
age      1              2                -0.09522   0.5591   437   -0.17     0.8648
age      1              3                -1.2596    0.5681   438   -2.22     0.0271
age      1              4                -1.7274    0.5749   437   -3.00     0.0028
age      1              5                -1.7757    0.5868   437   -3.03     0.0026
age      1              6                -1.3772    0.5564   436   -2.48     0.0137
age      2              3                -1.1643    0.2403   445   -4.85     <.0001
age      2              4                -1.6321    0.2626   447   -6.22     <.0001
age      2              5                -1.6805    0.2849   441   -5.90     <.0001
age      2              6                -1.2820    0.2206   443   -5.81     <.0001
age      3              4                -0.4678    0.2708   441   -1.73     0.0848
age      3              5                -0.5161    0.2926   439   -1.76     0.0784
age      3              6                -0.1176    0.2316   439   -0.51     0.6118
age      4              5                -0.04831   0.3073   439   -0.16     0.8752
age      4              6                 0.3502    0.2496   440    1.40     0.1613
age      5              6                 0.3985    0.2714   438    1.47     0.1428
sea          1                2           0.3428    0.2423   462    1.41     0.1578
sea          1                3           0.8803    0.2267   479    3.88     0.0001
sea          1                4           0.5853    0.2300   474    2.54     0.0113
sea          2                3           0.5375    0.2354   463    2.28     0.0229
sea          2                4           0.2425    0.2443   470    0.99     0.3215
sea          3                4          -0.2950    0.2161   470   -1.36     0.1729
```

```
                        Differences of Least Squares Means
Effect     age   sea   dim   _age   _sea   _dim   Adjustment      Adj P
age         1                 2                   Tukey-Kramer    1.0000
age         1                 3                   Tukey-Kramer    0.2319
age         1                 4                   Tukey-Kramer    0.0333
age         1                 5                   Tukey-Kramer    0.0312
age         1                 6                   Tukey-Kramer    0.1337
age         2                 3                   Tukey-Kramer    <.0001
age         2                 4                   Tukey-Kramer    <.0001
age         2                 5                   Tukey-Kramer    <.0001
age         2                 6                   Tukey-Kramer    <.0001
age         3                 4                   Tukey-Kramer    0.5142
age         3                 5                   Tukey-Kramer    0.4902
age         3                 6                   Tukey-Kramer    0.9959
```

(continued on next page)

Output 6.23 *(continued)*

```
          age          4             5            Tukey-Kramer    1.0000
          age          4             6            Tukey-Kramer    0.7252
          age          5             6            Tukey-Kramer    0.6847
          sea          1             2            Tukey-Kramer    0.4906
          sea          1             3            Tukey-Kramer    0.0007
          sea          1             4            Tukey-Kramer    0.0546
          sea          2             3            Tukey-Kramer    0.1033
          sea          2             4            Tukey-Kramer    0.7538
          sea          3             4            Tukey-Kramer    0.5221
```

In Output 6.23, the CS structure leads to the estimation of two parameters, the between-cows variance σ^2_1, 2.5259, and the residual variance σ^2, 2.4569. The **R** matrix shows a constant variance on the diagonal, given by the sum of the two covariance parameters (2.5259 + 2.4569 = 4.9828), and constant covariance of the diagonal represented by 2.5259. The average correlation among test day milk yields within cows is 2.5259/(2.5259 + 2.4569) = 0.506.

All fixed factors included in the analysis affected TD milk yields significantly. The TD fixed factor, peculiar to TD models, absorbs a relevant amount of the original variability. It accounts for the effects of the season in which a TD occurs. Daily milk yield increases with the age of animals until age class 5. The difference observed between summer and winter calvings (about 0.9 kg milk/day) can be explained mainly by the depressive effect of high temperatures at the start of lactation.

Differences observed in least squares means are confirmed by lactation curve shapes. The following program yields the plot of estimated lactation curves for different calving season and age at calving classes.

```
data sea;
set curves;
if effect ne 'dim(sea)' then delete;
run;
symbol1 interpol=join color=black value=star;
symbol2 interpol=join color=black value=none;
symbol3 interpol=join color=black value=diamond;
symbol4 interpol=join color=black value=plus;
proc gplot data=sea;
plot estimate*dim=sea;
run;
data age;
set curves;
if effect ne 'dim(age)' then delete;
run;
symbol1 interpol=join color=black value=star;
symbol2 interpol=join color=black value=none;
symbol3 interpol=join color=black value=diamond;
symbol4 interpol=join color=black value=plus;
symbol4 interpol=join color=black value=plus;
symbol5 interpol=join color=black value=triangle;
symbol6 interpol=join color=black value=square;
proc gplot data=age;
plot estimate*dim=age;
run;
```

Output 6.24 shows that the evolution over time of daily milk yield follows the typical pattern of dairy animals, with a first ascending phase to the lactation peak and a subsequent decrease toward the dry-off. Lactation curves of buffaloes calving in summer (SEA=3) are at a slightly lower level at the beginning of lactation.

Output 6.24 Estimated lactation curves of buffaloes calving in winter (sea=1), spring (sea=2), summer (sea=3), and fall (sea=4).

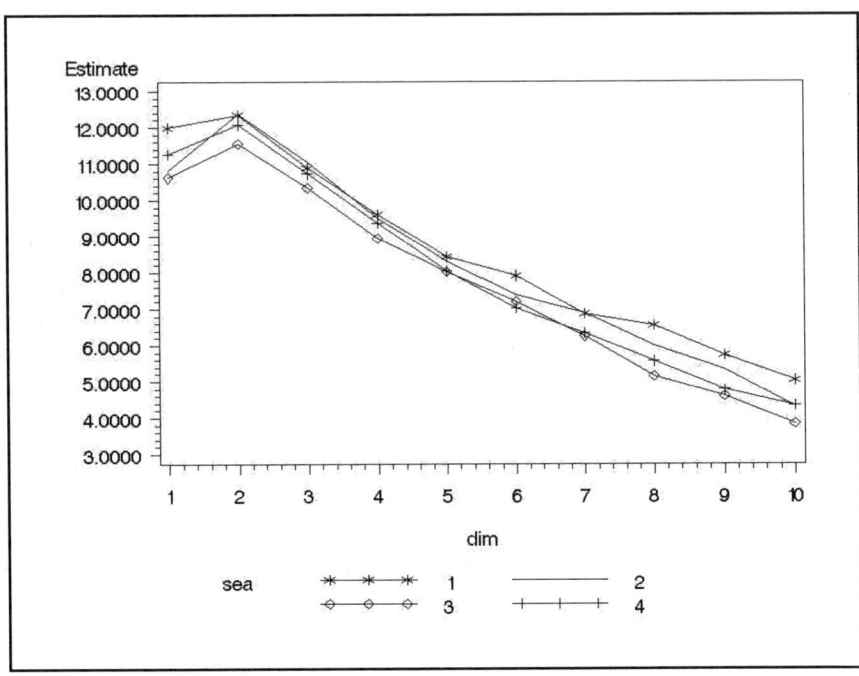

In Output 6.25, lactation curves of buffaloes of 2 to 3 years of age (age class 1) differ from those of older animals. Such a separation is evident in the first phase of lactation ($P<0.01$) and gradually decreases until the 5th–6th month of lactation. Peak yield occurs at around the 6th week of lactation in all age classes; lactation curves of buffalo cows aged 2 to 3 years are characterized by the highest persistency.

Output 6.25 Estimated lactation curves of buffaloes calving at different age classes (1 = two to three years; 2 = three to four years, ..., 6 = seven years).

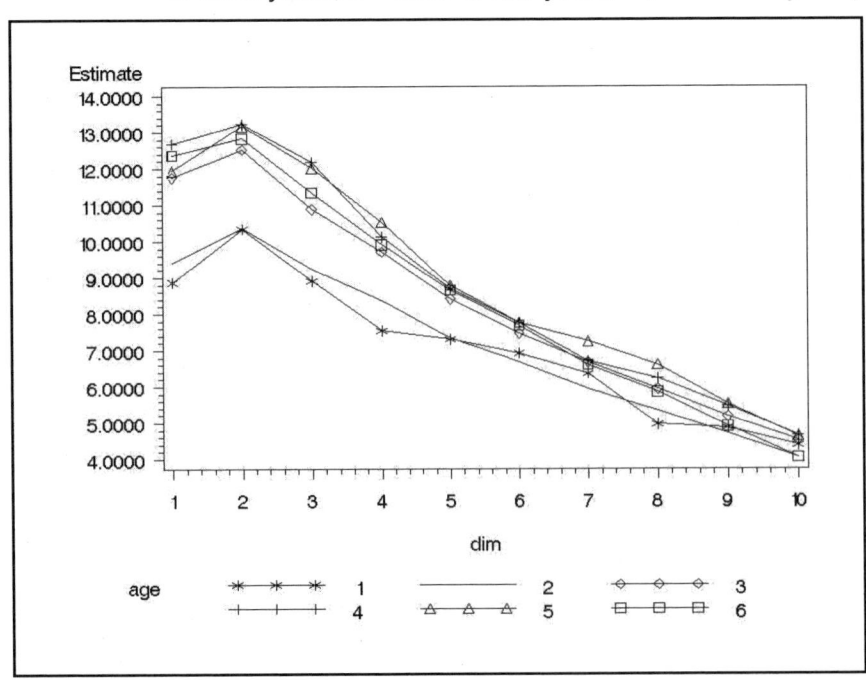

This model can be specified in PROC MIXED in two different but equivalent ways:

```
proc mixed data=buffaloes noclprint update info covtest;
class age testdate sea dim cow;
model milk=age sea dim testdate dim*sea dim*age/ddfm=kr;
random cow;
run;
```

or

```
proc mixed data=buffaloes noclprint update info covtest;
class age testdate sea dim cow;
model milk=age sea dim testdate dim*sea dim*age/ddfm=kr;
random intercept/sub=cow;
run;
```

The following SAS program fits the AR(1) structure:

```
proc mixed data=buffaloes noclprint update info covtest;
class age testdate sea dim cow;
model milk=age sea dim testdate dim*sea dim*age/ddfm=kr;
repeated dim/sub= cow type=AR(1) r rcorr;
lsmeans age sea/diff adjust=tukey;
lsmeans dim*sea dim*age;
ods output lsmeans=curves;
run;
```

Output 6.26 Results of mixed model analysis of BUFFALOES data set imposing a first-order autoregressive structure on the R (co)variance matrix.

```
                      The Mixed Procedure

                       Model Information
     Data Set                      WORK.BUFFALOES
     Dependent Variable            milk
     Covariance Structure          Autoregressive
     Subject Effect                cow
     Estimation Method             REML
     Residual Variance Method      Profile
     Fixed Effects SE Method       Prasad-Rao-Jeske-
                                   Kackar-Harville
     Degrees of Freedom Method     Kenward-Roger

                          Dimensions
               Covariance Parameters         2
               Columns in X                291
               Columns in Z                  0
               Subjects                    493
               Max Obs Per Subject          10
               Observations Used          4187
               Observations Not Used       578
               Total Observations         4765

                        Iteration History
   Iteration    Evaluations    -2 Res Log Like      Criterion
           0              1       17953.05791910
           1              2       16336.62744636    0.00005503
           2              1       16336.37523708    0.00000000
                     Convergence criteria met.
```

(continued on next page)

Output 6.26 *(continued)*

```
                        Estimated R Matrix for cow 1
Row     Col1      Col2      Col3      Col4      Col5      Col6      Col7      Col8
 1    4.9416    3.2238    2.1032    1.3721    0.8952    0.5840    0.3810    0.2486
 2    3.2238    4.9416    3.2238    2.1032    1.3721    0.8952    0.5840    0.3810
 3    2.1032    3.2238    4.9416    3.2238    2.1032    1.3721    0.8952    0.5840
 4    1.3721    2.1032    3.2238    4.9416    3.2238    2.1032    1.3721    0.8952
 5    0.8952    1.3721    2.1032    3.2238    4.9416    3.2238    2.1032    1.3721
 6    0.5840    0.8952    1.3721    2.1032    3.2238    4.9416    3.2238    2.1032
 7    0.3810    0.5840    0.8952    1.3721    2.1032    3.2238    4.9416    3.2238
 8    0.2486    0.3810    0.5840    0.8952    1.3721    2.1032    3.2238    4.9416

                   Estimated R Correlation Matrix for cow 1
Row     Col1      Col2      Col3      Col4      Col5      Col6      Col7      Col8
 1    1.0000    0.6524    0.4256    0.2777    0.1811    0.1182   0.07710   0.05030
 2    0.6524    1.0000    0.6524    0.4256    0.2777    0.1811    0.1182   0.07710
 3    0.4256    0.6524    1.0000    0.6524    0.4256    0.2777    0.1811    0.1182
 4    0.2777    0.4256    0.6524    1.0000    0.6524    0.4256    0.2777    0.1811
 5    0.1811    0.2777    0.4256    0.6524    1.0000    0.6524    0.4256    0.2777
 6    0.1182    0.1811    0.2777    0.4256    0.6524    1.0000    0.6524    0.4256
 7   0.07710    0.1182    0.1811    0.2777    0.4256    0.6524    1.0000    0.6524
 8   0.05030   0.07710    0.1182    0.1811    0.2777    0.4256    0.6524    1.0000

                      Covariance Parameter Estimates
                                      Standard      Z
        Cov Parm    Subject   Estimate   Error    Value    Pr Z
        AR(1)       cow        0.6524   0.01276   51.13   <.0001
        Residual               4.9416   0.1663    29.72   <.0001

                           Fit Statistics
             -2 Res Log Likelihood          16336.4
             AIC  (smaller is better)       16340.4
             AICC (smaller is better)       16340.4
             BIC  (smaller is better)       16348.8

                 Null Model Likelihood Ratio Test
                  DF    Chi-Square     Pr > ChiSq
                   1      1616.68        <.0001

                   Type 3 Tests of Fixed Effects
                        Num    Den
             Effect      DF    DF     F Value    Pr > F
             age          5    589     17.08    <.0001
             sea          3    648      4.57    0.0036
             dim          9   3347    135.72    <.0001
             testdate   169   3727      2.24    <.0001
             dim(sea)    27   3503      1.39    0.0883
             dim(age)    45   3509      1.86    0.0005

                         Least Squares Means
                                    Standard
Effect    age   sea   dim   Estimate    Error     DF    t Value    Pr > |t|
age        1                 6.9445    0.4601    584     15.09     <.0001
age        2                 7.0559    0.1797    647     39.27     <.0001
age        3                 8.1763    0.1963    633     41.66     <.0001
age        4                 8.6451    0.2088    627     41.40     <.0001
age        5                 8.6077    0.2279    621     37.78     <.0001
age        6                 8.3559    0.1818    628     45.96     <.0001
sea              1           8.2782    0.2044    640     40.51     <.0001
sea              2           8.0622    0.2150    643     37.49     <.0001
sea              3           7.5753    0.1847    670     41.02     <.0001
sea              4           7.9412    0.1906    652     41.66     <.0001

                   Differences of Least Squares Means
                                                Standard
Effect  age sea dim  _age  _sea  _dim   Estimate   Error    DF    t Value   Pr > |t|
age      1            2                 -0.1114   0.4602   577     -0.24    0.8088
age      1            3                 -1.2317   0.4686   579     -2.63    0.0088
age      1            4                 -1.7006   0.4733   578     -3.59    0.0004
age      1            5                 -1.6632   0.4838   578     -3.44    0.0006
age      1            6                 -1.4113   0.4578   576     -3.08    0.0022
```

(continued on next page)

Output 6.26 *(continued)*

```
age      2            3                  -1.1203   0.1993   595   -5.62   <.0001
age      2            4                  -1.5892   0.2177   596   -7.30   <.0001
age      2            5                  -1.5518   0.2354   592   -6.59   <.0001
age      2            6                  -1.2999   0.1827   595   -7.12   <.0001
age      3            4                  -0.4689   0.2242   593   -2.09   0.0369
age      3            5                  -0.4315   0.2416   591   -1.79   0.0746
age      3            6                  -0.1796   0.1916   590   -0.94   0.3489
age      4            5                   0.0373   0.2537   591    0.15   0.8829
age      4            6                   0.2893   0.2064   593    1.40   0.1616
age      5            6                   0.2519   0.2241   589    1.12   0.2615
sea          1        2                   0.2160   0.2063   637    1.05   0.2956
sea          1        3                   0.7029   0.1960   673    3.59   0.0004
sea          1        4                   0.3370   0.1979   631    1.70   0.0891
sea          2        3                   0.4870   0.1997   642    2.44   0.0150
sea          2        4                   0.1210   0.2110   653    0.57   0.5666
sea          3        4                  -0.3660   0.1846   657   -1.98   0.0479

                    Differences of Least Squares Means

            Effect    age   sea   dim   _age   _sea   _dim   Adjustment      Adj P
            age        1                  2                  Tukey-Kramer   0.9999
            age        1                  3                  Tukey-Kramer   0.0920
            age        1                  4                  Tukey-Kramer   0.0047
            age        1                  5                  Tukey-Kramer   0.0082
            age        1                  6                  Tukey-Kramer   0.0260
            age        2                  3                  Tukey-Kramer   <.0001
            age        2                  4                  Tukey-Kramer   <.0001
            age        2                  5                  Tukey-Kramer   <.0001
            age        2                  6                  Tukey-Kramer   <.0001
            age        3                  4                  Tukey-Kramer   0.2930
            age        3                  5                  Tukey-Kramer   0.4757
            age        3                  6                  Tukey-Kramer   0.9367
            age        4                  5                  Tukey-Kramer   1.0000
            age        4                  6                  Tukey-Kramer   0.7263
            age        5                  6                  Tukey-Kramer   0.8714
            sea              1                   2           Tukey-Kramer   0.7219
            sea              1                   3           Tukey-Kramer   0.0020
            sea              1                   4           Tukey-Kramer   0.3231
            sea              2                   3           Tukey-Kramer   0.0711
            sea              2                   4           Tukey-Kramer   0.9400
            sea              3                   4           Tukey-Kramer   0.1957
```

Output 6.26 clearly shows the different structure of the **R** matrix imposed by the AR(1) model, which is closer to the UN structure than to the CS structure. Akaike's information criterion (AIC) and the Bayesian information criterion (BIC, more conservative in terms of penalization for the number of parameters used) are lower for the AR(1) structure (AIC 16511 vs. 16340 and BIC 16519 vs. 16348 for CS and AR(1), respectively), indicating this structure as the best (Wada and Kashiwagi, 1990). Some differences between the two models can be observed at the level of statistical significance of pairwise comparisons.

Actually, most applications of these linear mixed models concern milk production data. However, they can be usefully adopted to estimate growth curves. The following SAS program (LAMBS.SAS) fits a linear mixed model with a UN structure to growth patterns of Sarda dairy breed lambs, reported in the data set LAMBS, subjected to a feeding treatment with three different levels of supplementation. The analysis is aimed at investigating the effect of the feeding treatment on the mean growth curve and at estimating average growth curves of each treatment group. The month factor models the effect of the growth pattern in analogy with the DIM factor in milk production analysis. Least squares means of month (group) factors allow for the reconstruction of growth curves pertaining to different levels of the feeding treatment.

```
proc mixed data=lambs noclprint update info covtest;
class lamb group month;
model weight = group month month(group)/ddfm=kr;
repeated month/sub=lamb type=UN r rcorr;
lsmeans group /diff adjust=tukey;
lsmeans month(group);
ods output lsmeans=curves;
run;

symbol1 interpol=join color=black value=none;
symbol2 interpol=join color=black value=star;
symbol3 interpol=join color=black value=diamond;
proc gplot data=curves;
plot estimate*month=group;
run;
quit;
```

Output 6.27 estimates an increasing trend in variances that can be observed on the diagonal of the **R** matrix. This is a peculiar feature of growth curve data, because weights may increase linearly with age and the variance of weights may increase quadratically with age. Thus the residual variance would be expected to increase in a manner similar to the phenotypic variance (Schaeffer, 2004). Moreover, the **R** correlation matrix shows the same decreasing trend for increasing lags observed for lactation curves, even if to a marked lower extent. This behavior can be ascribed to the nature of the trait examined. It is well known that growth traits have a higher genetic component and therefore a higher repeatability than milk production traits. The fit of a compound symmetry structure (not reported) leads to the calculation of an average correlation value for weights within animal of about 75%. Output 6.28 shows the average growth curves of the three levels of feeding treatment.

Output 6.27 Results of mixed model analysis of lamb growth patterns.

```
                        The Mixed Procedure

                         Model Information
            Data Set                     WORK.LAMBS
            Dependent Variable           Weight
            Covariance Structure         Unstructured
            Subject Effect               Lamb
            Estimation Method            REML
            Residual Variance Method     None
            Fixed Effects SE Method      Prasad-Rao-Jeske-
                                         Kackar-Harville
            Degrees of Freedom Method    Kenward-Roger

                             Dimensions
                   Covariance Parameters        45
                   Columns in X                 40
                   Columns in Z                  0
                   Subjects                     30
                   Max Obs Per Subject           9
                   Observations Used           265
                   Observations Not Used         6
                   Total Observations          271

                          Iteration History
       Iteration    Evaluations    -2 Res Log Like    Criterion
               0              1       1091.62146576
               1              3        613.87292451    0.03590614
               2              2        611.73793886    0.00789272
               3              1        610.93716688    0.00080803
               4              1        610.86152563    0.00001430
               5              1        610.86026988    0.00000001
```

(continued on next page)

Output 6.27 *(continued)*

```
                        Convergence criteria met.
                        Estimated R Matrix for Lamb 251
Row    Col1    Col2    Col3    Col4    Col5    Col6    Col7    Col8    Col9
  1   2.0733  2.2205  1.9837  1.8962  1.8524  1.9345  1.7562  1.8846  1.5275
  2   2.2205  2.9800  2.6714  2.6354  2.5442  2.5152  2.2380  2.5643  2.1112
  3   1.9837  2.6714  3.0144  2.8394  3.0117  3.3013  3.1273  3.3587  2.9754
  4   1.8962  2.6354  2.8394  3.5165  3.6285  3.8991  3.8096  3.8026  3.6157
  5   1.8524  2.5442  3.0117  3.6285  4.2002  4.6637  4.6441  4.7511  4.5943
  6   1.9345  2.5152  3.3013  3.8991  4.6637  5.7879  5.5643  5.5768  5.4700
  7   1.7562  2.2380  3.1273  3.8096  4.6441  5.5643  5.8514  5.7354  5.7285
  8   1.8846  2.5643  3.3587  3.8026  4.7511  5.5768  5.7354  6.7075  6.2561
  9   1.5275  2.1112  2.9754  3.6157  4.5943  5.4700  5.7285  6.2561  6.3650

                   Estimated R Correlation Matrix for Lamb 251
Row    Col1    Col2    Col3    Col4    Col5    Col6    Col7    Col8    Col9
  1   1.0000  0.8933  0.7935  0.7023  0.6277  0.5584  0.5042  0.5054  0.4205
  2   0.8933  1.0000  0.8913  0.8141  0.7191  0.6056  0.5359  0.5736  0.4848
  3   0.7935  0.8913  1.0000  0.8721  0.8464  0.7903  0.7446  0.7469  0.6793
  4   0.7023  0.8141  0.8721  1.0000  0.9441  0.8643  0.8398  0.7830  0.7642
  5   0.6277  0.7191  0.8464  0.9441  1.0000  0.9459  0.9368  0.8951  0.8885
  6   0.5584  0.6056  0.7903  0.8643  0.9459  1.0000  0.9561  0.8950  0.9012
  7   0.5042  0.5359  0.7446  0.8398  0.9368  0.9561  1.0000  0.9155  0.9387
  8   0.5054  0.5736  0.7469  0.7830  0.8951  0.8950  0.9155  1.0000  0.9575
  9   0.4205  0.4848  0.6793  0.7642  0.8885  0.9012  0.9387  0.9575  1.0000

                              Fit Statistics
                    -2 Res Log Likelihood         610.9
                    AIC  (smaller is better)      700.9
                    AICC (smaller is better)      722.4
                    BIC  (smaller is better)      763.9

                       Null Model Likelihood Ratio Test
                          DF    Chi-Square    Pr > ChiSq
                          44      480.76        <.0001

                         Type 3 Tests of Fixed Effects
                                Num     Den
                  Effect         DF     DF     F Value   Pr > F
                  Group           2     27        0.39   0.6835
                  Month           8     19.8    327.69   <.0001
                  Month(Group)   16     29.5      5.85   <.0001

                              Least Squares Means
                                         Standard
Effect       Group   Month    Estimate    Error     DF    t Value    Pr > |t|
Group          1              28.3808    0.5951     27     47.69     <.0001
Group          2              27.6796    0.5941     26.9   46.59     <.0001
Group          3              27.8283    0.5941     26.9   46.84     <.0001
Month                  1      16.8083    0.2629     27     63.94     <.0001

                        Differences of Least Squares Means
                                                    Standard
Effect    Group   Month   Group   Month   Estimate   Error    DF   t Value   Pr > |t|
Group       1              2               0.7012   0.8410    27     0.83    0.4117
Group       1              3               0.5525   0.8409    27     0.66    0.5167
Group       2              3              -0.1487   0.8402    26.9  -0.18    0.8608

                        Differences of Least Squares Means
        Effect    Group   Month   Group   Month   Adjustment     Adj P
        Group       1              2              Tukey-Kramer   0.6856
        Group       1              3              Tukey-Kramer   0.7900
        Group       2              3              Tukey-Kramer   0.9829
```

Output 6.28 Estimated average growth patterns for the three levels of feeding treatments.

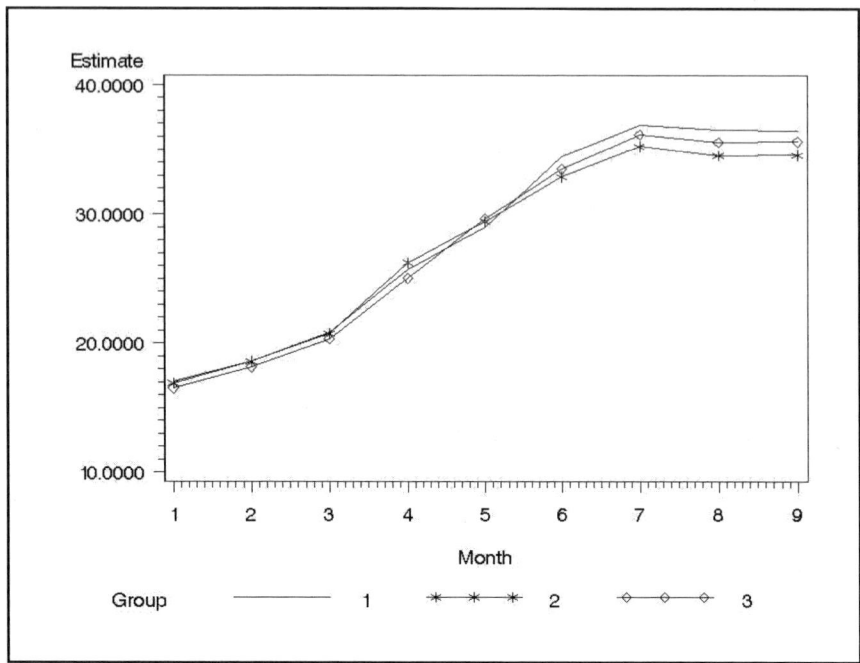

The heterogeneous (co)variances of growth traits can be modeled by a heterogeneous autoregressive structure of the first-order, ARH(1), similar to the AR(1) structure except for variances that vary on the diagonal. The following program fits the ARH(1) to growth data.

```
proc mixed data=lambs noclprint update info covtest;
class lamb group month;
model weight = group month month(group)/ddfm=kr;
repeated month/sub=lamb type=ARH(1) r rcorr;
lsmeans group /diff adjust=tukey;
lsmeans month(group);
ods output  lsmeans=curves;
run;
```

With only 10 covariance parameters, this model gives a −2 log likelihood of 672.6, as shown in Output 6.29. It is certainly an adequate fit compared to the unstructured model's 45 covariances and −2 log likelihood of 610.9 in Output 6.27.

Output 6.29 Results of mixed model analysis of lamb growth patterns with a heterogeneous first-order autoregressive structure, ARH(1).

```
                        The Mixed Procedure
                         Model Information
          Data Set                    WORK.LAMBS
          Dependent Variable          Weight
          Covariance Structure        Heterogeneous
                                      Autoregressive
          Subject Effect              Lamb
          Estimation Method           REML
          Residual Variance Method    None
          Fixed Effects SE Method     Prasad-Rao-Jeske-
                                      Kackar-Harville
          Degrees of Freedom Method   Kenward-Roger

                              Dimensions
                Covariance Parameters         10
                Columns in X                  40
                Columns in Z                   0
                Subjects                      30
                Max Obs Per Subject            9
                Observations Used            265
                Observations Not Used          6
                Total Observations           271

                          Iteration History
      Iteration    Evaluations    -2 Res Log Like     Criterion
          0             1           1091.62146576
          1             2            957.90379468     0.55224631
          2             1            774.44729536     0.33632012
          3             1            703.70457530     0.12696827
          4             1            683.18464661     0.03527998
          5             1            677.65859735     0.02059677
          6             1            674.75693028     0.01151187
          7             1            673.06723508     0.00348169
          8             1            672.60662798     0.00043578
          9             1            672.55212896     0.00000615
         10             1            672.55139935     0.00000000
                       Convergence criteria met.
```

```
                      Estimated R Matrix for Lamb 251
Row    Col1     Col2     Col3     Col4     Col5     Col6     Col7     Col8     Col9
 1    2.6013   2.9118   2.6718   2.4991   2.3426   2.4440   2.2560   2.2328   2.0077
 2    2.9118   3.8125   3.4984   3.2722   3.0672   3.2000   2.9538   2.9234   2.6288
 3    2.6718   3.4984   3.7550   3.5123   3.2923   3.4348   3.1705   3.1379   2.8217
 4    2.4991   3.2722   3.5123   3.8429   3.6022   3.7581   3.4690   3.4333   3.0873
 5    2.3426   3.0672   3.2923   3.6022   3.9497   4.1207   3.8037   3.7645   3.3851
 6    2.4440   3.2000   3.4348   3.7581   4.1207   5.0288   4.6419   4.5941   4.1311
 7    2.2560   2.9538   3.1705   3.4690   3.8037   4.6419   5.0122   4.9606   4.4607
 8    2.2328   2.9234   3.1379   3.4333   3.7645   4.5941   4.9606   5.7429   5.1642
 9    2.0077   2.6288   2.8217   3.0873   3.3851   4.1311   4.4607   5.1642   5.4320
```

```
                  Estimated R Correlation Matrix for Lamb 251
Row    Col1     Col2     Col3     Col4     Col5     Col6     Col7     Col8     Col9
 1    1.0000   0.9246   0.8549   0.7904   0.7308   0.6757   0.6248   0.5777   0.5341
 2    0.9246   1.0000   0.9246   0.8549   0.7904   0.7308   0.6757   0.6248   0.5777
 3    0.8549   0.9246   1.0000   0.9246   0.8549   0.7904   0.7308   0.6757   0.6248
 4    0.7904   0.8549   0.9246   1.0000   0.9246   0.8549   0.7904   0.7308   0.6757
 5    0.7308   0.7904   0.8549   0.9246   1.0000   0.9246   0.8549   0.7904   0.7308
 6    0.6757   0.7308   0.7904   0.8549   0.9246   1.0000   0.9246   0.8549   0.7904
 7    0.6248   0.6757   0.7308   0.7904   0.8549   0.9246   1.0000   0.9246   0.8549
 8    0.5777   0.6248   0.6757   0.7308   0.7904   0.8549   0.9246   1.0000   0.9246
 9    0.5341   0.5777   0.6248   0.6757   0.7308   0.7904   0.8549   0.9246   1.0000
```

```
                      Covariance Parameter Estimates
         Cov                                    Standard       Z
         Parm      Subject      Estimate         Error       Value       Pr Z
         Var(1)    Lamb          2.6013         0.6885        3.78      <.0001
         Var(2)    Lamb          3.8125         1.0013        3.81      <.0001
         Var(3)    Lamb          3.7550         0.9927        3.78      <.0001
         Var(4)    Lamb          3.8429         1.0095        3.81      <.0001
         Var(5)    Lamb          3.9497         1.0221        3.86      <.0001
```

(continued on next page)

Output 6.29 *(continued)*

```
            Var(6)     Lamb        5.0288      1.2845      3.91     <.0001
            Var(7)     Lamb        5.0122      1.2567      3.99     <.0001
            Var(8)     Lamb        5.7429      1.4089      4.08     <.0001
            Var(9)     Lamb        5.4320      1.3401      4.05     <.0001
            ARH(1)     Lamb        0.9246      0.01782    51.88     <.0001

                              Fit Statistics
                       -2 Res Log Likelihood           672.6
                       AIC (smaller is better)         692.6
                       AICC (smaller is better)        693.5
                       BIC (smaller is better)         706.6

                       Null Model Likelihood Ratio Test
                        DF     Chi-Square       Pr > ChiSq
                         9        419.07          <.0001

                       Type 3 Tests of Fixed Effects
                                Num      Den
                     Effect      DF       DF     F Value    Pr > F
                     Group        2     28.1        0.39    0.6776
                     Month        8      109      467.46    <.0001
                     Month(Group)16      132        3.78    <.0001

                            Least Squares Means
                                             Standard
   Effect     Group    Month    Estimate      Error      DF    t Value    Pr > |t|
   Group        1                28.3791     0.5887     28.3    48.21      <.0001
   Group        2                27.6775     0.5863     28      47.21      <.0001
   Group        3                27.8283     0.5862     27.9    47.47      <.0001

                       Differences of Least Squares Means
                                                       Standard
   Effect   Group  Month  Group  Month   Estimate       Error      DF   t Value   Pr > |t|
   Group      1            2              0.7015       0.8308     28.1    0.84    0.4056
   Group      1            3              0.5507       0.8308     28.1    0.66    0.5128
   Group      2            3             -0.1508       0.8291     28     -0.18    0.8570

                       Differences of Least Squares Means
              Effect    Group   Month   Group   Month   Adjustment       Adj P
              Group       1              2              Tukey-Kramer    0.6790
              Group       1              3              Tukey-Kramer    0.7866
              Group       2              3              Tukey-Kramer    0.9819
```

6.7 Prediction of Individual Test Day Data

A serious drawback of lactation and growth curve modeling by continuous functions of time and linear mixed models is that residuals must be treated only as a noise component. Predictions of missing or future TDs are therefore based on the deterministic component of variation. An alternative empirical procedure for predicting test day records can be based on time series models, whose fundamental property is the ability to improve regression analysis when residuals are serially correlated. In the case of milk yield, time series models are able to take account of the average lactation curve shape of homogeneous groups of animals but also of the (co)variance pattern of residuals.

A time series is usually a succession of real values of a variable Y ($Y_1, Y_2, ..., Y_n$) ordered by an index that refers to an interval of time. However, mathematical methods specifically developed for analyzing and modeling time series can be applied to any ordered succession of values regardless of the meaning of the ordering index (Hamilton, 1994). Thus, if there are N lactations of dairy ewes, each with n TD records, a succession of $N \times n$ TDs ($Y_1, Y_2, ... Y_k, ... Y_{N \times n}$) can be constructed where the index value (k) refers to a specific TD of a specific lactation (e.g., assuming that $N=100$ and $n=7$, there is a succession of 700 TDs in which at $k=74$ corresponds to the 4th TD record of the 11th lactation). The autoregressive moving average

(ARMA) seasonal model (Box and Jenkins, 1970) is able to describe a periodic succession as the one cited above, taking account both of the relationships among TD records within lactation (the non-seasonal part of the model) and of the relationships among correspondent TD records of different lactations of the succession (the seasonal part of the model) (Macciotta et al., 2000).

Box-Jenkins methodology for analyzing and modeling time series is characterized by three steps: (1) model identification, (2) parameter estimation, and (3) model validation.

Model identification defines the orders of the autoregressive (AR) and moving average (MA) components, both seasonal and non-seasonal. In this step, fundamental analytical tools are the spectral analysis of the Fourier transform of the original series and the autocorrelation functions. Results of spectral analysis are summarized in a periodogram function that plots the average squared amplitudes of the elementary waves in which the succession of analyzed values is decomposed. A periodogram that shows only isolated and discrete peaks underlines a deterministic process, whose future values can be predicted easily from the data; a periodogram nearly parallel to the frequency axis is evidence of processes that are random and unpredictable—i.e., white noise (WN) processes. The repeated occurrence of well-defined peaks at equally spaced intervals of frequency highlights the existence of a seasonal component.

The function of sample autocorrelation (ACF)—i.e., the linear correlation coefficient between Y_t and Y_{t-k}, calculated for $k = 0, 1, 2$—is equally important for the definition of the internal structure of the analyzed series and to evaluate the order of the two components (AR or MA). Moreover, the existence of a seasonal component of length s is underlined by the presence of a periodical pattern of period s in the ACF.

Goodness of fit of the ARMA model can be assessed by evaluating the statistical significance of parameter estimates, by analyzing the ACF of residuals, and by testing the white noise connotation of residuals, particularly by means of the Kolmogorov test, which compares the integrated periodogram (obtained as the normalized cumulative periodogram of residuals) with the theoretical periodogram of a WN process. In general, the rejection of the hypothesis that the series of residuals represents a WN process (i.e., completely random) also indicates whether to define an alternative model that will be subjected to identification, estimation, and validation steps. Moreover, by analogy with the determination coefficient of the multiple regression analysis, the general goodness of fit of an ARMA model can be tested by

$$R^2 = 1 - \frac{\mathrm{var}(\varepsilon)}{\mathrm{var}(Y)},$$

where var(ε) is the residual, or innovation, variance and var(Y) is the original variance of the time series.

An ARMA model that accurately fits the time series under analysis can identify and disentangle the different structural components of the series itself, and it can also be used to predict future values of the series, mainly for a limited horizon of prediction.

6.8 The SPECTRA and ARIMA Procedures for Time Series Prediction

The file EWES reports milk test day data of 100 Sarda dairy breed ewes, with seven TD records each, recorded every month. Data were organized as described in the previous section, with different ewes arranged sequentially (1, …, 100) and, within each ewe, TD records ordered according to their distance from lambing (1,…,7). Original data are plotted by the following statements, producing Output 6.30.

```
symbol1 interpol=join;
proc gplot data=ewes;
plot milk*N;
run;
```

Output 6.30 Plot of the ordered succession of milk test day data.

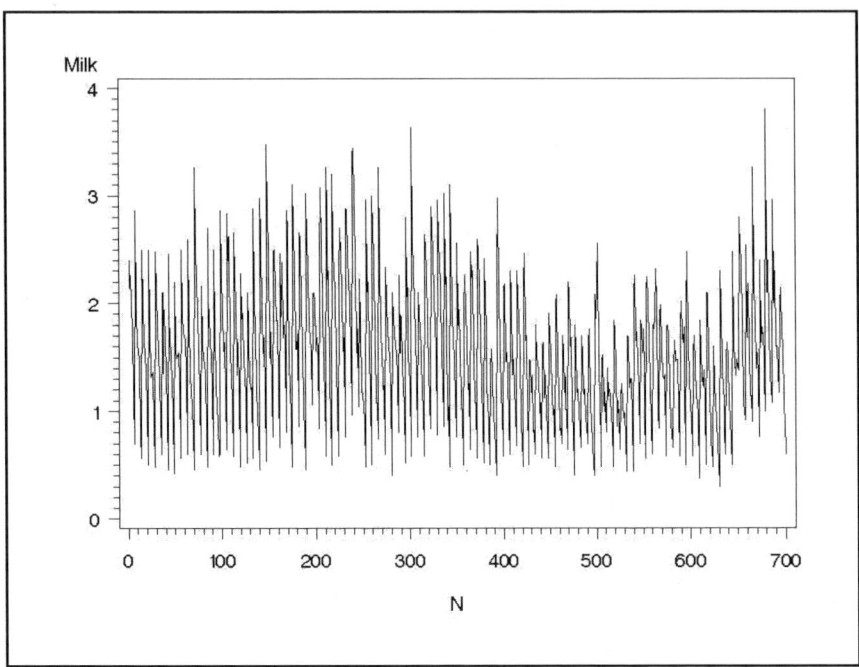

Procedures for fitting time series models are implemented in the SAS ETS module. PROC SPECTRA is first used to perform the Fourier spectral analysis on the variable MILK to evaluate the order of the MA and AR operators of the ARMA model. Then a plot of the periodogram of the milk yields against the frequency values is obtained with the GPLOT procedure.

```
proc spectra data=ewes out=per adjmean;
var milk;
run;
symbol1 interpol=join;
proc gplot data=per;
plot p_01*freq;
run;
```

The periodogram (Output 6.31) is characterized by well-defined peaks at regular intervals of the angular frequencies $n*2\pi/T$, where T is the period of 7 lags of the index variable. This pattern highlights a periodic deterministic component that occurs in all lactations and that can be identified with the amount of variability explained by the average lactation curve of animals. Moreover, several less-defined peaks can be observed at low angular frequencies, evidencing a residual deterministic linkage among TDs within each lactation.

Output 6.31 Plot of the integrated periodogram of the milk yields against the frequency values.

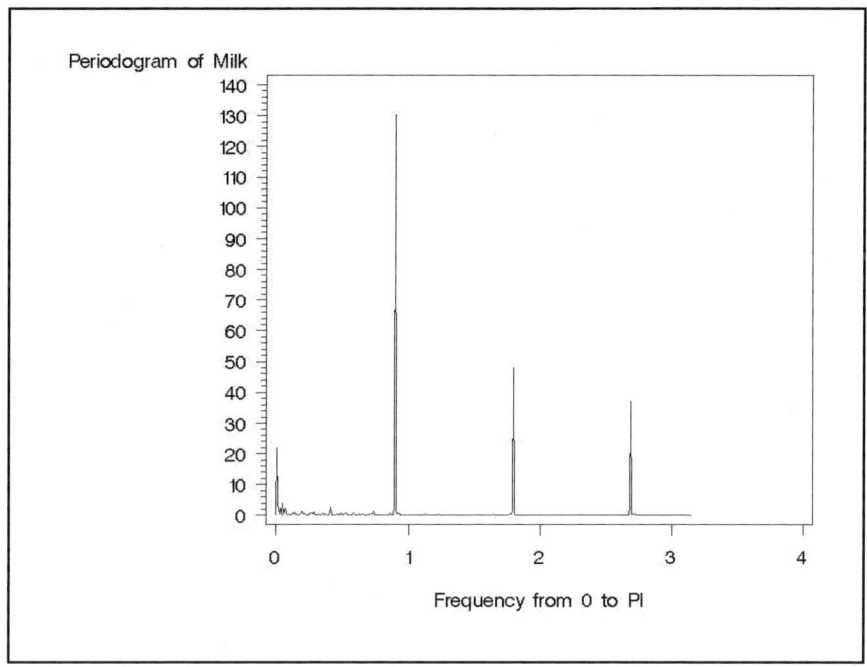

The following SAS program (EWES.SAS) fits an ARMA model with an AR non-seasonal parameter, an AR seasonal parameter, an MA non-seasonal parameter, and an MA seasonal parameter. In the conventional syntax of time series analysis, the model has the form $(1,1) \times (1,1)_7$. The model is fit to data from the first 50 ewes (WHERE N<351). Parameter estimates are then used to forecast missing TD records of the remaining 50 ewes.

```
proc arima data=ewes;
identify var=milk;
estimate p=(1) (7) q=(1) (7) method=ml;
where n<351;
run;
quit;
```

The IDENTIFY statement of the ARMA procedure reads the variable MILK and plots its autocorrelation function. Three autocorrelations are plotted: sample autocorrelation, partial autocorrelation, and inverse autocorrelation. In this section only the first will be discussed. However, the general criterion for reading autocorrelation plots is the same: autocorrelation patterns obtained from the series under study have to be compared with theoretical autocorrelation patterns typical of various AR and MA processes.

The ACF plot in Output 6.32 confirms the results of the spectral analysis, with values significantly different from zero (exceeding the dotted line) at lower lags (within-lactation relationships) and at lags that are multiples of eight (average lactation curve). The autocorrelation check for white noise tests the null hypothesis that none of the autocorrelation values of the series up to the defined lag are significantly different from zero. In this case they are different from zero; therefore the series does not represent a WN process and it can be modeled.

Output 6.32 Results of fitting a seasonal ARMA models to milk test day data from the EWES data set.

```
                              The ARIMA Procedure
                             Name of Variable = Milk
                         Mean of Working Series    1.689514
                         Standard Deviation        0.721651
                         Number of Observations       350

                                  Autocorrelations
 Lag    Covariance    Correlation  -1 9 8 7 6 5 4 3 2 1 0 1 2 3 4 5 6 7 8 9 1   Std Error
  0      0.520780      1.00000                               |********************|         0
  1      0.117572      0.22576                               |*****               |    0.053452
  2     -0.045510     -.08739                             **|.                    |    0.056111
  3     -0.145731     -.27983                         ******|.                    |    0.056498
  4     -0.152341     -.29253                         ******|.                    |    0.060328
  5     -0.071261     -.13684                            ***|.                    |    0.064253
  6      0.066988      0.12863                               |***                 |    0.065080
  7      0.425120      0.81631                               |****************    |    0.065803
  8      0.067725      0.13005                               |***.                |    0.090210
  9     -0.071103     -.13653                            . ***|.                   |    0.090744
 10     -0.155235     -.29808                         ******|.                    |    0.091329
 11     -0.154670     -.29700                         ******|.                    |    0.094068
 12     -0.071648     -.13758                            .***|.                    |    0.096710
 13      0.056325      0.10816                               |** .                |    0.097267
 14      0.408947      0.78526                               |****************    |    0.097610
 15      0.068933      0.13237                             . |***.                |    0.114243
 16     -0.068221     -.13100                            . ***|                    |    0.114680
 17     -0.148718     -.28557                         ******|                     |    0.115107
 18     -0.153352     -.29447                         ******|                     |    0.117113
 19     -0.076202     -.14632                            . ***|                    |    0.119210
 20      0.052766      0.10132                             . |** .                |    0.119722
 21      0.394090      0.75673                             . |***************     |    0.119967
 22      0.061432      0.11796                             . |** .                |    0.132907
 23     -0.074022     -.14214                            . ***|                    |    0.133206
 24     -0.152138     -.29214                         ******|.                    |    0.133638

                            "." marks two standard errors

                              Inverse Autocorrelations
          Lag    Correlation  -1 9 8 7 6 5 4 3 2 1 0 1 2 3 4 5 6 7 8 9 1
           1      -0.40587                  *******|.                    |
           2      -0.04647                       . *|.                    |
           3       0.06146                       . |*.                    |
           4      -0.05025                       . *|.                    |
           5       0.07769                       . |**                    |
           6       0.06589                       . |*.                    |
           7      -0.27948                  ******|.                      |
           8       0.15848                       . |***                   |
           9      -0.02736                       . *|.                    |
          10       0.03437                       . |*.                    |
          11       0.00171                       . |.                     |
          12      -0.09722                      **|.                      |
          13       0.14472                       . |***                   |
          14      -0.11459                      **|.                      |
          15       0.00008                       . |.                     |
          16       0.03534                       . |*.                    |
          17      -0.04782                       .*|.                     |
          18       0.00356                       . |.                     |
          19       0.05921                       . |*.                    |
          20      -0.00817                       . |.                     |
          21      -0.10011                      **|.                      |
          22       0.05285                       . |*.                    |
          23       0.00767                       . |.                     |
          24       0.00596                       . |.                     |
```

(continued on next page)

Output 6.32 *(continued)*

```
                              Partial Autocorrelations
        Lag   Correlation   -1 9 8 7 6 5 4 3 2 1 0 1 2 3 4 5 6 7 8 9 1
         1       0.22576     |                    . |*****              |
         2      -0.14579     |                  ***|  .                 |
         3      -0.24149     |                *****|  .                 |
         4      -0.20893     |                 ****|  .                 |
         5      -0.10074     |                   **|  .                 |
         6       0.07593     |                    . |**                 |
         7       0.78703     |                    . |****************   |
         8      -0.49592     |           **********|  .                 |
         9      -0.04820     |                    .*|  .                |
        10       0.00771     |                    . |  .                |
        11      -0.00790     |                    . |  .                |
        12       0.05796     |                    . |*.                 |
        13       0.04584     |                    . |*.                 |
        14       0.28107     |                    . |******             |
        15      -0.14520     |                  ***|  .                 |
        16      -0.04356     |                    .*|  .                |
        17       0.02839     |                    . |*.                 |
        18      -0.04202     |                    .*|  .                |
        19       0.00787     |                    . |  .                |
        20       0.11227     |                    . |**                 |
        21       0.07243     |                    . |*.                 |
        22      -0.11294     |                   **|  .                 |
        23      -0.02348     |                    . |  .                |
        24      -0.01040     |                    . |  .                |

                        Autocorrelation Check for White Noise
 To       Chi-              Pr >
 Lag     Square    DF       ChiSq    ------------------Autocorrelations------------------
   6      91.58     6      <.0001    0.226   -0.087   -0.280   -0.293   -0.137    0.129
  12     414.91    12      <.0001    0.816    0.130   -0.137   -0.298   -0.297   -0.138
  18     720.40    18      <.0001    0.108    0.785    0.132   -0.131   -0.286   -0.294
  24     991.73    24      <.0001   -0.146    0.101    0.757    0.118   -0.142   -0.292
```

WARNING: Estimates did not improve after a ridge was encountered in the objective function.
The iteration process has been terminated.
WARNING: Estimates may not have converged.

```
                        ARIMA Estimation Optimization Summary

        Estimation Method                                    Maximum Likelihood
        Parameters Estimated                                                  5
        Termination Criteria                 Maximum Relative Change in Estimates
        Iteration Stopping Value                                          0.001
        Criteria Value                                                  1.9E-13
        Maximum Absolute Value of Gradient                             22.04764
        R-Square Change from Last Iteration                             0.01565
        Objective Function                              Log Gaussian Likelihood
        Objective Function Value                                       -47.0497
        Marquardt's Lambda Coefficient                                     1E12
        Numerical Derivative Perturbation Delta                           0.001
        Iterations                                                            9
        Warning Message                          Estimates may not have converged.

                           Maximum Likelihood Estimation
                                    Standard                    Approx
           Parameter    Estimate      Error     t Value         Pr > |t|   Lag
           MU            1.64919      0.54166      3.04          0.0023      0
           MA1,1         0.04191      0.09618      0.44          0.6630      1
           MA2,1         0.86010      0.03126     27.51         <.0001       7
           AR1,1         0.57604      0.07830      7.36         <.0001       1
           AR2,1         0.99835      0.0011247  887.68         <.0001       7

                      Constant Estimate          0.001151
                      Variance Estimate          0.072901
                      Std Error Estimate         0.270001
                      AIC                      104.0993
                      SBC                      123.389
                      Number of Residuals          350
```

(continued on next page)

Output 6.32 *(continued)*

```
                          Correlations of Parameter Estimates
              Parameter        MU       MA1,1       MA2,1       AR1,1       AR2,1
              MU            1.000      -0.064       0.041      -0.099      -0.038
              MA1,1        -0.064       1.000       0.055       0.839       0.044
              MA2,1         0.041       0.055       1.000       0.061       0.576
              AR1,1        -0.099       0.839       0.061       1.000       0.053
              AR2,1        -0.038       0.044       0.576       0.053       1.000

                          Autocorrelation Check of Residuals
   To      Chi-                Pr >
   Lag    Square      DF      ChiSq   -------------------Autocorrelations------------------
    6      1.04        2     0.5954   -0.002    0.031   -0.030   -0.004   -0.010    0.030
   12      3.73        8     0.8810    0.006   -0.037    0.036   -0.006    0.028    0.062
   18     11.79       14     0.6235   -0.089   -0.056    0.062    0.046    0.065    0.028
   24     17.00       20     0.6530   -0.022   -0.024   -0.100    0.033    0.012   -0.040
   30     25.46       26     0.4928    0.069    0.077    0.063    0.065   -0.048    0.033
   36     34.80       32     0.3359   -0.038    0.049   -0.036    0.110    0.075   -0.032
   42     48.52       38     0.1179   -0.009   -0.035   -0.106    0.031   -0.046    0.137
   48     62.99       44     0.0315   -0.089    0.048   -0.025    0.094   -0.050   -0.116

                              Model for variable Milk
                              Estimated Mean     1.64919

                                Autoregressive Factors
                              Factor 1:  1 - 0.57604 B**(1)
                              Factor 2:  1 - 0.99835 B**(7)

                                Moving Average Factors
                              Factor 1:  1 - 0.04191 B**(1)
                              Factor 2:  1 - 0.8601 B**(7)
```

The warning message indicates a problem in the convergence of the estimate. Reasons can be found in the way the succession of data has been built up. Theoretically, values of the time series should be a random sample extracted from a population, whereas the series analyzed in this section is the result of a planned data arrangement. As a consequence, the statistical significance of parameters should not be strictly emphasized. However, the main aim of the analysis is not to study parameter estimates in their biological and asymptotic meaning but to find a model able to fit the original data well and to make reasonably accurate predictions of missing data.

Parameter estimates are all significant, and the general goodness of fit is evidenced by the test of autocorrelations among residuals that are not significantly different from zero. The variance estimate (0.072901) is the so-called innovation variance—i.e., the component of the original variance of the series that is not explained by the model. The ratio of the variance explained to the total variance gives a measure of the goodness of fit of the model: in this case it is equal to 0.86.

Parameter estimates are then used by the following SAS program (EWES.SAS) to make predictions for the remaining 50 ewes: a situation of incomplete lactation was simulated, with only the first four TD records available and the last three to be predicted.

The NOEST statement suppresses the estimation process except for estimation of the residual variance and allows for the use of parameter final values indicated by the MU=, AR=, and MA= options. The BACK option specifies the number of observations before the end of the data that the forecast is to begin. The LEAD option specifies the number of forecast values to compute: in this case the forecast is made individually for each ewe. The ID option indicates the index variable associated with the values of the time series.

```
proc arima data=ewes out=predictions;
identify var=milk;
estimate p=(1) (7) q=(1) (7)
noest mu=1.64919 ar=0.57604 0.99835 ma=0.04191 0.86010;
forecast back=346  lead=3 id=n noprint nooutall;
forecast back=339  lead=3 id=n noprint nooutall;
forecast back=332  lead=3 id=n noprint nooutall;
forecast back=325  lead=3 id=n noprint nooutall;
forecast back=318  lead=3 id=n noprint nooutall;
forecast back=311  lead=3 id=n noprint nooutall;
forecast back=304  lead=3 id=n noprint nooutall;
forecast back=297  lead=3 id=n noprint nooutall;
forecast back=290  lead=3 id=n noprint nooutall;
forecast back=283  lead=3 id=n noprint nooutall;
forecast back=276  lead=3 id=n noprint nooutall;
forecast back=269  lead=3 id=n noprint nooutall;
forecast back=262  lead=3 id=n noprint nooutall;
forecast back=255  lead=3 id=n noprint nooutall;
forecast back=248  lead=3 id=n noprint nooutall;
forecast back=241  lead=3 id=n noprint nooutall;
forecast back=234  lead=3 id=n noprint nooutall;
forecast back=227  lead=3 id=n noprint nooutall;
forecast back=220  lead=3 id=n noprint nooutall;
forecast back=213  lead=3 id=n noprint nooutall;
forecast back=206  lead=3 id=n noprint nooutall;
forecast back=199  lead=3 id=n noprint nooutall;
forecast back=192  lead=3 id=n noprint nooutall;
forecast back=185  lead=3 id=n noprint nooutall;
forecast back=178  lead=3 id=n noprint nooutall;
forecast back=171  lead=3 id=n noprint nooutall;
forecast back=164  lead=3 id=n noprint nooutall;
forecast back=157  lead=3 id=n noprint nooutall;
forecast back=150  lead=3 id=n noprint nooutall;
forecast back=143  lead=3 id=n noprint nooutall;
forecast back=136  lead=3 id=n noprint nooutall;
forecast back=129  lead=3 id=n noprint nooutall;
forecast back=122  lead=3 id=n noprint nooutall;
forecast back=115  lead=3 id=n noprint nooutall;
forecast back=108  lead=3 id=n noprint nooutall;
forecast back=101  lead=3 id=n noprint nooutall;
forecast back=94   lead=3 id=n noprint nooutall;
forecast back=87   lead=3 id=n noprint nooutall;
forecast back=80   lead=3 id=n noprint nooutall;
forecast back=73   lead=3 id=n noprint nooutall;
forecast back=66   lead=3 id=n noprint nooutall;
forecast back=59   lead=3 id=n noprint nooutall;
forecast back=52   lead=3 id=n noprint nooutall;
forecast back=45   lead=3 id=n noprint nooutall;
forecast back=38   lead=3 id=n noprint nooutall;
forecast back=31   lead=3 id=n noprint nooutall;
forecast back=24   lead=3 id=n noprint nooutall;
forecast back=17   lead=3 id=n noprint nooutall;
forecast back=10   lead=3 id=n noprint nooutall;
forecast back=3    lead=3 id=n noprint nooutall;
run;
quit;
proc print data=predictions;
run;
```

Predicted values are stored in the data predictions that contain the actual values of the series (MILK), the predicted values (FORECAST), the standard error, the confidence interval of predictions, and the residuals (actual-predicted). The first rows of the data set PREDICTIONS are reported in Output 6.33. As an example, for the position 364 of the time series analyzed (i.e., to the 7th TD record of the 52nd ewe of the succession), which corresponds to the actual value of milk yield of 0.64 kg, the model predicts 0.65 kg.

Output 6.33 First rows of the data set PREDICTIONS.

```
     Obs    N        Milk     FORECAST      STD         L95          U95       RESIDUAL
      1    355     1.06000    1.27862     0.30328     0.68421      1.87304    -0.21862
      2    356     0.98000    1.15165     0.34383     0.47775      1.82554    -0.17165
      3    357     0.50000    0.66207     0.35627    -0.03620      1.36034    -0.16207
      4    362     1.20000    1.29165     0.30328     0.69724      1.88606    -0.09165
      5    363     1.26000    1.15330     0.34383     0.47941      1.82719     0.10670
      6    364     0.64000    0.65544     0.35627    -0.04283      1.35371    -0.01544
      7    369     1.08000    1.20264     0.30328     0.60822      1.79705    -0.12264
      8    370     1.08000    1.12455     0.34383     0.45066      1.79845    -0.04455
      9    371     0.56000    0.62942     0.35627    -0.06885      1.32769    -0.06942
     10    376     1.24000    1.26842     0.30328     0.67401      1.86284    -0.02842
     11    377     0.90000    1.16650     0.34383     0.49260      1.84039    -0.26650
     12    378     0.52000    0.64871     0.35627    -0.04955      1.34698    -0.12871
     13    383     1.10000    1.02162     0.30328     0.42720      1.61603     0.07838
     14    384     0.88000    0.99018     0.34383     0.31629      1.66407    -0.11018
     15    385     0.50000    0.55177     0.35627    -0.14650      1.25004    -0.05177
```

6.9 References

Ali, T. E., and L. R. Schaeffer. 1987. Accounting for covariances among test day milk yields in dairy cows. *Can. J. Anim. Sci.* 67:637-644.

Box, G. E. P., and G. M. Jenkins. 1970. *Time Series Analysis: Forecasting and Control*. San Francisco: Holden Day.

Brody, S. 1945. *Bioenergetics and Growth*. New York: Reinhold Publishing.

Cappio-Borlino, A., B. Portolano, M. Todaro, N. P. P. Macciotta, P. Giaccone, and G. Pulina. 1997. Lactation curves of Valle del Belice dairy ewes for milk, fat, and protein estimated with test day models. *J. Dairy Sci.* 80:3023-3029.

Carvalheira, J. G. V., R. V. Blake, E. J. Pollak, R. L. Quaas, and C. V. Duran-Castro. 1998. Application of an autoregressive process to estimate genetic parameters and breeding values for daily milk yield in a tropical herd of Lucerna cattle and in United States Holstein herds. *J. Dairy Sci.* 81:2738-2751.

Catillo, G., N. P. P. Macciotta, A. Carretta, and A. Cappio-Borlino. 2002. Effects of age and calving season on lactation curve of milk production traits in Italian water buffalo. *J. Dairy Sci.* 5:1298-1306.

Ferguson, J. D., and R. Boston. 1993. Lactation curve analysis: Comparison of gamma function, polynomial and exponential methods toward a mechanistic model of milk production. *J. Dairy Sci.* 76 (Suppl. 1): 268, abstract.

Hamilton, J. D. 1994. *Time Series Analysis*. Princeton, NJ: Princeton University Press.

Kirkpatrick, M., and N. Heckman. 1989. A quantitative genetic model for growth, shape, reaction norms, and other infinite-dimensional characters. *J. Math. Biol.* 27:429-450.

Lawrence, T. L. J., and V. R. Fowler. 1997. *Growth of Farm Animals*. Wallingford, UK: CAB International.

Littell, R. C., P. R. Henry, and C. Ammermann. 1998. Statistical analysis of repeated measures data using SAS procedures. *J. Anim. Sci.* 76:1216-1231.

Lynch, M., and B. Walsh. 1998. *Genetics and Analysis of Quantitative Traits*. Sunderland, MA: Sinauer Associates.

Macciotta, N. P. P., A. Cappio-Borlino, and G. Pulina. 2000. Time series autoregressive integrated moving average modelling of test day milk yields of dairy ewes. *J. Dairy Sci.* 83:1094-1103.

Mepham, T. B. 1987. An analysis of lactation as a productive system. In *Physiology of Lactation*. Philadelphia: Open Univ. Press, Milton Keynes.

Olori, V. E., S. Brotherstone, W. G. Hill, and B. J. McGuirk. 1999. Fit of standard models of the lactation curve to weekly records of milk production of cows in a single herd. *Livest. Prod. Sci.* 58:55-63.

Pletcher, S. D., and C. J. Geyer. 1999. The genetic analysis of age-dependent traits: Modeling the character process. *Genetics* 151:825-835.

Ratkowsky, D. A. 1990. *Handbook of Nonlinear Regression Models.* New York: Marcel Dekker.

Rekaya, R., M. J. Caratano, and M. A. Toro. 2000. Bayesian analysis of lactation curves of Holstein Friesian cattle using a nonlinear model. *J. Dairy Sci.* 83:2691-2701.

Richards, F. J. 1959. A flexible growth function for empirical use. *J. Exp. Bot.* 10:290-300.

Schaeffer, L. R. 2004. Application of random regression models in animal breeding. *Livest. Prod. Sci.* 86 (1-3): 35-45.

Shanks, R. D., P. J. Berger, A. E. Freeman, and F. N. Dickinson. 1981. Genetic aspects of lactation curves. *J. Dairy Sci.* 64:1852-1860.

Van der Werf, J., and L. R. Schaeffer. 1997. Random regression in animal breeding. Course notes. CGIL Guelph, June 25–28, 1997.

Wada, Y., and N. Kashiwagi. 1990. Selecting statistical models with information statistics. *J. Dairy Sci.* 73:3575-3582.

Wade, K. M., R. L. Quaas, and L. D. Van Vleck. 1993. Estimation of the parameters involved in a first order autoregressive process for contemporary groups. *J. Dairy Sci.* 76:3033-3040.

Wood, P. D. P. 1967. Algebraic model of the lactation curve in cattle. *Nature* 216:164-165.

Chapter 7

Empirical Bayes Approaches to Mixed Model Inference in Quantitative Genetics

Robert J. Tempelman and Guilherme J. M. Rosa

7.1 Introduction 149
    7.1.1 Hierarchical Models 151
7.2 An Example of Linear Mixed Model Inference 153
7.3 Empirical Bayes Inference for Normally Distributed Data 157
    7.3.1 Empirical Bayes Analysis in the Linear Mixed Model 158
7.4 Generalized Linear Mixed Models 160
    7.4.1 Inference on Fixed and Random Effects 162
    7.4.2 Using the %GLIMMIX Macro for Quantitative Genetic Inference 163
    7.4.3 Empirical Bayes Inference in GLMM 166
7.5 Empirical Bayes vs. MCMC 169
7.6 Final Comments 174
7.7 References 175

7.1 Introduction

Mixed models have provided the foundational tools for quantitative genetic analyses since the expository work of Charles Henderson (Henderson, 1963; Henderson, 1973; Henderson et al., 1959) in animal breeding, with applications now being adapted to human genetics (Almasy and Blangero, 1998; Lange and Boehnke, 1983). Because mixed models in quantitative genetics are often highly parameterized, such that the number of parameters, including random effects, exceeds the number of observations, there has been a greater interest in adapting Bayesian inference to such models to help alleviate the large sample asymptotic assumptions required from a likelihood inference perspective.

It is not our purpose to give a comprehensive discussion of Bayesian inference for quantitative genetic analyses based on mixed effects models. Excellent expositions can be found in Gianola and Fernando (1986) and more recently in Sorensen and Gianola (2002). We highlight a few key concepts and illustrate how SAS software can be used for quantitative genetic inference, whether for normally or non-normally distributed characters.

Consider an unobservable vector θ and data $\mathbf{y}$ and their joint density $p(\theta,\mathbf{y})$. From standard probability theory,

$$p(\theta,\mathbf{y}) = p(\mathbf{y}\mid\theta)\, p(\theta) \tag{7.1}$$

and

$$p(\theta,\mathbf{y}) = p(\theta\mid\mathbf{y})\, p(\mathbf{y}), \tag{7.2}$$

where $p(\theta)$ and $p(\mathbf{y})$ are the *marginal densities* of θ and $\mathbf{y}$, respectively. Combining Equations 7.1 and 7.2, we can use the Bayes theorem to derive the conditional density of $\theta\mid\mathbf{y}$ as

$$p(\theta\mid\mathbf{y}) = \frac{p(\mathbf{y}\mid\theta)\, p(\theta)}{p(\mathbf{y})}.$$

Note that

$$p(\mathbf{y}) = \int_{R_\theta} p(\mathbf{y},\theta)\, d\theta = \int_{R_\theta} p(\mathbf{y}\mid\theta)\, p(\theta)\, d\theta = \mathop{\mathrm{E}}_{\theta}\left[p(\mathbf{y}\mid\theta) \right]$$

is merely a normalizing constant such that, as in likelihood inference, we can write the joint posterior density of $\theta\mid\mathbf{y}$ proportionately:

$$p(\theta\mid\mathbf{y}) \propto p(\mathbf{y}\mid\theta)\, p(\theta); \tag{7.3}$$

i.e., the *joint posterior density* $p(\theta\mid\mathbf{y})$ is *proportional* to the *likelihood function* $p(\mathbf{y}\mid\theta)$ times the *prior* density of θ or $p(\theta)$. Essentially, this joint posterior density in Equation 7.3 is then written as a two-stage model.

7.1.1 Hierarchical Models

Suppose that θ can be partitioned into two components, an unknown $m \times 1$ vector θ_1 and a known $s \times 1$ vector θ_2:

$$\theta = \begin{bmatrix} \theta_1 \\ \theta_2 \end{bmatrix}.$$

In such a situation, the posterior density for θ_1 might be derived as a hierarchical two-stage model:

$$p(\theta_1 \mid \mathbf{y}, \theta_2) \propto p(\mathbf{y} \mid \theta_1) p(\theta_1 \mid \theta_2).$$

Suppose, however, that θ_2 is not known. Then a third-stage prior density on θ_2 should be added to the hierarchy to derive the joint posterior density of θ:

$$p(\theta_1, \theta_2 \mid \mathbf{y}) \propto p(\mathbf{y} \mid \theta_1) p(\theta_1 \mid \theta_2) p(\theta_2), \tag{7.4}$$

where $p(\theta_2)$ represents the prior density of θ_2, the density itself presumably dependent on a set of known parameters, often referred to as *hyperparameters*. Suppose that θ_1 is of primary interest to the investigator such that θ_2 represents a vector of *nuisance* parameters. This uncertainty should be integrated out from the joint posterior specification in Equation 7.4 in order to derive the marginal posterior density of θ_1:

$$p(\theta_1 \mid \mathbf{y}) = \int_{R_{\theta_2}} p(\theta_1, \theta_2 \mid \mathbf{y}) d\theta_2. \tag{7.5}$$

This type of hierarchy characterizes the linear mixed effects model such that SAS PROC MIXED is often referred to as an example of Bayesian software (Carlin and Louis, 1996; Natarajan and Kass, 2000). For the example, the first stage of the linear mixed model (LMM) can be written as

$$\mathbf{y} = \mathbf{X}\boldsymbol{\beta} + \mathbf{Z}\mathbf{u} + \mathbf{e},$$

where $\boldsymbol{\beta}$ represents the unknown vector of fixed effects, and $\mathbf{u}$ represents the unobserved vector of random effects. Here $\mathbf{X}$ and $\mathbf{Z}$ are corresponding known incidence matrices that may represent either dummy variables for classification factors or covariates; for more details on fixed effects parameterizations in PROC MIXED, refer to Littell et al. (1996). Finally, $\mathbf{e}$ is a vector of multivariate normally distributed residuals with variance covariance matrix $\mathbf{R}$. Suppose that these residuals are iid normal, with $\mathbf{R} = \mathbf{I}\sigma_e^2$. Then the conditional first-stage density (or conditional likelihood) can be written as

$$p(\mathbf{y} \mid \boldsymbol{\beta}, \mathbf{u}, \sigma_e^2) = N(\mathbf{X}\boldsymbol{\beta} + \mathbf{Z}\mathbf{u}, \mathbf{I}\sigma_e^2).$$

A second-stage multivariate normal density is specified for **u** in the LMM, specifically,

$$\mathbf{u} \sim f(\mathbf{u}|\boldsymbol{\varphi}) = \mathrm{MVN}(\mathbf{0}, \mathbf{G}(\boldsymbol{\varphi})), \tag{7.6}$$

such that the variance covariance matrix $\mathbf{G}(\boldsymbol{\varphi})$ is a function of a vector of (co)variance components $\boldsymbol{\varphi}$. Let's presume, for now, that the dispersion parameters, specifically σ_e^2 and $\boldsymbol{\varphi}$, are known. From a Bayesian perspective, a prior specification is required for all unknown parameters, including $\boldsymbol{\beta}$. We might reasonably assume such a prior $p(\boldsymbol{\beta})$ to be independent *a priori* of **u**. Then the joint posterior density of $\boldsymbol{\beta}$ and **u** can be determined proportionately:

$$p(\boldsymbol{\beta}, \mathbf{u}|\mathbf{y}, \sigma_e^2, \boldsymbol{\varphi}) \propto p(\mathbf{y}|\boldsymbol{\beta}, \mathbf{u}, \sigma_e^2) p(\mathbf{u}|\mathbf{G}(\boldsymbol{\varphi})) p(\boldsymbol{\beta}). \tag{7.7}$$

From a classical mixed models perspective, a *flat* prior $p(\boldsymbol{\beta}) \propto 1$ is typically specified for $\boldsymbol{\beta}$. Then Equation 7.7 simplifies to $p(\boldsymbol{\beta}, \mathbf{u}|\mathbf{y}, \sigma_e^2, \boldsymbol{\varphi}) \propto p(\mathbf{y}|\boldsymbol{\beta}, \mathbf{u}, \sigma_e^2) p(\mathbf{u}|\mathbf{G}(\boldsymbol{\varphi}))$, which is multivariate normal with mean

$$\begin{bmatrix} \hat{\boldsymbol{\beta}} \\ \hat{\mathbf{u}} \end{bmatrix} = \mathrm{E}\begin{bmatrix} \boldsymbol{\beta} \\ \mathbf{u} \end{bmatrix} | \mathbf{y}, \sigma_e^2, \boldsymbol{\varphi} = \begin{bmatrix} \mathbf{C}_{\beta\beta} & \mathbf{C}_{\beta u} \\ \mathbf{C}_{u\beta} & \mathbf{C}_{uu} \end{bmatrix} \begin{bmatrix} \mathbf{X}'\mathbf{R}^{-1}\mathbf{y} \\ \mathbf{Z}'\mathbf{R}^{-1}\mathbf{y} \end{bmatrix}$$

and variance-covariance matrix **C** (Gianola et al., 1990; Sorensen and Gianola, 2002), where

$$\mathbf{C} = \begin{bmatrix} \mathbf{C}_{\beta\beta} & \mathbf{C}_{\beta u} \\ \mathbf{C}_{u\beta} & \mathbf{C}_{uu} \end{bmatrix} = \begin{bmatrix} \mathbf{X}'\mathbf{R}^{-1}\mathbf{X} & \mathbf{X}'\mathbf{R}^{-1}\mathbf{Z} \\ \mathbf{Z}'\mathbf{R}^{-1}\mathbf{X} & \mathbf{Z}'\mathbf{R}^{-1}\mathbf{Z} + (\mathbf{G}(\boldsymbol{\varphi}))^{-1} \end{bmatrix}^{-1}. \tag{7.8}$$

Readers acquainted with classical interpretations of mixed model inference might recognize that $\hat{\boldsymbol{\beta}}$ is the generalized least squares estimate of $\boldsymbol{\beta}$, having expectation $\boldsymbol{\beta}$ and covariance matrix $\mathbf{C}_{\beta\beta}$ over repeated sampling. Furthermore, $\hat{\mathbf{u}}$ represents the best linear unbiased predictor (BLUP) of **u** having covariance matrix of predictors errors $\mathrm{var}(\hat{\mathbf{u}} - \mathbf{u}) = \mathbf{C}_{uu}$ (Searle et al., 1992). From a Bayesian viewpoint, $\hat{\mathbf{u}} = \mathrm{E}(\mathbf{u}|\mathbf{y}, \sigma_e^2, \boldsymbol{\varphi})$ is a posterior or conditional mean estimator that has been shown to have optimal properties for ranking subjects for genetic merit based on truncation selection (Fernando and Gianola, 1986). From the perspective of reporting the relative accuracy of $\hat{\mathbf{u}}$, it has been further shown that $\mathrm{cov}(\hat{\mathbf{u}}, \mathbf{u}) = \mathbf{G}(\boldsymbol{\varphi}) - \mathbf{C}_{uu}$ (Searle et al., 1992) is maximized using $\hat{\mathbf{u}}$. Furthermore, the *i*th diagonal element of $\mathrm{cov}(\hat{\mathbf{u}}, \mathbf{u})$ divided by the *i*th diagonal element of $\mathbf{G}(\boldsymbol{\varphi})$ is typically reported in livestock genetic evaluation to be the accuracy $\rho(\hat{u}_i, u_i)$ of $\hat{u}_i$ (Van Vleck, 1993) with $\hat{u}_i$ and u_i being the *i*th elements of $\hat{\mathbf{u}}$ and **u**, respectively.

Now, as a property of the multivariate normal density, any linear combination, say, $\mathbf{K}'\boldsymbol{\beta} + \mathbf{M}'\mathbf{u}$, also has a multivariate normal posterior density with posterior mean $\mathbf{K}'\hat{\boldsymbol{\beta}} + \mathbf{M}'\hat{\mathbf{u}}$ and posterior variance-covariance matrix

$$\mathrm{var}\left(\mathbf{K}'\hat{\boldsymbol{\beta}} + \mathbf{M}'\hat{\mathbf{u}}|\mathbf{y},\boldsymbol{\varphi},\sigma_e^2\right) = \begin{bmatrix} \mathbf{K}' & \mathbf{M}' \end{bmatrix} \begin{bmatrix} \mathbf{C}_{\beta\beta} & \mathbf{C}_{\beta u} \\ \mathbf{C}_{u\beta} & \mathbf{C}_{uu} \end{bmatrix} \begin{bmatrix} \mathbf{K} \\ \mathbf{M} \end{bmatrix}. \qquad (7.9)$$

Of course, posterior probability statements (e.g., using key percentiles) can be readily derived from such a specification.

7.2 An Example of Linear Mixed Model Inference

Consider the simulated data set provided in its entirety in Meyer (1989). A mixed effects model was used to simulate one record for each of 282 animals deriving from either of two generations (1 and 2). We consider Model 1 as presented in Meyer (1989). In that model, the fixed effects $\boldsymbol{\beta}$ include effects for those of the two generations, whereas the random effects $\mathbf{u}$ are specified by the animals' additive genetic effects. The pedigrees for each of these 282 animals derive from an additional 24 base population (Generation 0) animals that do not have records of their own but, nevertheless, are of interest with respect to the inference on their own additive genetic values. Furthermore, it is presumed that these original 24 base animals are not related to each other. Therefore, the row dimension of $\mathbf{u}$ is 306 (282+24) even though the row dimension of the data vector $\mathbf{y}$ is only 282! That is, in the context of the second-stage prior for $\mathbf{u}$ in Equation 7.6, $\mathbf{G}(\boldsymbol{\varphi}) = \mathbf{A}\sigma_u^2$, where $\mathbf{A}$ is the known numerator relationship matrix of square dimension 306 as derived from a pedigree file, and $\boldsymbol{\varphi} \equiv \sigma_u^2$. SAS PROC INBREED can be used to compute $\mathbf{A}$ from a pedigree file that has three variables, corresponding to the identification of the animal, its sire, and its dam. The matrix $\mathbf{A}$ can then be imported into PROC MIXED to provide the joint posterior density of $\boldsymbol{\beta}$ and $\mathbf{u}$. Consider the following SAS program.

```
options ls = 75 ps = 60 nodate pageno=1 ;

/* Input pedigree file
   Animal id in first field, sire id in second field, dam id in third field. If
sire or dam is unknown, indicate with period '.' */

data pedigree;
   infile 'E:\meyer.ped';
   input animal sire dam;
run;

proc inbreed data=pedigree covar outcov=amatrix;
 var animal sire dam;
run;

/* Input data set with animal (1-306), generation (1 or 2) and y (data)
   Include base animals (1-24) with missing records (y = .) */

data meyer;
   infile 'E:\meyer.dat';
   input animal generation y;
run;
```

```
/* Row numbers needed to be provided for pipelining numerator relationship
matrix (A) from PROC INBREED into PROC MIXED */

data L2DATA;
  set amatrix;
  parm = 1;
  row = _n_;
run;

/* Posterior inference on fixed and random effects treating variance components
as known */

proc mixed data=meyer noprofile;
  class generation animal;
  model y = generation /solution noint covb;
  random animal /type=lin(1) LDATA=L2data solution;
  parms (40) (50) /noiter;
  ods listing exclude solutionr; ods output solutionr = solutionr;
run;

data solutionr;
  set solutionr;
  accuracy = (40-StderrPred**2)/40;

/* classical accuracy of genetic evaluation (40 = additive genetic variance) */

run;

/* Print off posterior means of random effects with posterior standard
deviations for first 30 animals */

title 'Predicted Genetic Effects';
proc print data=solutionr (obs=30);
run;
```

The SAS programming strategy above may be summarized briefly as follows. The pedigree file data set (PEDIGREE) containing the animals' identifications and their sires' and dams' identifications is read into PROC INBREED to create the numerator relationship matrix **A**. This matrix is pipelined into a SAS data set that we've labeled AMATRIX (using the OUTCOV= option). Before this data set can be imported into PROC MIXED to help determine the corresponding covariance structure for **u**, the rows need to be labeled from 1 to number of rows as done in the data set L2DATA. This SAS data set is read, along with the MEYER data file, which should include missing data for the 24 base animals, into PROC MIXED. The variance components, σ_u^2 and σ_e^2, are held constant at 40 and 50, respectively, to be consistent with values used in the simulation data by Meyer (1989). These constraints are indicated in the PARMS statement and held with the NOITER option along with the NOPROFILE option that precedes it. The Output Delivery System (ODS) in SAS is used to output the elements of $\hat{\mathbf{u}}$ to the SAS data set SOLUTIONR, which is then processed to compute accuracies of the elements of $\hat{\mathbf{u}}$. For the purpose of brevity, only the elements of $\hat{\mathbf{u}}$ pertaining to the first 30 animals (including the 24 base animals) are printed out. The corresponding output is given in Output 7.1.

Output 7.1 Posterior inference on fixed and random (genetic) effects in linear mixed model conditional on known dispersion parameters.

```
                         The INBREED Procedure

                  Number of Individuals    306

                         The Mixed Procedure

                         Model Information
         Data Set                      WORK.MEYER
         Dependent Variable            y
         Covariance Structures         Linear, Variance
                                       Components
         Estimation Method             REML
         Residual Variance Method      Parameter
         Fixed Effects SE Method       Model-Based
         Degrees of Freedom Method     Containment

                       Class Level Information
      Class         Levels    Values
      generation       2      1 2
      animal         306      1 2 3 4 5 6 7 8 9 10 11 12 13
                              14 15 16 17 18 19 20 21 22 23
                              ...
                              292 293 294 295 296 297 298
                              299 300 301 302 303 304 305
                              306

                              Dimensions
              Covariance Parameters              2
              Columns in X                       2
              Columns in Z                     306
              Subjects                           1
              Max Obs Per Subject              306
              Observations Used                282
              Observations Not Used             24
              Total Observations               306

                           Parameter Search
  CovP1        CovP2          Res Log Like     -2 Res Log Like
 40.0000      50.0000          -1016.9567          2033.9133

                            Fit Statistics
              -2 Res Log Likelihood         2033.9
              AIC (smaller is better)       2033.9
              AICC (smaller is better)      2033.9
              BIC (smaller is better)       2033.9

                    Solution for Fixed Effects    ❶
                                    Standard
  Effect       generation   Estimate    Error     DF   t Value   Pr > |t|
  generation       1         220.32    1.6534      0    133.25      .
  generation       2         236.74    1.9200      0    123.30      .

              Covariance Matrix for Fixed Effects  ❷
         Row    Effect        generation      Col1       Col2
          1     generation        1          2.7337     2.3814
          2     generation        2          2.3814     3.6865
```

(continued on next page)

Output 7.1 *(continued)*

```
                      Type 3 Tests of Fixed Effects
                          Num       Den
              Effect      DF        DF      F Value      Pr > F
              generation  2         0       9501.11      .

                      Predicted Genetic Effects  ❸
                                    StdErr
   Obs   Effect   animal   Estimate   Pred     DF   tValue   Probt   accuracy
    1    animal    1       -6.7673   4.4027    0    -1.54     .      0.51541
    2    animal    2       -4.2600   4.8008    0    -0.89     .      0.42381
    3    animal    3       -8.2827   4.7720    0    -1.74     .      0.43071
    4    animal    4        5.7754   4.9237    0     1.17     .      0.39394
    5    animal    5       -7.2904   4.3995    0    -1.66     .      0.51611
    6    animal    6        2.0610   4.8232    0     0.43     .      0.41841
    7    animal    7       -4.4713   5.0175    0    -0.89     .      0.37061
    8    animal    8       -4.8801   4.7283    0    -1.03     .      0.44107
    9    animal    9       -0.6550   4.4460    0    -0.15     .      0.50583
   10    animal   10       -4.3031   5.0284    0    -0.86     .      0.36788
   11    animal   11        4.9761   4.8642    0     1.02     .      0.40849
   12    animal   12       -1.3280   4.8651    0    -0.27     .      0.40828
   13    animal   13        5.2327   4.3491    0     1.20     .      0.52713
   14    animal   14        6.8234   4.7975    0     1.42     .      0.42460
   15    animal   15       -6.3013   4.7288    0    -1.33     .      0.44095
   16    animal   16        4.7106   4.6405    0     1.02     .      0.46164
   17    animal   17        5.5351   4.3388    0     1.28     .      0.52937
   18    animal   18       -2.7515   4.6522    0    -0.59     .      0.45893
   19    animal   19        7.2634   4.6082    0     1.58     .      0.46912
   20    animal   20        1.0232   4.8051    0     0.21     .      0.42278
   21    animal   21        3.9449   4.4066    0     0.90     .      0.51455
   22    animal   22        2.5831   4.7530    0     0.54     .      0.43522
   23    animal   23       -0.7235   4.9247    0    -0.15     .      0.39368
   24    animal   24        2.0854   4.8482    0     0.43     .      0.41237
   25    animal   25       -4.0305   4.3634    0    -0.92     .      0.52402
   26    animal   26       -6.3162   4.3634    0    -1.45     .      0.52402
   27    animal   27       -3.7447   4.3634    0    -0.86     .      0.52402
   28    animal   28       -7.7447   4.3634    0    -1.77     .      0.52402
   29    animal   29       -4.6019   4.3634    0    -1.05     .      0.52402
   30    animal   30       -9.4590   4.3634    0    -2.17     .      0.52402
```

Some key portions of Output 7.1 are annotated below.

❶ Solution for Fixed Effects

The posterior mean of β is given using the SOLUTION option to be

$$E(\beta \mid y, \sigma_e^2, \sigma_u^2) = \begin{bmatrix} 220.32 \\ 236.74 \end{bmatrix}.$$

Note that the NOINT option was used, meaning that for the case where there is only one fixed factor, the elements of β are expressed on a cell means basis (Searle, 1987) for the two generation effects. This is not a required specification since this example with the default parameterization would be a factor level effects model; further details are provided in Littell et al. (1996). Of interest, note further that the degrees of freedom (df) for all estimates and ANOVA terms are reported to be zero. Mixed effects models where the number of random effects exceeds the number of observations are, in this sense, pathological models relative to determination of df, at least from a classical statistics viewpoint.

Chapter 7: Empirical Bayes Approaches to Mixed Model Inference in Quantitative Genetics **157**

Zero degrees of freedom thereby lead to missing *p*-values under the PR>|T|, Pr>F, and PROBT columns in Output 7.1.

❷ Covariance Matrix for Fixed Effects

The output due to the option COVB in the MODEL statement determines the posterior variance-covariance matrix of $\boldsymbol{\beta} \mid \mathbf{y}, \sigma_e^2, \sigma_u^2$. That is,

$$\mathrm{var}\left(\boldsymbol{\beta} \mid \mathbf{y}, \sigma_e^2, \sigma_u^2\right) = \mathbf{C}_{\beta\beta} = \begin{bmatrix} 2.7337 & 2.3814 \\ 2.3814 & 3.6865 \end{bmatrix}.$$

Further, note that the square roots of the diagonal elements of $\mathbf{C}_{\beta\beta}$ are identical to the reported standard errors of the cell mean estimates for the two generations.

❸ Predicted Genetic Effects

The first 30 elements of $\mathrm{E}\left(\mathbf{u} \mid \mathbf{y}, \sigma_e^2, \sigma_u^2\right)$ are printed from the SAS data set SOLUTIONR. Recall that the first 24 animals have no records of their own but have descendants in generations 1 and 2 that have records. The STDERRPRED reported alongside each posterior mean represents the posterior standard deviation of the respective element, or the square root of the corresponding diagonal element of $\mathbf{C}_{uu}$. The accuracy of each genetic evaluation is computed using the expression from Van Vleck (1993), as previously noted. Note that the accuracies of genetic evaluations for some of the animals having no records of their own, namely animals 13 and 17, are actually higher than for some animals (i.e., 25 through 30) having their own records. This result is a function of $\mathbf{A}$, or specifically, how much information is available for each element of $\mathbf{u}$, based on the number of close relatives for the animal in question. Note that the posterior standard deviations and associated accuracies for animals 25 through 30 are identical since they each have one record and share the same sire and dam.

7.3 Empirical Bayes Inference for Normally Distributed Data

In the previous example, the variance components were treated as known and equal to the values that Meyer (1989) used to simulate the data. What if variance components are unknown, as is generally the case?

Recall our generic joint posterior density specification in Equation 7.4. Suppose that $\boldsymbol{\theta}_1$ is of particular interest (i.e., $\boldsymbol{\beta}$ and $\mathbf{u}$ in the mixed model), whereas $\boldsymbol{\theta}_2$ is not (i.e., φ and σ_e^2 in the mixed model). As indicated in Equation 7.5, it is then most appropriate to characterize the inference on $\boldsymbol{\theta}_1$ based on its marginal posterior density, which we reformulate below:

$$\begin{aligned} p(\boldsymbol{\theta}_1 \mid \mathbf{y}) &= \int_{R_{\theta_2}} p(\boldsymbol{\theta} \mid \mathbf{y}) d\boldsymbol{\theta}_2 = \int_{R_{\theta_2}} p(\boldsymbol{\theta}_1, \boldsymbol{\theta}_2 \mid \mathbf{y}) d\boldsymbol{\theta}_2 = \\ &= \int_{R_{\theta_2}} p(\boldsymbol{\theta}_1 \mid \boldsymbol{\theta}_2, \mathbf{y}) p(\boldsymbol{\theta}_2 \mid \mathbf{y}) d\boldsymbol{\theta}_2 = \mathop{\mathrm{E}}_{\boldsymbol{\theta}_2 \mid \mathbf{y}} p(\boldsymbol{\theta}_1 \mid \boldsymbol{\theta}_2, \mathbf{y}). \end{aligned} \quad (7.10)$$

Note that from Equation 7.10, the marginal posterior density is represented as a weighted average of $p(\boldsymbol{\theta}_1 | \boldsymbol{\theta}_2, \mathbf{y})$, with the weights being $p(\boldsymbol{\theta}_2 | \mathbf{y})$, the marginal posterior density of $\boldsymbol{\theta}_2$. If $p(\boldsymbol{\theta}_2 | \mathbf{y})$ is reasonably symmetric, then

$$p(\boldsymbol{\theta}_1 | \mathbf{y}) = p(\boldsymbol{\theta}_1 | \boldsymbol{\theta}_2 = \hat{\boldsymbol{\theta}}_2, \mathbf{y}),$$

where $\hat{\boldsymbol{\theta}}_2$ is the maximizer or joint posterior mode of $p(\boldsymbol{\theta}_2 | \mathbf{y})$. This type of inferential strategy is often referred to as *parametric empirical Bayes* (Carlin and Louis, 1996); we'll simply refer to it as empirical Bayes.

In the specific context of the LMM, the maximizer or the joint posterior mode of $p(\boldsymbol{\varphi}, \sigma_e^2 | \mathbf{y})$ happens to be the restricted maximum likelihood estimator (REML) of σ_e^2 and $\boldsymbol{\varphi}$ (Harville, 1974) under flat prior specifications ($p(\boldsymbol{\varphi}, \sigma_e^2) \propto 1$). Now REML is the default option in PROC MIXED! As a consequence, the reported estimates for $\boldsymbol{\beta}$ and $\mathbf{u}$ are then their respective posterior means based on variance components being equal to their "plugged in" (Natarajan and Kass, 2000) REML estimates for σ_e^2 and $\boldsymbol{\varphi}$.

7.3.1 Empirical Bayes Analysis in the Linear Mixed Model

We reconsider the same example and SAS code as presented previously. Our code for the empirical Bayes implementation is virtually identical to the previous code except for the PARMS statement, where now the NOITER option is removed to facilitate REML estimation of the variance components:

```
proc mixed data=meyer covtest asycov;
   class generation animal;
   model y = generation /solution noint covb;
   random animal /type=lin(1) LDATA=L2data solution ;
   parms (40) (50) ;
   ods listing exclude solutionr; ods output solutionr = solutionr;
run;
```

Part of the output relative to the execution of this code is given in Output 7.2.

Output 7.2 Empirical Bayes inference on fixed and random (genetic) effects conditional on REML estimates of dispersion parameters.

```
                         The Mixed Procedure

                              Dimensions
                    Covariance Parameters             2
                    Columns in X                      2
                    Columns in Z                    306
                    Subjects                          1
                    Max Obs Per Subject             306
                    Observations Used               282
                    Observations Not Used            24
                    Total Observations              306

                           Parameter Search
    CovP1          CovP2     Variance       Res Log Like    -2 Res Log Like
  40.0000        50.0000      52.2326         -1016.8211          2033.6423

                           Iteration History
           Iteration    Evaluations    -2 Res Log Like       Criterion
                   1              2      2033.61276908      0.00000040
                   2              1      2033.61246114      0.00000000
                       Convergence criteria met.
```

(continued on next page)

Output 7.2 *(continued)*

```
                    Covariance Parameter Estimates  ❶
                              Standard      Z
        Cov Parm      Estimate    Error    Value    Pr Z
        LIN(1)         43.9778   13.4410    3.27    0.0011
        Residual       50.9400    8.6835    5.87    <.0001

             Asymptotic Covariance Matrix of Estimates  ❷
             Row    Cov Parm        CovP1        CovP2
              1     LIN(1)         180.66      -81.4068
              2     Residual      -81.4068      75.4036

                         Fit Statistics
             -2 Res Log Likelihood              2033.6
             AIC (smaller is better)            2037.6
             AICC (smaller is better)           2037.7
             BIC (smaller is better)            2045.1

                 PARMS Model Likelihood Ratio Test
                   DF    Chi-Square    Pr > ChiSq
                    1        0.03        0.8630

                     Solution for Fixed Effects
                                   Standard
Effect         generation   Estimate    Error    DF   t Value   Pr > |t|
generation        1          220.32    1.7252     0   127.71       .
generation        2          236.69    1.9993     0   118.39       .

                 Covariance Matrix for Fixed Effects
        Row    Effect        generation       Col1       Col2
         1     generation        1           2.9762     2.6174
         2     generation        2           2.6174     3.9971

                    Type 3 Tests of Fixed Effects
                        Num     Den
        Effect          DF      DF     F Value    Pr > F
        generation       2       0     8698.51      .
```

Some key portions of Output 7.2 are annotated below.

❶ **Covariance Parameter Estimates**

The REML estimates of σ_u^2 and σ_e^2 are reported to be $REML(\sigma_u^2) = 43.9778$ and $REML(\sigma_e^2) = 50.9400$. That is, these estimates jointly maximize $p(\sigma_u^2, \sigma_e^2 \mid \mathbf{y})$ and exactly match the REML estimates reported by Meyer (1989) for this particular model. The COVTEST option from the code prints the approximate standard errors for these two estimates as being 13.4410 and 8.6835, respectively.

❷ **Asymptotic Covariance Matrix of Estimates**

The standard errors reported using the COVTEST option are actually derived from the observed REML information matrix $\dfrac{\partial^2 p(\boldsymbol{\sigma} \mid \mathbf{y})}{\partial \boldsymbol{\sigma} \partial \boldsymbol{\sigma}'}$ for $\boldsymbol{\sigma} = \begin{bmatrix} \sigma_u^2 & \sigma_e^2 \end{bmatrix}'$ such that the asymptotic covariance matrix of $\boldsymbol{\sigma} \mid \mathbf{y}$ is defined to be $\operatorname{asyvar}(\boldsymbol{\sigma} \mid \mathbf{y}) = \left[-\dfrac{\partial^2 p(\boldsymbol{\sigma} \mid \mathbf{y})}{\partial \boldsymbol{\sigma} \partial \boldsymbol{\sigma}'} \right]^{-1}$ as reported here. This output is due to the use of the ASYCOV option in the code. The reported asymptotic standard errors for $REML(\sigma_u^2)$ and $REML(\sigma_e^2)$ are then simply determined to be the square roots of the two diagonal elements 180.66 and 75.4036, as reported here for $\text{asyvar}(\sigma \mid y)$. The reported *p*-values for each of the two variance components are Wald test *p*-values (Pr Z) based on approximating the marginal density $p(\sigma \mid y)$ by a multivariate normal distribution with mean $REML(\sigma)$ and covariance matrix $\left[-\dfrac{\partial^2 p(\sigma \mid y)}{\partial \sigma \partial \sigma'} \right]^{-1}$. The *p*-values indicate that both $\sigma_u^2 > 0$ and $\sigma_e^2 > 0$, based on a Type I error rate of, say, 1%. Likelihood ratio tests can also be used for hypothesis testing on variance components with further details provided in Littell et al. (1996).

As before, random effect solutions can be output to a file (SOLUTIONR) using the ODS facility from which the accuracies can be computed. However, these accuracy computations should now be based on $REML(\sigma_u^2)$ if σ_u^2 is not known, as consistent with empirical Bayes inference.

7.4 Generalized Linear Mixed Models

Not all traits can be reasonably presumed to be conditionally normally distributed; in other words $p(y \mid \theta)$, or in our specific mixed model context, $p(y \mid \beta, u, \sigma_e^2)$ is not necessarily normal. Some examples include binary and ordered categorical responses, such as various disease and fitness measures, as well as count data, such as litter size. In that case, the empirical Bayes inference strategy for inference on fixed and random effects and variance components closely mirrors what was outlined for normally distributed characters in the previous section, except that $p(y \mid \beta, u, \sigma_e^2)$ is specified differently.

Generalized linear models were formalized for multifactorial analysis of non-normal characters (Nelder and Wedderburn, 1972) for specifications of $p(y \mid \theta)$ falling within the exponential family. Extensions allowing for both fixed and random effects were proposed concurrently by Gianola and Foulley (1983) and Harville and Mee (1984) for the specific case of mixed model analysis of binary and ordered categorical data. Subsequently, these models have been adapted by animal breeders for genetic evaluation of livestock for discrete responses on fitness and fertility (Tempelman, 1998).

In a generalized linear mixed model (GLMM), the conditional expectation $\mathrm{E}(y_i \mid \beta, u)$ for the *i*th element of **y** is *linked* to a function of a linear mixed model structure; that is,

$$h\big(\mathrm{E}(y_i \mid \beta, u)\big) = x_i' \beta + z_i' u,$$

where x_i' and z_i' represent the *i*th rows of the incidence matrices **X** and **Z**, respectively. The *link function* $h(.)$ is typically specified such that any *inverse link* function $h^{-1}(.)$ of $x_i'\beta + z_i'u$ leads to $\mathrm{E}(y_i \mid \beta, u) = h^{-1}(x_i'\beta + z_i'u)$ falling within the allowable parameter space. Depending on the nature of the data, the conditional first stage density or likelihood function $p(y \mid \beta, u)$ is specified accordingly as a product of conditionally independent components; i.e., $p(\mathbf{y}|\boldsymbol{\beta},\mathbf{u}) = \prod_{i=1}^{n} p(y_i|\boldsymbol{\beta},\mathbf{u})$, just as with the LMM when $\mathbf{R} = \mathbf{I}\sigma_e^2$.

For example, suppose that the responses of interest involve binary disease data. Then $p(\mathbf{y}|\boldsymbol{\beta},\mathbf{u})$ is specified as a product of conditionally independent Bernoulli distributions. Based on the *probit* link function $\Phi^{-1}(.)$, this conditional likelihood function can be specified explicitly as a function of fixed and random effects:

$$p(\mathbf{y}|\boldsymbol{\beta},\mathbf{u}) = \prod_{i=1}^{n} p(y_i|\boldsymbol{\beta},\mathbf{u}) = \prod_{i=1}^{n} \left(\Phi(\mathbf{x}_i'\boldsymbol{\beta} + \mathbf{z}_i'\mathbf{u})\right)^{y_i} \left(1 - \Phi(\mathbf{x}_i'\boldsymbol{\beta} + \mathbf{z}_i'\mathbf{u})\right)^{1-y_i}, \quad (7.11)$$

where $E(y_i|\boldsymbol{\beta},\mathbf{u}) = \Phi(\mathbf{x}_i'\boldsymbol{\beta} + \mathbf{z}_i'\mathbf{u})$ is specified such that any real-valued combination of $\mathbf{x}_i'\boldsymbol{\beta} + \mathbf{z}_i'\mathbf{u}$ falls with the allowable parameter space (0,1) for $E(y_i|\boldsymbol{\beta},\mathbf{u})$. Note that the inverse link function $\Phi(.)$ is simply the standard normal cumulative density function. Alternatively, a logistic link function can be used for binary data modeling, although there are both biological (Gianola and Foulley, 1983) and statistical (McCulloch, 1994) reasons that favor the use of the probit link function. Biologically, a probit link posits an underlying normally distributed latent variable that determines the measurement on the observed binary scale based on whether or not the latent variable exceeds a threshold. Subsequently, a probit link GLMM has often been referred to as a threshold mixed model (Foulley et al., 1990). Since the residual variance σ_e^2 is not identifiable or separately estimable from location parameters $\boldsymbol{\beta}$ and $\mathbf{u}$ in the model, it is typically constrained as $\sigma_e^2 = 1$. In a probit link GLMM, random effects variance and covariance components $\boldsymbol{\varphi}$ are also specified with respect to this underlying latent scale. The latent scale specification has been useful for fully Bayesian inference approaches using Markov chain Monte Carlo (MCMC) procedures (Albert and Chib, 1993; Sorensen et al., 1995).

Notice that observations are conditionally independent, as is evident from the joint density specification being the product of individual components in Equation 7.11; of course, once the conditioning on $\mathbf{u}$ is dropped in formulating a *marginal likelihood*

$$p(\mathbf{y}|\boldsymbol{\beta},\mathbf{G}(\boldsymbol{\varphi})) = \int p(\mathbf{y}|\boldsymbol{\beta},\mathbf{u}) p(\mathbf{u}|\mathbf{G}(\boldsymbol{\varphi})) d\mathbf{u},$$

then observations are dependent through $\mathbf{G}(\boldsymbol{\varphi})$. However, unlike the case when the conditional likelihood $p(\mathbf{y}|\boldsymbol{\beta},\mathbf{u})$ is multivariate normal, it is computationally intractable to formulate this marginal likelihood for most non-normal data distributions. This problem has inspired Bayesian approaches to inference in GLMM.

Given the usual second-stage assumptions, such as that for $\mathbf{u}$ in Equation 7.6 and a flat prior for $\boldsymbol{\beta}$, the joint posterior density for a GLMM is specified similarly as for LMM in Equation 7.7:

$$p(\boldsymbol{\beta},\mathbf{u}|\mathbf{y},\boldsymbol{\varphi}) \propto p(\mathbf{y}|\boldsymbol{\beta},\mathbf{u}) p(\mathbf{u}|\mathbf{G}(\boldsymbol{\varphi})).$$

Note that for the binary likelihood function, the conditional mean and variance are both specified by the fixed and random effects; i.e., $E(y_i|\mu_i) = \mu_i$ and $\text{var}(y_i|\mu_i) = \mu_i(1-\mu_i)$ for $\mu_i = \Phi(\mathbf{x}_i'\boldsymbol{\beta} + \mathbf{z}_i'\mathbf{u})$. For some field data, as is often true in quantitative genetic studies, the data variability may be greater than can be characterized with an exponential family distribution. This characterizes a situation of *overdispersion*. Overdispersion parameters can be introduced to accommodate this extra variability. This is particularly true when a Poisson likelihood function is adopted for the mixed model analysis of count data; in that specific case, overdispersion can be accommodated by alternative specifications including the use of a negative binomial mixed effects model (Tempelman and Gianola, 1996). For the analysis of binary data using a probit link GLMM, this overdispersion parameter is typically constrained to be equal to 1 for reasons provided in Gianola and Foulley (1983); however, for binomial data where each response is based on the number of successes relative to a total number (>1) of trials on each subject, it may be possible to additionally specify an overdispersion parameter.

7.4.1 Inference on Fixed and Random Effects

Statistical inference in GLMM from a Bayesian perspective does not conceptually differ from that for LMM presented earlier. Consider the case where all (co)variance components are presumed known, assuming no overdispersion. Then Equation 7.7 applies in determining the joint posterior density of the fixed and random effects for GLMM just as it does for LMM. However, unlike the case for multivariate normal conditional likelihood using PROC MIXED, it is very difficult to analytically determine the posterior mean and variance covariance matrix using a non-normal GLMM specification. The joint posterior mode has been proposed as an approximation to the posterior mean $E(\beta, u|y, \varphi)$ of $p(\beta, u|y, \varphi)$ (Fahrmeir and Tutz, 2001; Gianola and Foulley, 1983), with standard errors based on the information matrix of $p(\beta, u|y, \varphi)$. Subsequent inference on β and u is then based on approximating $p(\beta, u|y, \varphi)$ with a multivariate normal density having as mean vector the joint posterior mode $\begin{bmatrix} \tilde{\beta} \\ \tilde{u} \end{bmatrix}$ and a covariance matrix $\tilde{C}$ closely resembling C used for LMM in Equation 7.8:

$$\tilde{C} = \begin{bmatrix} \tilde{C}_{\beta\beta} & \tilde{C}_{\beta u} \\ \tilde{C}_{u\beta} & \tilde{C}_{uu} \end{bmatrix} = \begin{bmatrix} -\dfrac{\partial^2 p(\beta, u|y, \varphi)}{\partial \beta \partial \beta'} & -\dfrac{\partial^2 p(\beta, u|y, \varphi)}{\partial \beta \partial u'} \\ -\dfrac{\partial^2 p(\beta, u|y, \varphi)}{\partial u \partial \beta'} & -\dfrac{\partial^2 p(\beta, u|y, \varphi)}{\partial u \partial u'} \end{bmatrix}^{-1} = \begin{bmatrix} X'WX & X'WZ \\ Z'WX & Z'WZ + (G(\varphi))^{-1} \end{bmatrix}^{-1}.$$

Here W is a diagonal matrix with the ith diagonal element being $-\dfrac{\partial^2 p(\beta, u|y, \varphi)}{\partial (\mu_i)^2}$. Note that $\tilde{C}$, which is the inverse of the information matrix of $p(\beta, u|y, \varphi)$, depends on knowledge of β and u such that $\tilde{C}$ is typically evaluated at the joint posterior mode $\begin{bmatrix} \tilde{\beta} \\ \tilde{u} \end{bmatrix}$. Unlike the solving equations for LMM in PROC MIXED, the solving equations for GLMM are iterative, being based on Newton-Raphson or Fisher's scoring; the only difference is that if Fisher's scoring is used, then $E_y(\tilde{C}^{-1})$, the expected value of $\tilde{C}^{-1}$ with respect to y, is evaluated instead of $\tilde{C}^{-1}$. Nevertheless, the system of equations resembles classical mixed model equations such that the %GLIMMIX macro makes extensive calls to PROC MIXED. Since $p(\beta, u|y, \varphi)$ is asymptotically multivariate normal with mean $\begin{bmatrix} \tilde{\beta} \\ \tilde{u} \end{bmatrix}$ and variance-covariance $\tilde{C}$, then $p(\beta|y, \varphi)$ can be approximated by a multivariate normal density with mean $\tilde{\beta}$ and covariance matrix $\tilde{C}_{\beta\beta}$, whereas $p(u|y, \varphi)$ can be approximated by a multivariate normal density with mean $\tilde{u}$ and covariance matrix $\tilde{\mathbf{C}}_{uu}$. Similarly, linear combinations (i.e., $\mathbf{K}'\boldsymbol{\beta}+\mathbf{M}'\mathbf{u}$) can be approximated by a multivariate normal density in a manner analogous to LMM in Equation 7.9.

7.4.2 Using the %GLIMMIX Macro for Quantitative Genetic Inference

The macro %GLIMMIX, written by Russ Wolfinger at SAS, is based on a pseudo-likelihood approach developed by Wolfinger and O'Connell (1993); however, their inference strategy for $\boldsymbol{\beta}$ and $\mathbf{u}$ is not at all different from determining the joint posterior mode of $p(\boldsymbol{\beta},\mathbf{u}|\mathbf{y},\boldsymbol{\varphi})$. Further details on use of the %GLIMMIX macro are given in Littell et al. (1996). Let's revisit the example of Meyer (1989), except that the continuous responses from the simulated data vector $\mathbf{y}$ are discretized to binary data $\mathbf{z}$ such that any response y_i less than, say, 235 is coded as 0; otherwise it is coded as 1. We analyze this binary data (response labeled as Z in the MODEL statement) using the %GLIMMIX macro as illustrated below:

```
%glimmix(data=meyer,
  stmts=%str(
  class generation animal;
  model z = generation /solution noint covb;
  random animal /type=lin(1) LDATA=L2data solution;
  parms (0.8) (1) /eqcons = 1,2;
  ods output solutionf=solutionf;
  ods listing exclude solutionr; ods output solutionr = solutionr;
  ),error=binomial, link=probit
)
run;

data solutionr;
  set solutionr;
  accuracy = (0.8-StderrPred**2)/0.8;

/* classical accuracy of genetic evaluation (0.8 = additive genetic variance) */

run;

/* Print off posterior means of random effects with posterior standard
deviations for first 30 animals */

title 'Predicted Genetic Effects';
proc print data= solutionr (obs=30);
run;
```

You might notice that this %GLIMMIX macro analysis of Z differs little from the PROC MIXED specification of Y given previously. In the %GLIMMIX macro, it is required that the overdispersion parameter be specified as the last dispersion parameter in the PARMS statement. Here we choose not to specify such a parameter by simply constraining the last (co)variance parameter in the PARMS statement as equal to 1, as indicated by the second term in parentheses and constrained using the EQCONS= 2 option. The string enclosure within the STMTS= provides the same fixed and random effect specifications for the GLMM analysis of Z as we saw previously for the LMM analysis of Y from this same data set. Note that the binary likelihood specification (ERROR = BINOMIAL) and the link specification (LINK = PROBIT) specify the particular GLMM. Given the underlying normally distributed latent variable specification inherent with probit mixed effects models, the additive genetic variance (σ_u^2) is specified relative to the residual variance $(\sigma_e^2 = 1)$ on this latent scale. To be consistent with the ratio $\sigma_u^2/\sigma_e^2 = 40/50$ specified in the LMM analysis of the normally distributed data using PROC MIXED earlier, we constrain $\sigma_u^2 = 0.8$ as indicated by the first value in the PARMS statement and the EQCONS= 1 specification. The output is provided in Output 7.3.

Output 7.3 Approximate posterior inference on fixed and random (genetic) effects using generalized linear mixed model conditional on known dispersion parameters.

```
                        The Mixed Procedure
                         Parameter Search
     CovP1         CovP2         Res Log Like       -2 Res Log Like
     0.8000        1.0000          -614.4253           1228.8506

                         Iteration History
      Iteration    Evaluations    -2 Res Log Like      Criterion
          1            1            1228.85064018      0.00000000
                      Convergence criteria met.

                    Covariance Parameter Estimates
                       Cov Parm       Estimate
                       LIN(1)          0.8000
                       Residual        1.0000

                            Fit Statistics
                 -2 Res Log Likelihood        1228.9
                 AIC (smaller is better)      1228.9
                 AICC (smaller is better)     1228.9
                 BIC (smaller is better)      1228.9

                  PARMS Model Likelihood Ratio Test
                    DF      Chi-Square      Pr > ChiSq
                     0         0.00           1.0000

                    Solution for Fixed Effects   ❶
                                   Standard
   Effect       generation   Estimate    Error      DF    t Value    Pr > |t|
   generation       1         -2.1137    0.3112      0     -6.79        .
   generation       2          0.09390   0.2988      0      0.31        .

               Covariance Matrix for Fixed Effects   ❷
        Row    Effect       generation       Col1         Col2

          1    generation       1           0.09686      0.05141
          2    generation       2           0.05141      0.08926

                    Type 3 Tests of Fixed Effects
                        Num     Den
           Effect        DF      DF      F Value    Pr > F
           generation     2       0       34.99        .

                      GLIMMIX Model Statistics
               Description                        Value
               Deviance                          141.3499
               Scaled Deviance                   141.3499
               Pearson Chi-Square                112.4891
               Scaled Pearson Chi-Square         112.4891
               Extra-Dispersion Scale              1.0000
                    Predicted genetic effects   ❸
                                       StdErr
   Obs   Effect   animal   Estimate    Pred      DF   tValue   Probt   accuracy
    1    animal      1     -0.8924    0.7760      0   -1.15      .     0.24732
    2    animal      2     -0.5002    0.8357      0   -0.60      .     0.12699
    3    animal      3     -0.3754    0.8070      0   -0.47      .     0.18601
    4    animal      4     -0.06822   0.8432      0   -0.08      .     0.11132
    5    animal      5     -0.5071    0.7282      0   -0.70      .     0.33718
    6    animal      6      0.07596   0.7709      0    0.10      .     0.25716
    7    animal      7     -0.08116   0.8388      0   -0.10      .     0.12052
```

(continued on next page)

Output 7.3 *(continued)*

```
    8    animal     8   -0.5703    0.8093    0    -0.70    .    0.18130
    9    animal     9    0.05737   0.7305    0     0.08    .    0.33297
   10    animal    10   -0.1207    0.8275    0    -0.15    .    0.14406
   11    animal    11    0.2543    0.7598    0     0.33    .    0.27843
   12    animal    12   -0.1471    0.8195    0    -0.18    .    0.16052
   13    animal    13    0.8264    0.6326    0     1.31    .    0.49977
   14    animal    14    0.7524    0.7017    0     1.07    .    0.38460
   15    animal    15   -0.1816    0.7459    0    -0.24    .    0.30447
   16    animal    16    0.1965    0.6940    0     0.28    .    0.39796
   17    animal    17    0.6454    0.6391    0     1.01    .    0.48942
   18    animal    18    0.005852  0.7143    0     0.01    .    0.36217
   19    animal    19    0.7060    0.6842    0     1.03    .    0.41477
   20    animal    20   -0.1390    0.7589    0    -0.18    .    0.28004
   21    animal    21    0.06744   0.7166    0     0.09    .    0.35810
   22    animal    22    0.2788    0.7428    0     0.38    .    0.31031
   23    animal    23   -0.1351    0.8229    0    -0.16    .    0.15353
   24    animal    24   -0.1482    0.8190    0    -0.18    .    0.16157
   25    animal    25   -0.6994    0.8291    0    -0.84    .    0.14074
   26    animal    26   -0.6994    0.8291    0    -0.84    .    0.14074
   27    animal    27   -0.6994    0.8291    0    -0.84    .    0.14074
   28    animal    28   -0.6994    0.8291    0    -0.84    .    0.14074
   29    animal    29   -0.6994    0.8291    0    -0.84    .    0.14074
   30    animal    30   -0.6994    0.8291    0    -0.84    .    0.14074
```

The key portions of Output 7.3 are annotated below, each of which is analogous to the same key portion from Output 7.1 for LMM inference on **y**.

❶ Solution for Fixed Effects

This portion is analogous in interpretation to the similarly titled portion from Output 7.1, except that the solutions $\tilde{\boldsymbol{\beta}}$ for $\boldsymbol{\beta}$ are simply the elements of the joint posterior mode of $p(\boldsymbol{\beta},\mathbf{u}|\mathbf{y},\boldsymbol{\varphi})$ with the standard errors determined by the square roots of the diagonal elements of $\tilde{\mathbf{C}}_{\beta\beta}$. Here

$$\tilde{\boldsymbol{\beta}} = \begin{bmatrix} \tilde{\beta}_1 \\ \tilde{\beta}_2 \end{bmatrix} = \begin{bmatrix} -2.1137 \\ 0.09390 \end{bmatrix}.$$

Now, the conditional expectation $E(\mathbf{y}_i | \boldsymbol{\beta}, \mathbf{u}) = \Phi(\mathbf{x}_i'\boldsymbol{\beta} + \mathbf{z}_i'\mathbf{u})$ is expressed on a *subject-specific* basis; this expression is appropriate for inferences involving specific linear combinations of **u** (e.g., how well an animal will perform in a certain environment). If fixed effects, $\boldsymbol{\beta}$, are of primary interest, however, then a *population-averaged* inference is appropriate such that one determines a different conditional mean but with a higher level of marginalization. That is, the marginal expectation of $\mathbf{y}_i$ involves an expectation over **u** (Gilmour et al., 1985; Zeger et al., 1988):

$$E(\mathbf{y}_i | \boldsymbol{\beta}) = \underset{\mathbf{u}}{E}(\mathbf{y}_i | \boldsymbol{\beta}, \mathbf{u}) = \underset{\mathbf{u}}{E}\big(\Phi(\mathbf{x}_i'\boldsymbol{\beta} + \mathbf{z}_i'\mathbf{u})\big) = \Phi\left(\frac{\mathbf{x}_i'\boldsymbol{\beta}}{\sqrt{1+\mathbf{z}_i'\mathbf{G}(\boldsymbol{\varphi})\mathbf{z}_i}}\right) = \Phi\left(\frac{\mathbf{x}_i'\boldsymbol{\beta}}{\sqrt{1+\sigma_u^2}}\right)$$

In the context of our example, then, the estimated marginal expectation of each observation in Generation 1 is defined to be

$$\Phi\left(\frac{\tilde{\beta}_1}{\sqrt{1+\sigma_u^2}}\right) = \Phi\left(\frac{-2.1137}{\sqrt{1+0.8}}\right) = 0.058,$$

whereas the marginal expectation of each observation in Generation 2 is

$$\Phi\left(\frac{\tilde{\beta}_2}{\sqrt{1+\sigma_u^2}}\right) = \Phi\left(\frac{0.09390}{\sqrt{1+0.8}}\right) = 0.528.$$

That is, the estimated marginal probability of individuals with z-scores equal to 1 is 0.058 in Generation 1, whereas it is 0.528 in Generation 2.

❷ Covariance Matrix for Fixed Effects

This portion due to the COVB option is $\tilde{\mathbf{C}}_{\beta\beta}$ from the inverse information matrix of $p(\boldsymbol{\beta}, \mathbf{u}|\mathbf{y}, \boldsymbol{\varphi})$ and is analogous to the similarly titled portion from Output 7.1.

❸ Predicted Genetic Effects

The first 30 elements of the **u**-components of the joint posterior mode of β and **u** are analogous to what was provided in the similarly titled portion in Output 7.1. The standard errors of prediction are also reported for each animal, being determined as the square roots of the respective diagonal elements of $\tilde{\mathbf{C}}_{\mathbf{uu}}$. Notice that the accuracies reported here for the GLMM analysis of the binarized **z** are appreciably less than those from the LMM analysis of continuous normal **y** in Output 7.1; this is anticipated since there is generally far less information in binary data than in continuous data for the prediction of genetic merit of individuals.

Now these genetic predictions can also be expressed on a subject-specific basis as discussed earlier. For example, for animal 25 (or any of animals 26–30 since each has the same **u**-component estimate), the predicted response (or genetic merit) relative to a Generation 1 baseline is

$$\Phi(\tilde{\beta}_1 + \tilde{u}_{25}) = \Phi(-2.1137 - 0.6994) = 0.0024.$$

7.4.3 Empirical Bayes Inference in GLMM

As noted earlier in the LMM section, the random effects (co)variances components $\boldsymbol{\varphi}$ are generally not known and need to be estimated. The strategy for inferring upon variance components in GLMM conceptually differs little from the REML strategy used for LMM. That is, inference is based on the marginal density $p(\boldsymbol{\varphi}|\mathbf{y})$ with the joint posterior mode being the most commonly reported point estimate. The resulting estimates of $\boldsymbol{\varphi}$ have been generally labeled by animal breeders to be *marginal maximum likelihood* (MML) estimates (Foulley et al., 1990) to distinguish GLMM-based from LMM-based inference on $p(\boldsymbol{\varphi}|\mathbf{y})$. However, it is computationally intractable to analytically evaluate $p(\boldsymbol{\varphi}|\mathbf{y})$ for most non-Gaussian GLMM, so various MML approximations have been proposed (Harville and Mee, 1984; Stiratelli et al., 1984) that lead to estimates of φ being equivalent to those provided in %GLIMMIX (Wolfinger and O'Connell, 1993). As with LMM, the empirical Bayes estimates for β and $\mathbf{u}$ in a GLMM analysis would then be the joint posterior mode of $p(\beta, \mathbf{u}|\mathbf{y}, \varphi = \text{MML}(\varphi))$.

Reconsider the same binary data derived from Meyer (1989) considered previously, except that now we use the %GLIMMIX macro to provide the MML estimate of σ_u^2, labeled MML(σ_u^2), and empirical Bayes estimates of β and $\mathbf{u}$ based on the joint maximizer of $p(\beta, \mathbf{u}|\mathbf{y}, \sigma_u^2 = \text{MML}(\sigma_u^2))$. The SAS code follows:

```
%glimmix(data=meyer,
   procopt=noprofile covtest,
   stmts=%str(
   class generation animal;
   model b_235 = generation /solution noint covb;
   random animal /type=lin(1) LDATA=L2data solution;
   parms (0.8) (1) / eqcons=2 ;
   ods listing exclude solutionr; ods output solutionr = solutionr;
   ),error=binomial, link=probit
)
run;
```

Notice that this time the constraint is removed from σ_u^2 in the PARMS statement. The output is given in Output 7.4.

Output 7.4 Empirical Bayes inference on fixed and random (genetic) effects conditional on MML estimates of dispersion parameters.

```
                    The Mixed Procedure
                     Parameter Search
   CovP1         CovP2         Res Log Like    -2 Res Log Like
   0.4020        1.0000          -596.3083         1192.6166

                     Iteration History
   Iteration    Evaluations    -2 Res Log Like      Criterion
       1            1            1192.61656705     0.00000000
                Convergence criteria met.

                Covariance Parameter Estimates  ❶
                              Standard         Z
   Cov Parm     Estimate       Error        Value      Pr Z
   LIN(1)        0.4020        0.1768        2.27     0.0230
   Residual      1.0000          0            .         .

                       Fit Statistics
        -2 Res Log Likelihood            1192.6
        AIC (smaller is better)          1194.6
        AICC (smaller is better)         1194.6
        BIC (smaller is better)          1198.3

               PARMS Model Likelihood Ratio Test
             DF      Chi-Square       Pr > ChiSq
              1         0.00            1.0000
```

(continued on next page)

Output 7.4 *(continued)*

```
                      Solution for Fixed Effects  ❷
                                Standard
Effect       generation   Estimate     Error      DF    t Value   Pr > |t|    .
generation    1           -1.9922     0.2565       0    -7.77       .
generation    2            0.1503     0.2288       0     0.66       .

               Covariance Matrix for Fixed Effects  ❸
     Row    Effect         generation       Col1          Col2
      1    generation       1              0.06579       0.02575
      2    generation       2              0.02575       0.05235

                  Type 3 Tests of Fixed Effects
                      Num      Den
        Effect         DF       DF      F Value    Pr > F
        generation      2        0       40.39       .

                     GLIMMIX Model Statistics
         Description                          Value
         Deviance                           169.4140
         Scaled Deviance                    169.4140
         Pearson Chi-Square                 153.4077
         Scaled Pearson Chi-Square          153.4077
         Extra-Dispersion Scale               1.0000
```

The key items from Output 7.4 are as follows.

❶ Covariance Parameter Estimates

The approximate MML estimate of σ_u^2 is given to be 0.4020 with a standard error of 0.1768 based on an approximate determination of the information matrix of $p(\sigma_u^2 | \mathbf{y})$. The associated Wald test indicates evidence of $\sigma_u^2 > 0$ ($p=0.0230$); nevertheless, a likelihood ratio test may be preferable (Littell et al., 1996).

❷ Solution for Fixed Effects

The empirical Bayes estimate of $\boldsymbol{\beta}$ is taken to be the $\boldsymbol{\beta}$-component, $\tilde{\boldsymbol{\beta}}$, of the joint posterior mode of $p(\boldsymbol{\beta}, \mathbf{u}|\mathbf{y}, \sigma_u^2 = \text{MML}(\sigma_u^2))$. This is reported from the output to be $\tilde{\boldsymbol{\beta}} = \begin{bmatrix} -1.9922 \\ 0.1503 \end{bmatrix}$. The reported standard errors (or posterior standard deviations) are simply the square roots of the diagonal elements of $\tilde{\mathbf{C}}_{\beta\beta}$, evaluated at $\sigma_u^2 = \text{MML}(\sigma_u^2)$.

❸ Covariance Matrix for Fixed Effects

This is simply $\tilde{\mathbf{C}}_{\beta\beta}$, evaluated at $\sigma_u^2 = \text{MML}(\sigma_u^2)$.

Empirical Bayes predictors of $\mathbf{u}$, along with standard errors and associated accuracies, can also be computed as implied by the SOLUTION option in the RANDOM statement followed by the ODS creation of SOLUTIONR. Note that here standard errors and the corresponding accuracies for elements of $\tilde{\mathbf{u}}$ would be evaluated at $\sigma_u^2 = \text{MML}(\sigma_u^2)$.

7.5 Empirical Bayes vs. MCMC

Empirical Bayes procedures for GLMM are based on rather important asymptotic (i.e., large sample) assumptions that may not apply well to the data at hand. First, the MML estimates of φ are based on an approximation to $p(\varphi \mid \mathbf{y})$ or at least to its joint posterior mode (Harville and Mee, 1984; Stiratelli et al., 1984; Wolfinger and O'Connell, 1993). Second, the joint posterior mode of $p(\boldsymbol{\beta}, \mathbf{u} \mid \mathbf{y}, \varphi)$ may not be a good approximation to the posterior mean $E(\boldsymbol{\beta}, \mathbf{u} \mid \mathbf{y}, \varphi)$ in GLMM, although both sets of estimators are indeed equivalent using LMM for normally distributed characters. Finally, inference using

$$p(\boldsymbol{\beta}, \mathbf{u} \mid \mathbf{y}, \boldsymbol{\sigma} = \text{REML}(\boldsymbol{\sigma}))$$

for LMM or

$$p(\boldsymbol{\beta}, \mathbf{u} \mid \mathbf{y}, \varphi = \text{MML}(\varphi))$$

for GLMM may substantially understate the uncertainty—that is, the posterior standard deviations of $\boldsymbol{\beta}$ and $\mathbf{u}$—relative to the use of $p(\boldsymbol{\beta}, \mathbf{u} \mid \mathbf{y})$. In the latter case, Bayesian uncertainty on $\boldsymbol{\sigma}$ or φ is explicitly integrated out (Carlin and Louis, 2000).

Increasingly over the last decade, Bayesian analyses have been based on the use of stochastic Markov chain Monte Carlo (MCMC) procedures that render fully Bayesian inference (Carlin and Louis, 1996; Sorensen and Gianola, 2002). That is, exact small sample inference on $p(\varphi \mid \mathbf{y})$, $p(\boldsymbol{\beta} \mid \mathbf{y})$, and $p(\mathbf{u} \mid \mathbf{y})$ is possible within the limits of Monte Carlo error that can always be controlled by simply increasing the number of MCMC samples drawn. However, Carlin and Louis (2000) have emphasized that MCMC is by no means a "plug and play" procedure. MCMC convergence is very difficult to determine with any of the commonly used diagnostic methods because they each have serious flaws (Cowles and Carlin, 1996). Furthermore, as Carlin and Louis (2000) point out, the use of MCMC methods has tempted investigators to consider hierarchical structures more complex than can be supported by the data, especially if noninformative prior specifications (e.g., $p(\boldsymbol{\beta}) \propto 1$) are used. Such specifications may lead to improper joint posterior densities and unstable inference (Natarajan and Kass, 2000). Hence, we firmly believe that the empirical Bayes approaches to LMM and GLMM inference using PROC MIXED and the %GLIMMIX macro, respectively, will continue to be important and computationally feasible inference tools and generally somewhat safer than MCMC procedures, at least for most quantitative geneticists.

If the variance components are known, the classical mixed model inference strategy based on PROC MIXED is identical to fully Bayesian inference using MCMC (see Wang et al., 1994) on $\boldsymbol{\beta}$ and $\mathbf{u}$; however, the same may not be true for GLMM, as the joint posterior mode is substituted for the posterior mean. How much might these two sets of solutions differ? Consider the same binary response scores $\mathbf{z}$ derived from the data set in Meyer (1989), as previously. Here we also used MCMC to provide posterior means for each of the 306 animals, conditional on $\sigma_u^2 = 0.8$. Our (SAS PROC IML) implementation was based on saving every 10 MCMC samples from 50,000 samples after MCMC convergence, for a total of 5,000 samples. In Output 7.5 we plot the $\mathbf{u}$-components of the joint posterior mode $p(\boldsymbol{\beta}, \mathbf{u} \mid \mathbf{z}, \sigma_u^2 = 0.8)$, using %GLIMMIX, as presented previously, against the posterior mean $E(\mathbf{u} \mid \mathbf{z}, \sigma_u^2 = 0.8)$.

Output 7.5 The **u**-components of the joint posterior mode of $p(\boldsymbol{\beta}, \mathbf{u} | \mathbf{z}, \sigma_u^2 = 0.8)$ vs. $E(\mathbf{u} | \mathbf{z}, \sigma_u^2 = 0.8)$ for discretized data from Meyer (1989). A reference line (intercept 0, slope 1) is superimposed.

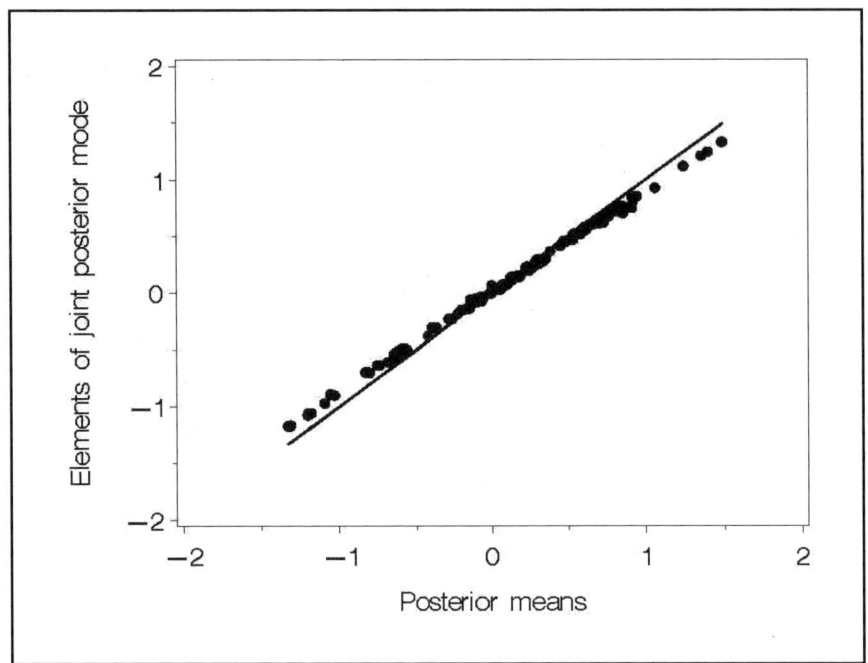

From Output 7.5, it appears that the estimates of **u** from both sets of solutions are practically identical; however, there is evidence of slighter greater shrinkage to 0 of the joint posterior modal solutions relative to the posterior means. This discrepancy hints that the marginal posterior densities of various elements of **u** deviate from normality.

The asymptotic standard errors of the joint posterior estimates of **u**, using the square roots of the diagonal elements of $\tilde{\mathbf{C}}_{\mathbf{uu}}$, are plotted against the corresponding posterior standard deviations of the MCMC samples of **u** in Output 7.6. Recall these standard errors are used for assessing the accuracy of $\tilde{\mathbf{u}}$ as predictors of **u**.

Output 7.6 Asymptotic standard errors (based on the information matrix of $p\left(\boldsymbol{\beta},\mathbf{u}|\mathbf{z},\sigma_u^2=0.8\right)$) vs. posterior standard deviations of $\boldsymbol{\beta},\mathbf{u}|\mathbf{z},\sigma_u^2=0.8$. A reference line (intercept 0, slope 1) is superimposed.

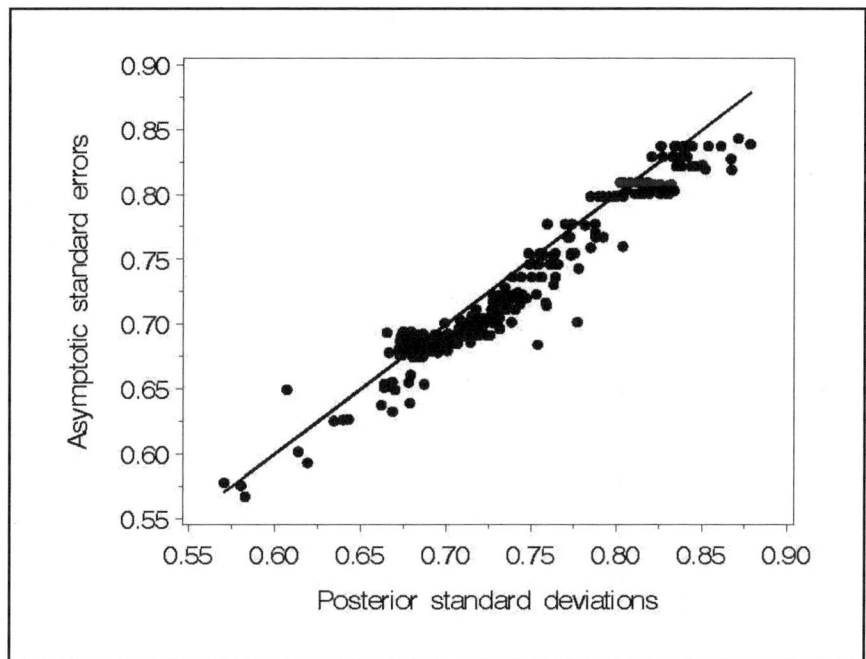

Again, there appears to be very little difference between the two sets of posterior standard deviations in Output 7.6, suggesting that the joint posterior modal inference with standard errors based on the conditional information matrix of $p\left(\boldsymbol{\beta},\mathbf{u}|\mathbf{z},\sigma_u^2\right)$ may be pragmatically satisfactory. You should note from Output 7.6 that some clusters of animals share the same asymptotic standard errors for $\tilde{\mathbf{u}}$ since they derive from the same full-sib family; however, their corresponding MCMC standard deviations slightly differ. This is likely attributable to a small amount of Monte Carlo error that could be further controlled by simply drawing more MCMC samples from $p\left(\boldsymbol{\beta},\mathbf{u}|\mathbf{z},\sigma_u^2\right)$.

Now, what if φ is not known? Some researchers have advocated that MML inference on φ be discouraged in situations characterized by low data information, particularly for binary data when there are few measurements per subject (Breslow and Lin, 1995). The data from Meyer (1989) involve more animals than data such that on average the number of measurements per subject is less than 1! However, because of the correlation matrix **A** used for **u**, the *effective population size* (Walsh and Lynch, 1998) or effective number of subjects is actually somewhat less than the 306 provided. Using %GLIMMIX, we have an approximate MML estimate of 0.40. How does this compare with representative values from the posterior density of σ_u^2 using MCMC? Based on the 5000 MCMC samples saved, we empirically derive the posterior density in Output 7.7.

Output 7.7 Posterior density of $p(\sigma_u^2|\mathbf{z})$ based on 5000 MCMC samples. Plot is truncated for $\sigma_u^2 > 4.00$.

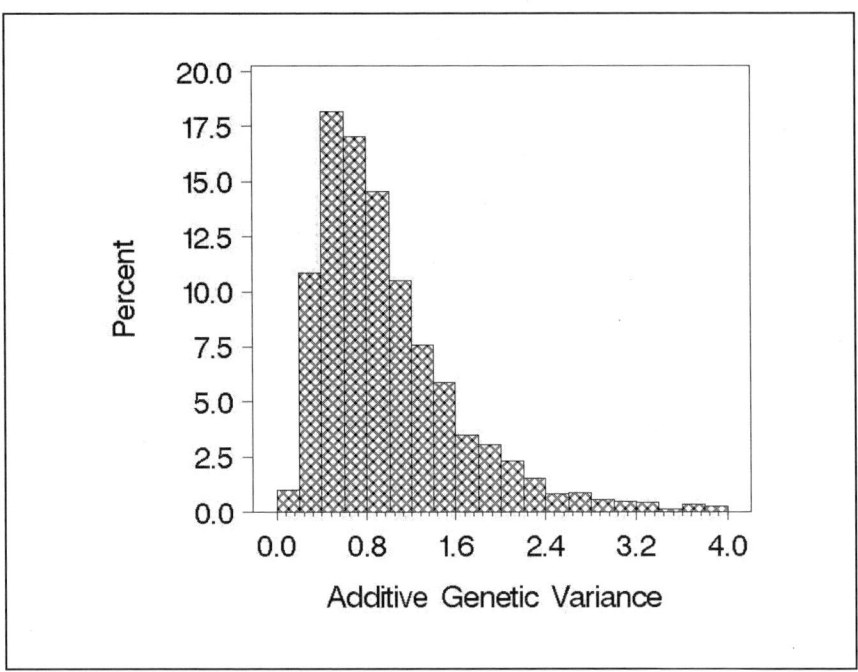

The modal estimate using %GLIMMIX does not differ much qualitatively from the mode derived from Output 7.7, lying somewhere between 0.40 and 0.60 (based on the bar with midpoint 0.5 in Output 7.7). Because of the observed skewness in Output 7.7, both the posterior median (0.83) and posterior mean (1.02) substantially exceed the posterior mode, obviously raising the question of which point estimate is most appropriate. This issue is beyond the scope of this chapter but depends on the specification of loss functions (Carlin and Louis, 1996).

Naturally, you might wonder whether the posterior mean of **u** for the case for unknown σ_u^2 (using MCMC) is different from the empirical Bayes estimate of **u** based on the **u**-components of the joint posterior mode of $p(\boldsymbol{\beta}, \mathbf{u}|\mathbf{z}, \sigma_u^2 = \text{MML}(\sigma_u^2))$ as illustrated previously. The scatterplot of the two sets of solutions is provided in Output 7.8.

Output 7.8 The **u**-components of the joint posterior mode of $p\left(\boldsymbol{\beta}, \mathbf{u} | \mathbf{z}, \sigma_u^2 = \mathrm{MML}\left(\sigma_u^2\right)\right)$ vs. $\mathrm{E}\left(\mathbf{u}|\mathbf{z}\right)$ for discretized data from Meyer (1989). A reference line (intercept 0, slope 1) is superimposed.

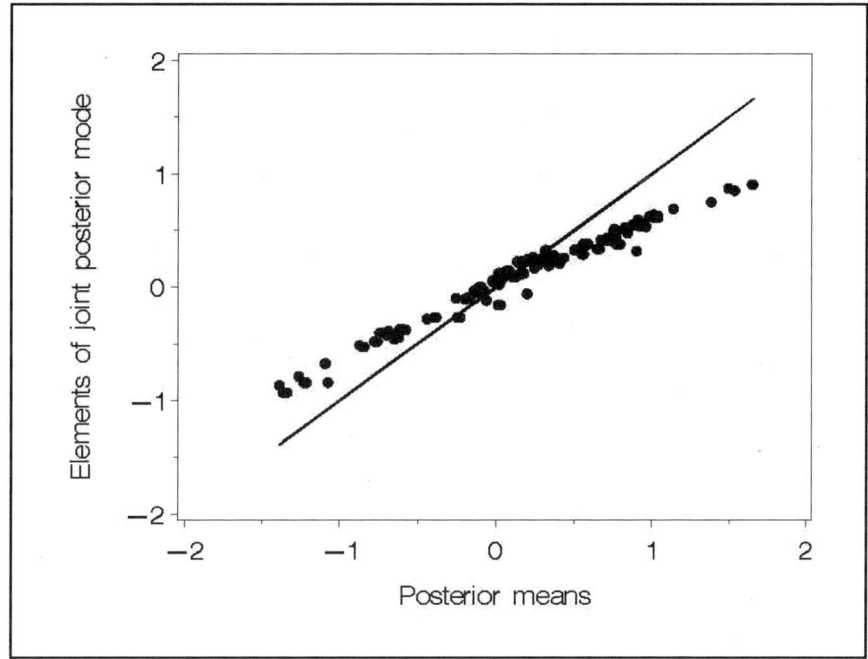

One might note that the nature of the shrinkage of the empirical Bayes estimates relative to the posterior mean in Output 7.8 is slightly more accentuated than in Output 7.5. This discrepancy is most likely due to the biased downward MML estimate of σ_u^2 that in turns further shrinks **u**-components of the joint posterior mode of $p\left(\boldsymbol{\beta}, \mathbf{u} | \mathbf{z}, \sigma_u^2 = \mathrm{MML}\left(\sigma_u^2\right)\right)$ toward zero.

Output 7.9 Asymptotic standard errors (based on the information matrix of $p\left(\boldsymbol{\beta},\mathbf{u}|\mathbf{z},\sigma_u^2 = \hat{\sigma}_u^2\right)$ vs. posterior standard deviations of $\mathbf{u}|\mathbf{z}$. A reference line (intercept 0, slope 1) is superimposed.

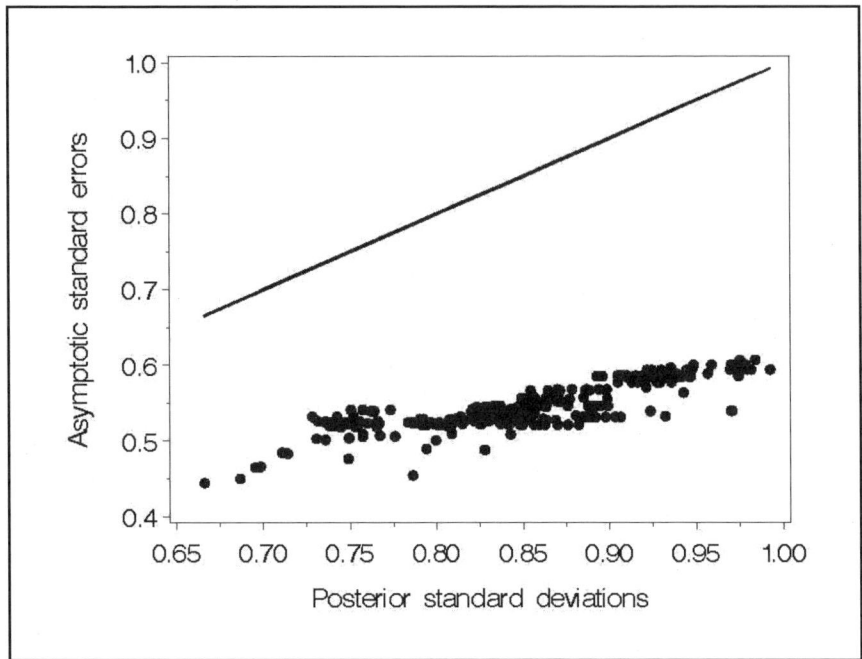

The asymptotic standard errors of the empirical Bayes estimates are plotted against the corresponding posterior means of **u** in Output 7.9. Notice that although qualitatively the standard errors (deviations) rank similarly between the two inference procedures, the empirical Bayes procedure substantially understates the standard errors, on average by over 50%, most likely due to the biased MML estimate of σ_u^2. This downward bias consequently leads to an overstatement of accuracies of genetic merit predictions using the empirical Bayes results. It should be noted, however, that these differences qualitatively depend upon the size of the data set, the population structure, and the amount of information (phenotypes and relatives) per animal. For example, Kizilkaya et al. (2002) found substantially smaller differences between MCMC and empirical Bayes methods for inference on **u** and σ_u^2 based on a GLMM analysis of ordinally scored calving ease observed in daughters of Italian Piedmontese sires.

7.6 Final Comments

We have presented a brief overview of linear and generalized linear mixed effects model implementations for quantitative genetics research from primarily an empirical Bayes perspective. Furthermore, our data sources have concentrated only on observed phenotypes and pedigrees of subjects. Molecular markers are becoming increasingly available such that our SAS coding strategies could be readily adapted to mixed model inferences for those quantitative trait loci mapping and genetic linkage analysis applications as well (Fernando and Grossman, 1989).

7.7 References

Albert, J. H., and S. Chib. 1993. Bayesian analysis of binary and polychotomous response data. *Journal of the American Statistical Association* 88:669-679.

Almasy, L., and J. Blangero. 1998. Multipoint quantitative trait linkage analysis in general pedigrees. *American Journal of Human Genetics* 62:1198-1211.

Breslow, N. E., and X. Lin. 1995. Bias correction in generalized linear mixed models with a single component of dispersion. *Biometrika* 82:81-91.

Carlin, B. P., and T. A. Louis. 1996. *Bayes and Empirical Bayes Methods for Data Analysis*. London: Chapman & Hall.

Carlin, B. P., and T. A. Louis. 2000. Empirical Bayes: Past, present, and future. *Journal of the American Statistical Association* 95:1286-1289.

Cowles, M. K., and B. P. Carlin. 1996. Markov chain Monte Carlo convergence diagnostics: A comparative review. *Journal of the American Statistical Association* 91:883-904.

Fahrmeir, L., and G. Tutz. 2001. *Multivariate Statistical Modeling Based on Generalized Linear Models*. 2d ed. New York: Springer-Verlag.

Fernando, R. L., and D. Gianola. 1986. Optimal properties of the conditional mean as a selection criterion. *Theoretical and Applied Genetics* 72:822-825.

Fernando, R. L., and M. Grossman. 1989. Marker assisted selection using best linear unbiased prediction. *Genetics, Selection, Evolution* 21:467-477.

Foulley, J. L., D. Gianola, and S. Im. 1990. Genetic evaluation for discrete polygenic traits in animal breeding. In *Advances in Statistical Methods for the Genetic Improvement of Livestock*, ed. D. Gianola and K. Hammond, 361-409. New York: Springer-Verlag.

Gianola, D., and R. L. Fernando. 1986. Bayesian methods in animal breeding theory. *Journal of Animal Science* 63:217-244.

Gianola, D., and J. L. Foulley. 1983. Sire evaluation for ordered categorical data with a threshold model. *Genetics, Selection, Evolution* 15:201-224.

Gianola, D., S. Im, and F. W. Macedo. 1990. A framework for prediction of breeding values. In *Advances in Statistical Methods for the Genetic Improvement of Livestock*, ed. D. Gianola and K. Hammond, 210-238. New York: Springer-Verlag.

Gilmour, A. R., R. D. Anderson, and A. L. Rae. 1985. The analysis of binomial data by a generalized linear mixed model. *Biometrika* 72:593-599.

Harville, D. A. 1974. Bayesian inference of variance components using only error contrasts. *Biometrika* 61:383-385.

Harville, D. A., and R. W. Mee. 1984. A mixed model procedure for analyzing ordered categorical data. *Biometrics* 40:393-408.

Henderson, C. R. 1963. Selection index and expected genetic advance. In *Statistical Genetics and Plant Breeding*, ed. W. D. Hanson and H. F. Robinson, 141-163. Washington: National Academy of Sciences and National Research Council.

Henderson, C. R. 1973. Sire evaluation and genetic trends. *Proceedings of the Animal Breeding and Genetics Symposium in Honor of Jay L. Lush*, Champaign, IL, 10-41.

Henderson, C. R., O. Kempthorne, S. R. Searle, and C. N. VonKrosig. 1959. Estimation of environmental and genetic trends from records subject to culling. *Biometrics* 15:192-218.

Kizilkaya, K., B. D. Banks, P. Carnier, A. Albera, G. Bittante, and R. J. Tempelman. 2002. Bayesian inference strategies for the prediction of genetic merit using threshold models with an application to calving ease scores in Italian Piemontese cattle. *Journal of Animal Breeding and Genetics* 119:209-220.

Lange, K., and M. Boehnke. 1983. Extensions to pedigree analysis. IV. Covariance components models for multivariate traits. *American Journal of Medical Genetics* 14:513-524.

Littell, R. C., G. A. Milliken, W. W. Stroup, and R. D. Wolfinger. 1996. *SAS System for Mixed Models*. Cary, NC: SAS Institute Inc.

McCulloch, C. E. 1994. Maximum likelihood variance components estimation for binary data. *Journal of the American Statistical Association* 89:330-335.

Meyer, K. 1989. Restricted maximum likelihood to estimate variance components for animal models with several random effects using a derivative-free algorithm. *Genetics, Selection, Evolution* 21:317-340.

Natarajan, R., and R. E. Kass. 2000. Reference Bayesian methods for generalized linear models. *Journal of the American Statistical Association* 95:227-237.

Nelder, J. A., and R. W. M. Wedderburn. 1972. Generalized linear models. *Journal of the Royal Statistical Society A* 135:370-384.

Searle, S. R. 1987. *Linear Models for Unbalanced Data*. New York: John Wiley & Sons.

Searle, S. R., G. Casella, and C. E. McCulloch. 1992. *Variance Components*. New York: John Wiley & Sons.

Sorensen, D., and D. Gianola. 2002. *Likelihood, Bayesian, and MCMC Methods in Quantitative Genetics*. New York: Springer-Verlag.

Sorensen, D. A., S. Andersen, D. Gianola, and I. Korsgaard. 1995. Bayesian inference in threshold models using Gibbs sampling. *Genetics, Selection, Evolution* 27:227-249.

Stiratelli, R. N., N. Laird, and J. H. Ware. 1984. Random-effects models for serial observations with binary response. *Biometrics* 40:961-971.

Tempelman, R. J. 1998. Generalized linear mixed models in dairy cattle breeding. *Journal of Dairy Science* 81:1428-1444.

Tempelman, R. J., and D. Gianola. 1996. A mixed effects model for overdispersed count data in animal breeding. *Biometrics* 52:265-279.

Van Vleck, L. D. 1993. *Selection Index and Introduction to Mixed Model Methods*. Boca Raton, FL: CRC Press.

Walsh, B., and M. Lynch. 1998. *Genetics and Analysis of Quantitative Traits*. Sunderland, MA: Sinauer Associates.

Wang, C. S., J. J. Rutledge, and D. Gianola. 1994. Bayesian analysis of mixed linear models via Gibbs sampling with an application to litter size in Iberian pigs. *Genetics, Selection, Evolution* 26:91-115.

Wolfinger, R. D., and M. O'Connell. 1993. Generalized linear models: A pseudo-likelihood approach. *Journal of Statistical Computation and Simulation* 48:233-243.

Zeger, S. L., K. Y. Liang, and P. S. Albert. 1988. Models for longitudinal data: A generalized estimating equation approach. *Biometrics* 44:1049-1060.

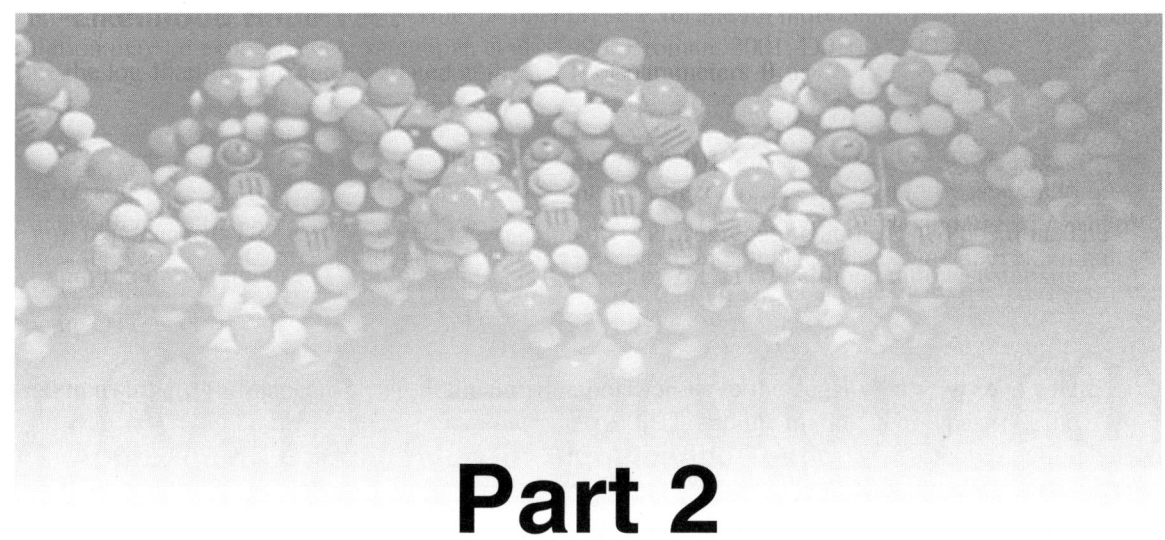

Part 2

Molecular Genetics

Chapter 8   Gene Frequencies and Linkage Disequilibrium   **179**

Chapter 9   The Extended Sib-Pair Method for Mapping QTL   **201**

Chapter 10   Bayesian Mapping Methodology   **225**

Chapter 11   Gene Expression Profiling Using Mixed Models   **251**

Chapter 8

Gene Frequencies and Linkage Disequilibrium

Wendy Czika and Xiang Yu

    8.1 Introduction 179
    8.2 Single-Locus Frequencies 180
    8.3 Hardy-Weinberg Proportions 184
    8.4 Multiple-Locus Frequencies 188
    8.5 Marker-Trait Association Tests 191
        8.5.1 Analyzing Samples of Unrelated Individuals 192
        8.5.2 Analyzing Family Data 195
    8.6 References 200

8.1 Introduction

In a non-experimental, or natural, population of a digametic species, association techniques can be more powerful than linkage studies for identifying genes of modest effect on a disease (Risch and Merikangas, 1996). Such methods use samples of unrelated individuals or nuclear families genotyped at particular genetic markers or polymorphisms, typically either from a genome scan or from a dense map within a previously identified candidate gene. Properties of the genetic markers used for mapping can themselves be examined using random samples of individuals prior to the association testing. Among the marker characteristics of interest are allele, genotype, and haplotype frequencies; significance of the test for Hardy-Weinberg proportions at each marker; and linkage disequilibrium (LD) between pairs or among multiple markers. This chapter will demonstrate how procedures from SAS/Genetics and SAS/STAT can be used for these calculations as well as the association mapping of a quantitative trait in a non-experimental population.

8.2 Single-Locus Frequencies

At a single marker locus, estimates of population allele and genotype frequencies may be of interest. Conveniently, maximum likelihood estimates (MLEs) of genotype frequencies are given by the observed sample genotype frequencies: $\tilde{P}_{uv} = n_{uv}/n$, where n_{uv} is the number of individuals with genotype M_u/M_v in the sample of size n. The same is true for allele frequencies only when Hardy-Weinberg proportions are in effect (described further in the following section): $\tilde{p}_u = n_u/(2n)$, where n_u is the number of M_u alleles in the sample. Using the multinomial distribution of the genotypes, the variance for allele and genotype frequencies can be obtained (Weir, 1996) as

$$\text{var}(\tilde{p}_u) = \frac{1}{2n}(p_u + P_{uu} - 2p_u^2)$$

$$\text{var}(\tilde{P}_{uv}) = \frac{P_{uv}(1 - P_{uv})}{n}$$

with estimates of variances calculated by substituting the parameters with their estimates. While the normal approximation can then be used to calculate confidence intervals, bootstrap confidence intervals are preferred when dealing with proportions.

Various measures of heterozygosity can be used to summarize genotype frequencies for individual markers. The polymorphism information content (PIC) (Botstein et al., 1980), observed heterozygosity, and expected heterozygosity (also called allelic diversity) give an indication of how informative a marker would be for association or linkage studies on the sample collected.

We look at these measures of marker informativeness, estimated genotype and allele frequencies, and test statistics for Hardy-Weinberg equilibrium (HWE) on genotype data from three population groups: African Americans, U.S. Caucasians, and southwestern Hispanics (Budowle and Moretti, 1999). We focus here on the seven loci HLA-DQA1, LDLR, GYPA, HBGG, D7S8, Gc, and D1S80. The following code can be used to first read in random samples of 25 individuals from each of the three groups, combine the three groups into a single data set, and then calculate the measures discussed above for the individual groups.

```
/**************************************************************/
/* Read in random sample of African American individuals      */
/**************************************************************/
data g1;
  input id $ (a1-a14) ($);
  group='Afr Amer';
  datalines;
B0610 24 24 2 3 B B B B A B A A C C
B0611 17 20 2 4.2/4.3 B B A B A A A B B B
B0616 21 25 1.3 4.1 B B A B B C A B B B
B0642 22 22 2 2 B B B B A B A B B B
B0651 18 31 1.2 4.2/4.3 A B B B A C A A B B
B0652 >41 22 1.1 4.1 A B A B A A A B C C
B0665 24 28 1.1 3 A B A B C C A A A C
B0671 0 0 2 3 B B A B A C A A B B
B0673 20 28 4.1 4.2/4.3 B B B B B C A A B B
B0675 28 29 1.2 4.1 B B A B A A A A B B
B0681 22 22 4.1 4.1 B B A B A B A B A B
B0692 21 24 1.2 4.2/4.3 B B B B A C A B B B
B0703 25 28 1.1 3 B B A A A A A B C
B0713 17 18 1.1 3 B B A B A A A B A C
B0718 24 25 1.2 3 B B B B A C A B B C
```

```
B0741 21 28 1.2 2     B B A B A A A B B B
B0744 25 28 1.2 3     A B A A A C B B B B
B0757 24 24 1.2 4.1   A B A A B C A B B B
B0760 0 0 1.2 4.1     B B A B A A B B A B
B0762 0 0 1.2 3       A B A B B C A B B B
B0773 28 32 1.2 1.3   A B A B A A A B C C
B0775 21 24 3 4.1     A B A B A C A A B C
B0791 28 31 1.2 4.2/4.3 A B A B B C A B B B
B0796 18 21 1.1 3     A B B B A C A A A B
B0798 21 21 2 3       B B A A B B A B A B
;

/**************************************************************/
/* Read in random sample of Caucasian individuals            */
/**************************************************************/
data g2;
 input id $ (a1-a14) ($);
 group='Cauc';
 datalines;
C008 18 37 1.3 2       A B A A B B A B C C
C009 22 29 1.2 4.2/4.3 A B A B A B A B A C
C010 18 28 3 3         B B B B A A A A C C
C013 24 24 3 4.1       A B A B B B A B B C
C033 22 24 1.2 3       A A A A A A A A B C
C048 29 31 3 4.1       B B A B A B A B B C
C050 18 24 2 4.1       B B A B B B B B C C
C061 16 24 2 4.1       A B A A B B A A C C
C062 18 29 2 4.1       A B A A A A A B A A
C064 18 23 1.1 2       B B A A A A A B A A
C084 18 24 1.1 4.1     A B B B A B A A A C
C087 18 18 1.2 3       A B A A A A A B A C
C099 24 31 1.3 2       B B A B A A A B B C
C112 24 29 2 3         A B A B B B B B A C
C114 24 24 1.1 1.2     A B A B A B A A A C
C119 20 34 1.1 4.1     A B A A A A A B A C
C124 18 24 1.2 2       A B A A A A A A C C
C127 18 26 4.1 4.1     A B A A A B B A A
C131 24 24 1.1 1.2     A B A B A B A B C C
C136 18 21 1.1 1.2     B B A A A B A A C C
C139 28 31 1.2 3       A B B B A A A B C
C155 18 31 4.1 4.2/4.3 B B A B B B B B B C
C196 24 29 1.2 4.2/4.3 A B A A A B A B C C
C198 22 24 1.2 2       B B A A A B A A C C
C199 24 24 4.1 4.2/4.3 A B A B A B A A A B
;

/**************************************************************/
/* Read in random sample of Southwest Hispanic individuals */
/**************************************************************/
data g3;
 input id $ (a1-a14) ($);
 group='SW Hisp';
 datalines;
H317 30 31 4.2/4.3 4.2/4.3 A B A B B B A B A C
H320 18 31 4.1 4.1     A A A B A B A A B C
H329 18 24 3 4.2/4.3   A B A A A B A B B C
H339 18 28 1.2 1.2     A B B B A B A B C C
H362 18 25 1.1 4.2/4.3 B B A A A B A B A C
H363 24 28 2 3         A B A A B B A B B
H376 18 24 1.1 4.2/4.3 A B A B A B A B B B
H384 22 31 3 4.1       A B A B A A B B B B
H393 24 25 3 3         A B B B A B A A B C
H400 18 25 1.3 4.1     A A A A A B A A A C
```

```
H409 18 31 4.1 4.2/4.3 B B A B B C A A B C
H415 17 24 1.1 3     A A A B B A A A B
H423 18 24 1.1 3     A A A B A A A B B C
H426 24 24 4.1 4.2/4.3 A B A A A B A B A B
H429 18 31 3   4.2/4.3 A A A B B B A A A C
H430 18 23 4.1 4.2/4.3 B B B B A A B B B B
H433 24 27 3 3       A B A B A A A B A C
H435 18 18 1.2 3     A B A B A B B B C C
H445 0  0  1.2 4.1   A A A B B B A B A B
H452 21 30 2   4.2/4.3 A B A A B B A B A C
H461 24 29 1.2 2     A A A A B B A B A C
H463 25 25 1.2 4.1   A B A A B B B B C C
H480 24 28 3   4.1   A B A B A A A A A A
H492 24 30 2   4.2/4.3 B B B B B A B C C
H501 24 24 1.1 4.1   B B A B B B A B A C
;

/***********************************************************/
/* Combine groups into a single data set and replace 0s    */
/* with blanks to represent missing alleles                */
/***********************************************************/
data all_gps;
 set g1 g2 g3;
 array a{14} $;
 do i=1 to 14;
   if a{i}='0' then a{i}=' ';
 end;
 drop i;
run;

/***********************************************************/
/* Create data set with names of marker loci               */
/***********************************************************/
data markers;
 input name $ @@;
 datalines;
D1S80 HLA-DQA1 LDLR GYPA HBGG D7S8 Gc
;

/***********************************************************/
/* Calculate marker summary statistics for loci by group   */
/***********************************************************/
proc allele data=all_gps ndata=markers nofreq;
 var a1-a14;
 by group notsorted;
run;
```

This code produces a "Marker Summary" table for each of the population groups, shown in Output 8.1.

Output 8.1 Marker summary statistics from PROC ALLELE.

```
---------------------------------- group=Afr Amer ------------------------------

                              The ALLELE Procedure

                                 Marker Summary
                                                      --------Test for HWE--------
              Number    Number
                of        of             Hetero-   Allelic    Chi-              Pr >
    Locus     Indiv   Alleles    PIC    zygosity  Diversity  Square    DF      ChiSq
    D1S80       22       12    0.8625    0.7727    0.8750   65.5003   66      0.4942
    HLA-DQA1    25        7    0.8069    0.9200    0.8296   24.9670   21      0.2486
    LDLR        25        2    0.2688    0.4000    0.3200    1.5625    1      0.2113
    GYPA        25        2    0.3714    0.5600    0.4928    0.4649    1      0.4954
    HBGG        25        3    0.5409    0.6000    0.6112    3.3678    3      0.3383
    D7S8        25        2    0.3546    0.5600    0.4608    1.1586    1      0.2818
    Gc          25        3    0.4455    0.3600    0.5016    6.2672    3      0.0993


---------------------------------- group=Cauc ----------------------------------

                              The ALLELE Procedure

                                 Marker Summary
                                                      --------Test for HWE--------
              Number    Number
                of        of             Hetero-   Allelic    Chi-              Pr >
    Locus     Indiv   Alleles    PIC    zygosity  Diversity  Square    DF      ChiSq
    D1S80       25       13    0.7802    0.8000    0.8024   76.0960   78      0.5399
    HLA-DQA1    25        7    0.8086    0.9200    0.8312   24.8616   21      0.2532
    LDLR        25        2    0.3546    0.6400    0.4608    3.7809    1      0.0518
    GYPA        25        2    0.3405    0.4000    0.4352    0.1635    1      0.6859
    HBGG        25        2    0.3685    0.3600    0.4872    1.7041    1      0.1918
    D7S8        25        2    0.3602    0.4400    0.4712    0.1096    1      0.7406
    Gc          25        3    0.4874    0.5200    0.5528    3.0617    3      0.3822


--------------------------------- group=SW Hisp---------------------------------

                              The ALLELE Procedure

                                 Marker Summary
                                                      --------Test for HWE--------
              Number    Number
                of        of             Hetero-   Allelic    Chi-              Pr >
    Locus     Indiv   Alleles    PIC    zygosity  Diversity  Square    DF      ChiSq
    D1S80       24       12    0.7994    0.8333    0.8203   46.2563   66      0.9691
    HLA-DQA1    25        7    0.7883    0.8000    0.8144   12.7772   21      0.9162
    LDLR        25        2    0.3734    0.5200    0.4968    0.0545    1      0.8154
    GYPA        25        2    0.3648    0.4800    0.4800    0.0000    1      1.0000
    HBGG        25        3    0.3908    0.4000    0.4952    1.8082    3      0.6131
    D7S8        25        2    0.3714    0.5600    0.4928    0.4649    1      0.4954
    Gc          25        3    0.5798    0.6400    0.6536    3.0751    3      0.3802
```

Allele and genotype frequencies and related statistics can be calculated for each of the markers; in the analysis below, a single marker is examined for the African American group only.

```
ods listing exclude MarkerSumm;

/**************************************************************/
/* Calculate allele and gentoype frequencies for African       */
/* American population group at marker locus D1S80             */
/**************************************************************/
proc allele data=all_gps ndata=markers(obs=1) boot=1000 seed=71;
 var a1-a2;
 where group='Afr Amer';
run;
```

The "Allele Frequencies" table in Output 8.2 displays the 12 alleles observed at marker D1S80 in this sample. Of the 78 possible genotypes, only the 18 that are observed are shown in the "Genotype Frequencies" table also included in the output.

Output 8.2 Allele and genotype frequencies from PROC ALLELE.

```
                      The ALLELE Procedure

                       Allele Frequencies
                                    Standard      95% Confidence
         Locus    Allele   Frequency   Error         Limits
         D1S80      17      0.0455    0.0306    0.0000    0.1136
         D1S80      18      0.0682    0.0366    0.0000    0.1364
         D1S80      20      0.0455    0.0306    0.0000    0.1136
         D1S80      21      0.1591    0.0591    0.0682    0.2955
         D1S80      22      0.1136    0.0637    0.0000    0.2500
         D1S80      24      0.1818    0.0685    0.0682    0.3182
         D1S80      25      0.0909    0.0411    0.0227    0.1818
         D1S80      28      0.1818    0.0513    0.0909    0.2955
         D1S80      29      0.0227    0.0222    0.0000    0.0682
         D1S80      31      0.0455    0.0306    0.0000    0.1136
         D1S80      32      0.0227    0.0222    0.0000    0.0682
         D1S80     >41      0.0227    0.0222    0.0000    0.0682

                       Genotype Frequencies
                                    HWD     Standard    95% Confidence
  Locus    Genotype   Frequency    Coeff     Error         Limits
  D1S80     17/18      0.0455    -0.0196    0.0199    -0.0620    0.0041
  D1S80     17/20      0.0455    -0.0207    0.0203    -0.0635    0.0021
  D1S80     18/21      0.0455    -0.0119    0.0188    -0.0553    0.0145
  D1S80     18/31      0.0455    -0.0196    0.0199    -0.0620    0.0031
  D1S80     20/28      0.0455    -0.0145    0.0182    -0.0537    0.0108
  D1S80     21/21      0.0455     0.0201    0.0349    -0.0331    0.0945
  D1S80     21/24      0.0909    -0.0165    0.0263    -0.0723    0.0279
  D1S80     21/25      0.0455    -0.0083    0.0190    -0.0496    0.0207
  D1S80     21/28      0.0455     0.0062    0.0199    -0.0403    0.0372
  D1S80     22/22      0.0909     0.0780    0.0480    -0.0021    0.1756
  D1S80     22/>41     0.0455    -0.0201    0.0193    -0.0635    0.0000
  D1S80     24/24      0.0909     0.0579    0.0438    -0.0186    0.1400
  D1S80     24/25      0.0455    -0.0062    0.0191    -0.0517    0.0227
  D1S80     24/28      0.0455     0.0103    0.0208    -0.0398    0.0465
  D1S80     25/28      0.0909    -0.0289    0.0245    -0.0811    0.0108
  D1S80     28/29      0.0455    -0.0186    0.0179    -0.0558    0.0000
  D1S80     28/31      0.0455    -0.0145    0.0182    -0.0553    0.0103
  D1S80     28/32      0.0455    -0.0186    0.0179    -0.0573    0.0000
```

8.3 Hardy-Weinberg Proportions

When working with outbred populations such as humans where many of the usual requirements for HWE can be assumed, it is still of interest to test for HWE at individual markers, often as a means of detecting genotyping errors. A Pearson chi-square statistic can be formed using the observed genotype counts and the product of the sample size and the frequencies of the two alleles comprising the genotype (multiplied by two for heterozygous genotypes) as the expected counts (Weir, 1996):

$$X^2 = \sum_u \frac{(n_{uu} - n\tilde{p}_u^2)^2}{n\tilde{p}_u^2} + \sum_u \sum_{v>u} \frac{(n_{uv} - 2n\tilde{p}_u\tilde{p}_v)^2}{2n\tilde{p}_u\tilde{p}_v}$$

This statistic has an asymptotic chi-square distribution with $m(m-1)/2$ degrees of freedom (df) for a marker with m alleles. Alternatively, when small genotype counts compromise the validity of the asymptotic distribution, a permutation test can be used to calculate exact *p*-values (Guo and Thompson, 1992). The probability of observing the genotype counts in the sample given the allele counts can be calculated by using the multinomial distribution of the genotype and allele counts, assuming HWE. Alleles in the sample are permuted holding the allele counts fixed, while creating new genotype counts. Since the number of possible permutations can quickly become unfeasible, a Monte Carlo version of this procedure can be implemented, where for P permutations performed, the exact p-value is estimated as the proportion of the genotype count probabilities from the P permuted samples that equal or exceed the probability of the original sample.

Data provided for GAW 12 (Wijsman et al., 2001) were read into a single SAS data set containing both phenotypic and genotypic information. Individuals were genotyped at a sequence containing 30 single nucleotide polymorphisms (SNPs) from candidate gene 6. The SAS data set GAWSNPS appears as in Table 8.1.

Table 8.1 GAW 12 data.

pedigree	id	father	mother	household_id	affection	q1	q2	q3	q4	q5	a1
1	678	1	460	1	A	22.12	20.58	27.2	17.82	39.64	1
1	11			2	U	10.41	15.6	17.89	22.73	30.9	1
1	1131	680	16	2	U	19.99	18.55	23.39	24.79	35.69	1
1	1418	11	1131	2	U	17.99	16.91	22.57	26.1	28.05	1
1	12			3	A	12.86	12.43	18.91	16.59	43.36	1
1	1133	680	16	3	A	21.02	24.25	25.65	25.26	37.7	1
1	1419	12	1133	3	U	16.32	20.01	23.92	22.36	28.19	1
1	13			4	U	15.63	16.17	18.72	20.46	35.25	1
1	684	462	3	4	U	19.48	22.85	23.62	19.07	30.31	1
1	1118	13	684	4	U	15.17	17.74	16.33	12.85	33	1
1	1119	13	684	4	A	20.11	21.32	20.6	16.36	33.71	1
1	14			5	U	17.52	18.45	25.81	24.16	28.45	1
1	681	5	463	5	A	17.53	19.27	23.21	19.5	39.64	1
1	1120	14	681	5	U	16.52	20.67	19.38	15.72	25.94	1
1	1121	14	681	5	U	16.04	17.76	19.01	16.18	27.33	1
1	15			6	U	21.63	23.07	24.96	25.81	37.16	1
1	685	462	3	6	A	22.13	24.88	21.31	23.76	33.49	1

The first 11 columns of this data set contain identifiers and phenotypic information including five quantitative covariates (Q1–Q5), and the remaining columns A1–A60 (not all shown) contain individuals' alleles at the markers.

In order to test for HWE, a sample of unrelated individuals is required. Thus, a subset of the data is created containing only the founders of the pedigrees.

```
data founders;
  set gawsnps;
  if father=. and mother=.;
run;
```

A map data set is also created, containing two columns: one with the SNP number (numbered consecutively) and the second containing the position of the SNP on the chromosome in base pairs. The 30 SNPs and their respective positions are read into the SAS data set MAP as follows:

```
data map;
  input name $ pos @@;
  datalines;
SNP1   993    SNP2   1748
SNP3   1868   SNP4   1987
SNP5   4411   SNP6   4848
SNP7   5007   SNP8   5663
SNP9   5762   SNP10  5782
SNP11  5901   SNP12  6805
SNP13  7073   SNP14  7332
SNP15  7987   SNP16  8067
SNP17  8226   SNP18  9497
SNP19  9616   SNP20  10054
SNP21  10955  SNP22  11782
SNP23  11981  SNP24  12408
SNP25  12716  SNP26  13710
SNP27  13869  SNP28  14544
SNP29  15021  SNP30  16849
;
run;
```

The following code can be run to test for HWE at each of the 30 SNPs, with significance evaluated by both an asymptotic chi-square test and a Monte Carlo permutation procedure to estimate the exact p-value. Note the SAS data set option RENAME= is used for the NDATA= data set so the chromosome position is used as the label for the marker loci in the ODS table that is created.

```
proc allele data=founders ndata=map(rename=(name=oldname pos=name))
        exact=10000 nofreq seed=123;
  var a1-a60;
run;
```

The "Marker Summary" table that is produced shows both the asymptotic chi-square p-values and the approximate exact p-values using 10,000 permutations of the alleles in the sample.

Output 8.3 Marker summary table for subset of GAW data.

```
                              The ALLELE Procedure

                                Marker Summary
                                                   -------------Test for HWE------------
        Number   Number
          of       of                 Hetero-   Allelic     Chi-                Pr >      Prob
Locus    Indiv   Alleles    PIC      zygosity  Diversity   Square    DF        ChiSq     Exact
  993     165       2      0.2449    0.2606    0.2858     1.2808     1        0.2577    0.2654
 1748     165       2      0.2449    0.2606    0.2858     1.2808     1        0.2577    0.2769
 1868     165       1      0.0000    0.0000    0.0000     0.0000     0          .         .
 1987     165       2      0.2449    0.2606    0.2858     1.2808     1        0.2577    0.2682
 4411     165       2      0.2421    0.2545    0.2818     1.5434     1        0.2141    0.2537
 4848     165       2      0.3566    0.4303    0.4644     0.8916     1        0.3450    0.4018
 5007     165       2      0.1516    0.1818    0.1653     1.6500     1        0.1990    0.3614
 5663     165       2      0.0060    0.0061    0.0060     0.0015     1        0.9689    1.0000
 5762     165       2      0.0060    0.0061    0.0060     0.0015     1        0.9689    1.0000
 5782     165       2      0.1557    0.1879    0.1702     1.7736     1        0.1829    0.3658
 5901     165       2      0.0060    0.0061    0.0060     0.0015     1        0.9689    1.0000
 6805     165       2      0.1516    0.1818    0.1653     1.6500     1        0.1990    0.3618
 7073     165       2      0.1516    0.1818    0.1653     1.6500     1        0.1990    0.3712
 7332     165       2      0.1516    0.1818    0.1653     1.6500     1        0.1990    0.3733
 7987     165       2      0.0060    0.0061    0.0060     0.0015     1        0.9689    1.0000
 8067     165       2      0.1516    0.1818    0.1653     1.6500     1        0.1990    0.3657
 8226     165       2      0.1557    0.1879    0.1702     1.7736     1        0.1829    0.3603
 9497     165       2      0.0179    0.0182    0.0180     0.0139     1        0.9062    1.0000
 9616     165       2      0.2852    0.3333    0.3446     0.1750     1        0.6757    0.6493
10054     165       2      0.2874    0.3394    0.3479     0.0990     1        0.7530    0.8178
10955     165       2      0.2784    0.3273    0.3343     0.0719     1        0.7885    0.8151
```

(continued on next page)

Output 8.3 *(continued)*

```
11782    165    2    0.3275    0.3879    0.4126    0.5906    1    0.4422    0.4471
11981    165    2    0.2177    0.2788    0.2486    2.4339    1    0.1187    0.2032
12408    165    2    0.1258    0.1333    0.1349    0.0216    1    0.8832    0.5849
12716    165    2    0.1346    0.1576    0.1452    1.2069    1    0.2719    0.6051
13710    165    2    0.0179    0.0182    0.0180    0.0139    1    0.9062    1.0000
13869    165    2    0.3275    0.3879    0.4126    0.5906    1    0.4422    0.4484
14544    165    2    0.1302    0.1394    0.1400    0.0035    1    0.9530    1.0000
15021    165    2    0.3599    0.5273    0.4706    2.3915    1    0.1220    0.1411
16849    165    2    0.0294    0.0303    0.0298    0.0391    1    0.8433    1.0000
```

Another use of the HWE test is as a test for association between a marker and a binary trait such as disease status, as a gene-mapping tool (Nielsen et al., 1999). This can be accomplished by testing only those affected with the disease for HWE at a marker; the power of this test is based on the square of the linkage disequilibrium measure between the marker and disease locus. The ALLELE procedure uses the WHERE statement to restrict the data that are analyzed.

```
proc allele data=founders ndata=map(rename=(name=oldname pos=name))
            exact=10000 nofreq seed=72;
  var a1-a60;
  where affection='A';
run;
```

In the "Marker Summary" table shown in Output 8.4, there are no significant *p*-values at the 0.05 level. Because linkage disequilibrium decays more slowly than its square, other association tests may have higher power for detecting linkage disequilibrium.

Output 8.4 Testing for marker-trait association with the test for HWE.

```
                          The ALLELE Procedure

                            Marker Summary
                                                    -------------Test for HWE------------
         Number    Number
         of        of                 Hetero-   .Allelic    Chi-                Pr >      Prob
Locus    Indiv     Alleles    PIC     zygosity  Diversity   Square     DF       ChiSq     Exact
  993    45        2          0.3081  0.3778    0.3805      0.0023     1        0.9618    1.0000
 1748    45        2          0.3081  0.3778    0.3805      0.0023     1        0.9618    1.0000
 1868    45        1          0.0000  0.0000    0.0000      0.0000     0        .         .
 1987    45        2          0.3081  0.3778    0.3805      0.0023     1        0.9618    1.0000
 4411    45        2          0.3012  0.3556    0.3694      0.0631     1        0.8017    0.6966
 4848    45        2          0.3146  0.4444    0.3911      0.8368     1        0.3603    0.4779
 5007    45        2          0.2166  0.2889    0.2472      1.2827     1        0.2574    0.5703
 5663    45        2          0.0217  0.0222    0.0220      0.0057     1        0.9399    1.0000
 5762    45        2          0.0217  0.0222    0.0220      0.0057     1        0.9399    1.0000
 5782    45        2          0.2166  0.2889    0.2472      1.2827     1        0.2574    0.5716
 5901    45        2          0.0217  0.0222    0.0220      0.0057     1        0.9399    1.0000
 6805    45        2          0.2166  0.2889    0.2472      1.2827     1        0.2574    0.5690
 7073    45        2          0.2166  0.2889    0.2472      1.2827     1        0.2574    0.5645
 7332    45        2          0.2166  0.2889    0.2472      1.2827     1        0.2574    0.5626
 7987    45        1          0.0000  0.0000    0.0000      0.0000     0        .         .
 8067    45        2          0.2166  0.2889    0.2472      1.2827     1        0.2574    0.5696
 8226    45        2          0.2166  0.2889    0.2472      1.2827     1        0.2574    0.5748
 9497    45        2          0.0217  0.0222    0.0220      0.0057     1        0.9399    1.0000
 9616    45        2          0.2166  0.2444    0.2472      0.0054     1        0.9412    1.0000
10054    45        2          0.2166  0.2444    0.2472      0.0054     1        0.9412    1.0000
10955    45        2          0.2044  0.2667    0.2311      1.0651     1        0.3021    1.0000
11782    45        2          0.2776  0.3333    0.3331      0.0000     1        0.9960    1.0000
11981    45        2          0.2392  0.2889    0.2778      0.0720     1        0.7884    1.0000
12408    45        2          0.1332  0.1111    0.1435      2.2877     1        0.1304    0.2196
12716    45        2          0.1167  0.1333    0.1244      0.2296     1        0.6318    1.0000
13710    45        1          0.0000  0.0000    0.0000      0.0000     0        .         .
13869    45        2          0.2776  0.3333    0.3331      0.0000     1        0.9960    1.0000
14544    45        2          0.1638  0.2000    0.1800      0.5556     1        0.4561    1.0000
15021    45        2          0.3496  0.5111    0.4516      0.7813     1        0.3767    0.5057
16849    45        2          0.0425  0.0444    0.0435      0.0232     1        0.8788    1.0000
```

8.4 Multiple-Locus Frequencies

We might want to know how frequently alleles at several different marker loci on the same chromosome occur together. When gametic phase is known in the collection of the marker data, then these haplotype frequencies can be estimated using observed frequencies, essentially treating haplotypes as alleles. Thus the observed estimates are the MLEs when HWE holds at the markers. However, it is more likely that gametic phase is unknown when collecting random samples of individuals from a population. When trying to estimate just two-locus haplotype frequencies at biallelic markers like SNPs, a cubic equation can be solved to obtain the MLE, assuming HWE (Weir, 1996). An analytical solution is no longer tractable when estimating frequencies for multiple (more than two) loci at once or a pair of multiallelic markers. It then becomes necessary to apply an iterative method such as the expectation-maximization (EM) algorithm (Excoffier and Slatkin, 1995; Hawley and Kidd, 1995; Long et al., 1995) to identify the set of haplotype probabilities that maximize the likelihood based on the multinomial distribution of the multi-locus genotypes.

First, we look at two-locus haplotypes.

```
ods listing select LDMeasures;

proc allele data=founders ndata=map haplo=est maxdist=1
        corrcoeff delta dprime propdiff yulesq;
  var a1-a60;
run;
```

With this code, only the "Linkage Disequilibrium Measures" table is produced in the ODS output. The MAXDIST=1 option indicates that only consecutive pairs of loci are examined, as opposed to all possible pairs of markers. Maximum likelihood estimates of haplotype frequencies $\tilde{p}_{uv}$ between allele M_u at locus $\mathbf{M}$ and N_v at locus $\mathbf{N}$ are calculated assuming HWE since the HAPLO=EST option is specified. Given the haplotype frequency, the LD coefficient between a pair of alleles at different loci can be calculated using the MLE of the haplotype frequency as $\hat{d}_{uv} = \hat{p}_{uv} - \tilde{p}_u \tilde{p}_v$. In addition to the LD coefficient, all five available LD measures are included in the ODS table since all were listed in the PROC ALLELE statement: the correlation coefficient r, population attributable risk δ, Lewontin's D', the proportional difference d, and Yule's Q (Devlin and Risch, 1995). A subset of the table is shown in Output 8.5.

Output 8.5 LD measures for GAW data.

```
                         The ALLELE Procedure

                     Linkage Disequilibrium Measures
                                 LD        Corr             Lewontin's     Prop     Yule's
Locus1  Locus2  Haplotype  Frequency   Coeff       Coeff    Delta      D'            Diff      Q
SNP1    SNP2    1-1         0.8273     0.1429      1.0000   1.0000     1.0000     1.0000    1.0000
SNP1    SNP2    1-2         0.0000    -0.1429     -1.0000      .      -1.0000    -1.0000   -1.0000
SNP1    SNP2    2-1         0.0000    -0.1429     -1.0000      .      -1.0000    -1.0000   -1.0000
SNP1    SNP2    2-2         0.1727     0.1429      1.0000   1.0000     1.0000     1.0000    1.0000
SNP2    SNP3    1-2         0.8273    -0.0000         .        .          .          .         .
SNP2    SNP3    2-2         0.1727     0.0000         .        .          .          .         .
SNP3    SNP4    2-1         0.1727     0.0000         .        .          .       0.0000       .
SNP3    SNP4    2-2         0.8273    -0.0000         .        .          .      -0.0000       .
SNP4    SNP5    1-1         0.1666     0.1373      0.9679   0.9819     0.9783    0.9747    0.9997
SNP4    SNP5    1-2         0.0061    -0.1373     -0.9679  -54.285    -0.9783   -0.9747   -0.9997
SNP4    SNP5    2-1         0.0030    -0.1373     -0.9679  -133.17    -0.9783   -0.9747   -0.9997
SNP4    SNP5    2-2         0.8242     0.1373      0.9679   0.9925     0.9783    0.9747    0.9997
SNP5    SNP6    1-1         0.1655     0.0580      0.3209   0.2528     0.9329    0.2500    0.9369
SNP5    SNP6    1-2         0.0042    -0.0580     -0.3209  -0.3384    -0.9329   -0.2500   -0.9369
SNP5    SNP6    2-1         0.4678    -0.0580     -0.3209  -21.940    -0.9329   -0.2500   -0.9369
SNP5    SNP6    2-2         0.3625     0.0580      0.3209   0.9564     0.9329    0.2500    0.9369
```

(continued on next page)

Output 8.5 *(continued)*

```
SNP6    SNP7    1-1    0.5424   -0.0333   -0.2406    .         -1.0000   -0.4033   -1.0000
SNP6    SNP7    1-2    0.0909    0.0333    0.2406    1.0000     1.0000    0.4033    1.0000
SNP6    SNP7    2-1    0.3667    0.0333    0.2406    0.4033     1.0000    0.4033    1.0000
SNP6    SNP7    2-2    0.0000   -0.0333   -0.2406   -0.6760    -1.0000   -0.4033   -1.0000
SNP7    SNP8    1-1    0.0000   -0.0028   -0.1743  -10.345     -1.0000   -0.9119   -1.0000
SNP7    SNP8    1-2    0.9091    0.0028    0.1743    0.9119     1.0000    0.9119    1.0000
SNP7    SNP8    2-1    0.0030    0.0028    0.1743    1.0000     1.0000    0.9119    1.0000
SNP7    SNP8    2-2    0.0879   -0.0028   -0.1743    .         -1.0000   -0.9119   -1.0000
SNP8    SNP9    1-1    0.0030    0.0030    1.0000    1.0000     1.0000    1.0000    1.0000
SNP8    SNP9    1-2    0.0000   -0.0030   -1.0000    .         -1.0000   -1.0000   -1.0000
SNP8    SNP9    2-1    0.0000   -0.0030   -1.0000    .         -1.0000   -1.0000   -1.0000
SNP8    SNP9    2-2    0.9970    0.0030    1.0000    1.0000     1.0000    1.0000    1.0000
SNP9    SNP10   1-1    0.0000   -0.0027   -0.1712   -0.0333    -1.0000   -0.0323   -1.0000
SNP9    SNP10   1-2    0.0030    0.0027    0.1712    0.0323     1.0000    0.0323    1.0000
SNP9    SNP10   2-1    0.9061    0.0027    0.1712    1.0000     1.0000    0.0323    1.0000
SNP9    SNP10   2-2    0.0909   -0.0027   -0.1712    .         -1.0000   -0.0323   -1.0000
SNP10   SNP11   1-1    0.0000   -0.0027   -0.1712   -9.9667    -1.0000   -0.9088   -1.0000
SNP10   SNP11   1-2    0.9061    0.0027    0.1712    0.9088     1.0000    0.9088    1.0000
SNP10   SNP11   2-1    0.0030    0.0027    0.1712    1.0000     1.0000    0.9088    1.0000
```

Additionally, an output data set can be created that contains chi-square test statistics for linkage disequilibrium between each pair of markers across all possible haplotypes at the two markers. The *p*-values from the tests for disequilibrium, both Hardy-Weinberg and linkage, can be represented graphically using the %TPLOT macro in SAS/Genetics to create an LD map. This macro requires another data set containing a *p*-value for each marker, typically from a marker-trait association test, but since we are not examining a trait at this point in the analysis, we can instead create a "dummy" data set as shown in this code.

```
/**********************************************************/
/* Create output data set with LD test statistics         */
/**********************************************************/
proc allele data=founders ndata=map noprint
            haplo=est outstat=ldtest;
 var a1-a60;
run;

/**********************************************************/
/* Create data set with marker names and dummy p-values   */
/**********************************************************/
data dummy;
 set map;
 rename name=locus;
 prob=1;
run;

%tplot(ldtest, dummy, prob)
```

From Output 8.6, we can determine, at a glance, that all markers have genotypes in Hardy-Weinberg proportions (since all triangles on the diagonal are white) and also identify patterns of LD among the 30 markers, with black squares representing a pair of markers with highly significant LD (level of 0.01), and gray representing significance of LD at the 0.05 level.

Output 8.6 %TPLOT results for GAW data.

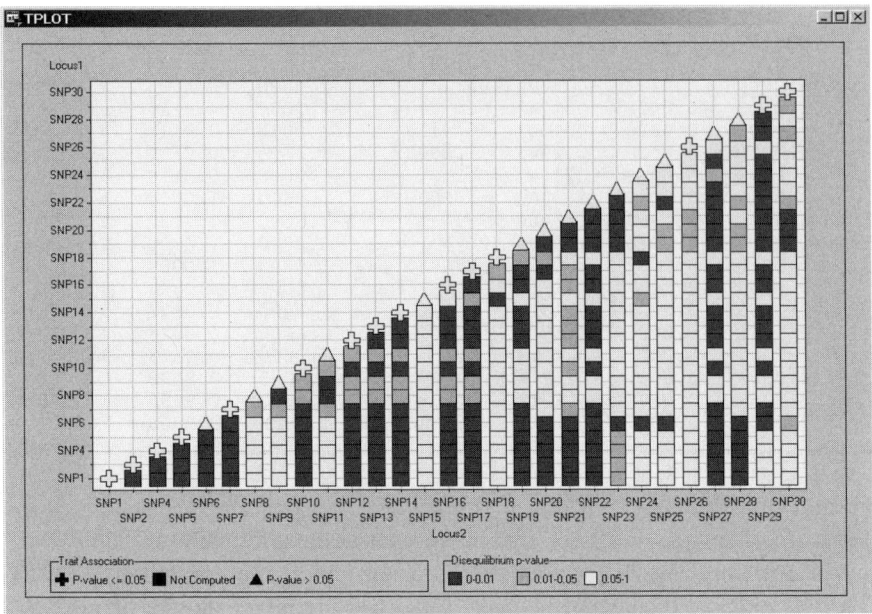

In order to examine more than two loci at a time, PROC HAPLOTYPE can be used to implement the EM algorithm for estimating haplotype frequencies assuming HWE. A likelihood ratio test can test the hypothesis of no LD among any of the marker loci (Zhao et al., 2000). This includes pair-wise LD as well as higher-order LD. A macro is shown here that estimates haplotype frequencies and tests for LD at all non-overlapping sets of five consecutive loci:

```
%macro fivelocushaps;
  %do first=1 %to 51 %by 10;
    %let last=%eval(&first+9);
    %let firstm=%eval((&first+1)/2);

    proc haplotype data=founders ndata=map(firstobs=&firstm) ld;
      var a&first-a&last;
    run;

  %end;
%mend;

%fivelocushaps
```

Results for the first set of five-locus haplotypes are shown in Output 8.7.

Output 8.7 PROC HAPLOTYPE results.

```
                        The HAPLOTYPE Procedure

                          Analysis Information
         Loci Used                         SNP1 SNP2 SNP3 SNP4 SNP5
         Number of Individuals                              165
         Number of Starts                                     1
         Convergence Criterion                          0.00001
         Iterations Checked for Conv.                         1
         Maximum Number of Iterations                       100
         Number of Iterations Used                            5
         Log Likelihood                              -136.62881
         Initialization Method                Linkage Equilibrium
         Standard Error Method                           Binomial
         Haplotype Frequency Cutoff                           0

       Algorithm converged.

                          Haplotype Frequencies
                                         Standard      95% Confidence
   Number    Haplotype    H0 Freq   H1 Freq    Error        Limits
      1      1-1-2-1-1    0.02006   0.00000   0.00000   0.00000   0.00000
      2      1-1-2-1-2    0.09815   0.00000   0.00000   0.00000   0.00000
      3      1-1-2-2-1    0.09608   0.00305   0.00304   0.00000   0.00900
      4      1-1-2-2-2    0.47009   0.82423   0.02098   0.78310   0.86535
      5      1-2-2-1-1    0.00419   0.00000   0.00000   0.00000   0.00000
      6      1-2-2-1-2    0.02049   0.00000   0.00000   0.00000   0.00000
      7      1-2-2-2-1    0.02006   0.00000   0.00000   0.00000   0.00000
      8      1-2-2-2-2    0.09815   0.00000   0.00000   0.00000   0.00000
      9      2-1-2-1-1    0.00419   0.00000   0.00000   0.00000   0.00000
     10      2-1-2-1-2    0.02049   0.00000   0.00000   0.00000   0.00000
     11      2-1-2-2-1    0.02006   0.00000   0.00000   0.00000   0.00000
     12      2-1-2-2-2    0.09815   0.00000   0.00000   0.00000   0.00000
     13      2-2-2-1-1    0.00087   0.16665   0.02055   0.12638   0.20692
     14      2-2-2-1-2    0.00428   0.00608   0.00428   0.00000   0.01448
     15      2-2-2-2-1    0.00419   0.00000   0.00000   0.00000   0.00000
     16      2-2-2-2-2    0.02049   0.00000   0.00000   0.00000   0.00000

                       Test for Allelic Associations
                                                       Chi-       Pr >
   Hypothesis                     DF       LogLike     Square     ChiSq
   H0: No Association              4    -487.33817
   H1: Allelic Associations       15    -136.62881    701.4187   <.0001
```

In the "Haplotype Frequencies" table shown in Output 8.7, there are two columns containing estimates of haplotype frequencies since the LD option was specified. The "H0 Freq" column displays estimates that assume no LD among the five markers, and thus haplotype frequencies are simply the product of the allele frequencies. The "H1 Freq" column contains the haplotype frequency estimates obtained from the EM algorithm that takes LD into account. The highly significant *p*-value from the likelihood ratio test displayed in the "Test for Allelic Associations" table indicates that there is significant LD, and thus it should be accounted for when estimating haplotype frequencies.

8.5 Marker-Trait Association Tests

Once properties about the markers have been studied and understood, the markers can be tested for association with the trait of interest in hopes that any significant association found is due to the physical proximity of the marker to the trait gene. The procedures in SAS/Genetics can handle binary traits only: PROC CASECONTROL performs contingency-table type tests on random samples of individuals, PROC FAMILY performs the TDT and other similar tests on nuclear families, and PROC HAPLOTYPE can test haplotypes for association with a binary trait using various chi-square tests. These procedures accommodate data sets structured like FOUNDERS and GAWSNPS with the trait variable AFFECTION (representing disease status). Testing for association of marker genotypes, alleles, or haplotypes with a quantitative trait requires either binning the trait into two classes and treating it as a binary variable or taking its continuous nature into account. We show in this section how SAS/STAT procedures can be applied to analyze quantitative traits in their original state and genotypic data from non-experimental populations.

8.5.1 Analyzing Samples of Unrelated Individuals

Continuing to use a random sample of individuals, we can perform an analysis of variance (ANOVA) to test for an effect of the marker genotypes on the continuous trait (Nielsen and Weir, 1999) in the FOUNDERS data set. Since these markers are biallelic, the genotypes can be represented uniquely as the number of "2" alleles the genotype comprises. When all three genotypes appear in the sample for a marker, orthogonal polynomial contrasts with coefficients (–1, 0, 1) and (1, –2, 1) can partition the marker effect into additive and quadratic (dominance) components, respectively.

```
/**********************************************************/
/* Create 1 variable per marker containing genotype        */
/* coded as the number of "2" alleles in the genotype      */
/**********************************************************/
data founders_g;
 set founders;
 array g{30};
 array a{60};
 do i=1 to 30;
   g{i}=a{2*i-1}+a{2*i}-2;
 end;
run;

/**********************************************************/
/* Perform an ANOVA on each individual marker              */
/**********************************************************/
%macro genanova;
  %do i=1 %to 30;
    title "SNP &i";
    proc glm data=founders_g;
      class g&i;
      model q1 = g&i;
    run;
  %end;
%mend;
%genanova
```

The ANOVA results for SNP10, the base pair with a true genetic effect on the quantitative trait Q1, are displayed in Output 8.8. Note that since only two distinct genotypes are present in the sample of founders for this marker, there is only 1 df for the genotype effect.

Output 8.8 ANOVA for GAW data.

```
                                  SNP 10

                            The GLM Procedure

Dependent Variable: q1
                                  Sum of
   Source                  DF    Squares      Mean Square    F Value    Pr > F
   Model                    1    443.112259    443.112259     45.04    <.0001
   Error                  163   1603.735211      9.838866
   Corrected Total        164   2046.847470

              R-Square    Coeff Var    Root MSE    q1 Mean
              0.216485    16.95306    3.136697    18.50224

   Source                  DF    Type I SS    Mean Square    F Value    Pr > F
   g10                      1    443.1122593   443.1122593    45.04    <.0001

   Source                  DF    Type III SS  Mean Square    F Value    Pr > F
   g10                      1    443.1122593   443.1122593    45.04    <.0001
```

An alternative testing approach not shown here is to use a regression model instead of an ANOVA. With the same coding as above, the marker genotypes can be treated as quantitative regressor variables in PROC REG. Thus for each marker, there is a 1 df test for an additive effect of the "2" allele. To test a dominant or recessive disease model in terms of allele "2," the heterozygous genotype can be coded as a 2 or 0, respectively.

In this next example, we use the output from PROC HAPLOTYPE to test the quantitative trait Q1 for association with haplotypes by applying the haplotype trend regression (HTR) method (Zaykin et al., 2002), again just using the FOUNDERS data set containing a sample of unrelated individuals. This method uses regressor columns, one for each possible haplotype, with values representing each individual's probability of having the haplotype. The OUT= option of PROC HAPLOTYPE creates an output data set containing, for each individual, a row for each possible haplotype pair given that individual's multi-locus genotype. This data set can be transformed, as shown in the code below, to be in the form for modeling the quantitative trait on the individuals' haplotype probabilities. Note that if the phase were known with certainty, these probabilities would all be either 0, ½, or 1, with probability of 1 occurring only for individuals homozygous at all loci.

```
proc haplotype data=gawsnps ndata=map out=outhap;
  var a11-a20;
run;

title 'Original PROC HAPLOTYPE OUT= Data Set';
proc print data=outhap;
  where id lt 20;
run;

data out1;
  set outhap;
  haplotype=tranwrd(haplotype1,'-','_');
run;

data out2;
  set outhap;
  haplotype=tranwrd(haplotype2,'-','_');
run;

data outnew;
  set out1 out2;
run;
```

```
proc sort data=outnew;
 by haplotype;
run;

data outnew2;
   set outnew;
   lagh=lag(haplotype);
   if haplotype ne lagh then num+1;
   hapname="H"||trim(left(num));
   call symput('numhaps', trim(left(num)));
run;

proc sort data=outnew2;
   by id haplotype;
run;

data outt;
   set outnew2;
   by id haplotype;
   if first.haplotype then totprob=prob/2;
   else totprob+prob/2;
   if last.haplotype;
run;

proc transpose data=outt out=outreg(drop=_NAME_) ;
   id hapname;
   idlabel haplotype;
   var totprob;
   by id;
run;

data outmissto0(drop=i);
   set outreg;
   array h{&numhaps};
   do i=1 to &numhaps;
     if h{i}=. then h{i}=0;
   end;
run;

data htr;
   merge outmissto0 gawsnps;
run;

title 'Haplotype Trend Regression';
proc reg data=htr;
   model q1 =  h2-h&numhaps ;
run;
```

The original OUT= data set from PROC HAPLOTYPE is shown in Output 8.9. After this data set is manipulated with the code above, the transformed data set HTR can be analyzed using PROC REG. Note that since there is a linear dependency among the haplotype probabilities, one of the haplotype columns (H1 is chosen here) must be omitted from the analysis.

Output 8.9 OUT= data set from PROC HAPLOTYPE and HTR results.

```
                      Original PROC HAPLOTYPE OUT= Data Set

ID  a11 a12 a13 a14 a15 a16 a17 a18 a19 a20  HAPLOTYPE1  HAPLOTYPE2    PROB
 1   1   2   1   1   2   2   2   2   1   1   1-1-2-2-1   2-1-2-2-1  1.00000
 2   1   2   1   1   2   2   2   2   1   1   1-1-2-2-1   2-1-2-2-1  1.00000
 3   1   2   1   1   2   2   2   2   1   1   1-1-2-2-1   2-1-2-2-1  1.00000
 4   1   1   1   1   2   2   2   2   1   1   1-1-2-2-1   1-1-2-2-1  1.00000
 5   1   1   1   1   2   2   2   2   1   1   1-1-2-2-1   1-1-2-2-1  1.00000
 6   1   1   1   1   2   2   2   2   1   1   1-1-2-2-1   1-1-2-2-1  1.00000
 7   1   1   1   1   2   2   2   2   1   1   1-1-2-2-1   1-1-2-2-1  1.00000
 8   1   2   1   1   2   2   2   2   1   1   1-1-2-2-1   2-1-2-2-1  1.00000
 9   1   1   1   2   2   2   2   2   1   2   1-1-2-2-1   1-2-2-2-2  1.00000
10   1   1   1   2   2   2   2   2   1   2   1-1-2-2-1   1-2-2-2-2  1.00000
11   1   1   1   2   2   2   2   2   1   2   1-1-2-2-1   1-2-2-2-2  1.00000
12   1   1   1   1   2   2   2   2   1   1   1-1-2-2-1   1-1-2-2-1  1.00000
13   2   2   1   1   2   2   2   2   1   1   2-1-2-2-1   2-1-2-2-1  1.00000
14   1   2   1   1   2   2   2   2   1   1   1-1-2-2-1   2-1-2-2-1  1.00000
15   1   2   1   1   2   2   2   2   1   1   1-1-2-2-1   2-1-2-2-1  1.00000
16   1   2   1   1   2   2   2   2   1   1   1-1-2-2-1   2-1-2-2-1  1.00000
17   1   1   1   2   2   2   2   2   1   2   1-1-2-2-1   1-2-2-2-2  1.00000
18   1   1   1   1   2   2   2   2   1   1   1-1-2-2-1   1-1-2-2-1  1.00000
19   1   1   1   2   2   2   2   2   1   2   1-1-2-2-1   1-2-2-2-2  1.00000
20   1   2   1   2   2   2   2   2   1   2   1-1-2-2-1   2-2-2-2-2  0.00022
20   1   2   1   2   2   2   2   2   1   2   1-2-2-2-2   2-1-2-2-1  0.99978


                         Haplotype Trend Regression

                             The REG Procedure
                             Model: MODEL1
                          Dependent Variable: q1

                            Analysis of Variance
                                 Sum of         Mean
     Source              DF     Squares        Square      F Value    Pr > F
     Model                5   2652.81188     530.56238       44.85    <.0001
     Error              994     11760         11.83059
     Corrected Total    999     14412

                Root MSE            3.43956    R-Square    0.1841
                Dependent Mean     17.86629    Adj R-Sq    0.1800
                Coeff Var          19.25169

                            Parameter Estimates
                                   Parameter     Standard
    Variable   Label        DF     Estimate       Error     t Value   Pr > |t|
    Intercept  Intercept     1     17.10430      0.18313      93.40    <.0001
    H2         1_2_1_1_2     1     10.99140      6.88887       1.60    0.1109
    H3         1_2_2_2_2     1      7.23292      0.60794      11.90    <.0001
    H4         2_1_2_2_1     1     -0.03462      0.36247      -0.10    0.9239
    H5         2_1_2_2_2     1     12.94602      6.88390       1.88    0.0603
    H6         2_2_2_2_2     1   -121.30965   4664.25856      -0.03    0.9793
```

8.5.2 Analyzing Family Data

The family information in the GAWSNPS data set can be used by applying family-based tests of association to the marker data. Note that nuclear families are treated as independent, though they may be related in this particular sample. We demonstrate here how to implement the TDT_{Q5} (Allison, 1997) and a slight modification of it, the $QTDT_M$ (Gauderman, 2003), again using the GLM procedure in SAS/STAT. These tests use samples of family trios: two parents and a single offspring, all genotyped at the particular marker. The mating types take on the values 1, ..., 6, representing the mating types of the parents: 1/1 × 1/1, 1/1 × 1/2, 1/1 × 2/2, 1/2 × 1/2, 1/2 × 2/2, and 2/2 × 2/2, respectively. For the quantitative trait Y, the ANOVA model being fit can be expressed as $Y_{ijk} = \mu + \alpha_i + \beta_j + e_{ijk}$, where the α_i represent the fixed effect of mating type i, the β_j represent the fixed genotype effect for genotype j, and $k=1,..., n_{ij}$. The test for both additive and quadratic effects of the marker on the trait is facilitated by the F-statistic for the βs after fitting the mating type parameters (as in the random sample case, the genotype codes could instead be treated as quantitative input variables in a regression with the heterozygous genotype coded appropriately for the model being tested). For the TDT_{Q5}, only offspring with a parental mating type of 2, 4, or 5 (at least one parent heterozygous) are included in the model, while the $QTDT_M$ includes all offspring. Extreme sampling can be applied to increase the power of these tests, where a subset of the sample of families to be analyzed is created based on the quantitative trait for the child being greater than a particular upper bound or less than a lower bound (Allison, 1997).

In this example, the quantitative trait being analyzed is Q1. This macro puts the data contained in GAWSNPS in the correct form to produce the *F*-statistic for testing the hypothesis of significant association and linkage between the trait locus and each individual marker locus. The %TPLOT macro can be implemented again, now including the *p*-values from the TDT_{Q5}, as a way of summarizing these results.

```
/***********************************************************/
/* Organize gawsnps into nuclear families with parents     */
/* listed first                                            */
/***********************************************************/
proc sort data=gawsnps;
 by pedigree household_id id;
run;

/***********************************************************/
/* Create data set with children only to determine         */
/* the cutoffs Zl and Zu for extreme sampling              */
/***********************************************************/
data offspring;
 set gawsnps;
 by pedigree household_id;
 if first.household_id then withinfam=0;
 withinfam+1;
 if withinfam <= 2 then delete;
 drop withinfam;
run;

/***********************************************************/
/* Determine 10% and 90% quantiles of Q1                   */
/***********************************************************/
ods output quantiles=q;
proc univariate data=offspring;
 var q1;
run;

data q2;
 set q;
 if quantile="10%" then call symput('zl',estimate);
 else if quantile="90%" then call symput('zu',estimate);
run;

/***********************************************************/
/* Create data set containing parents and children that    */
/* meet criteria for extreme sampling (Zl< or >Zu).        */
/***********************************************************/
data extreme;
 set gawsnps;
 by pedigree household_id;
 if first.household_id then withinfam=0;
 withinfam+1;
 if withinfam > 2 and q1 <= &zu and q1 >= &zl then delete;
 drop withinfam;
run;

%macro qtdt;
```

```sas
 /** Define fstats_all data set **/
 data fstats_all;
  length locus $5;
  delete;
 run;

 /** Loop over each marker **/
 %do one=1 %to 59 %by 2;
  %let two=%eval(&one+1);
  %let marker=%eval(&two/2);

  /***********************************************/
  /* Create variables containing children's marker */
  /* genotype and the parental mating type.        */
  /* Keep 1 child per family with valid mating type */
  /***********************************************/
  data extremeglm;
   set extreme;
   by pedigree household_id;
   if first.household_id then withinfam=0;
   withinfam+1;
   la1=lag(a&one);
   la2=lag(a&two);
   l2a1=lag2(a&one);
   l2a2=lag2(a&two);
   if la1+la2 > l2a1+l2a2 then
    pargeno=compress((la1+la2)||(l2a1+l2a2));
   else pargeno=compress((l2a1+l2a2)||(la1+la2));
   chigeno=a&one+a&two;
   if withinfam=3;
   * Use following statement for TDT_Q5, not QTDT_M;
   if pargeno in ('32', '33', '43');
  run;

  title "SNP &marker";

  ods output ModelANOVA=fstats;
  ods listing close;
  proc glm data=extremeglm;
   class pargeno chigeno;
   model q1=pargeno chigeno/ ss1;
  run;

  /***********************************************/
  /* Create data set containing F-statistics for  */
  /* the TDT_Q5.  This data set will be used in   */
  /* %TPLOT macro so Locus variable must be defined. */
  /***********************************************/
  data fstats_all;
   set fstats_all fstats(in=new);
   if Source='chigeno';
   if new then Locus="SNP&marker";
  run;

  data fstats;
   set fstats;
   probf=1;
  run;

 %end;
 ods listing;
%mend;

%qtdt

%tplot(ldtest, fstats_all, ProbF)
```

Output 8.10 %TPLOT results using TDTQ5 *p*-values.

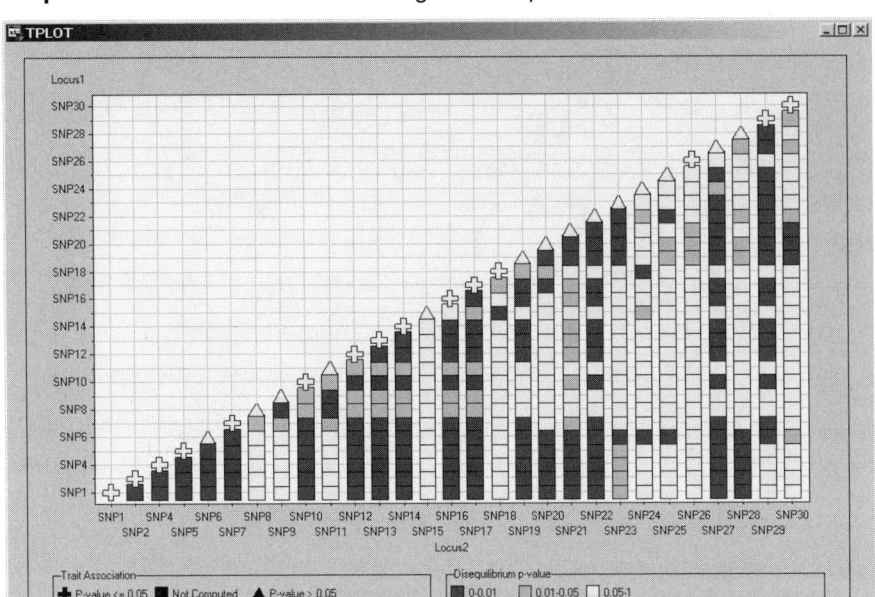

Output 8.10 is identical to Output 8.6, but now there are two different symbols shown on the diagonal of the plot. The symbols correspond to the *p*-value from the TDT_{Q5}, with the plus sign representing a significant *p*-value at level 0.05 and the triangle representing a *p*-value greater than 0.05. Thus, the results (in terms of the *p*-values for the *F*-statistics) of the 30 PROC GLMs that were run are all shown on this single plot. A model adjusting the effect of the offspring's genotype on phenotype for other covariates, such as age or gender, can easily be accommodated by adding these factors to the MODEL statement of PROC GLM (and CLASS statement if a variable is qualitative), with the covariates listed before the offspring genotype variable.

If parental genotype information is unavailable, a sibling test of linkage and association can be performed on Q1 (Allison et al., 1999). For this analysis, all the offspring can be included from sibships with at least two siblings. The mixed effects model for the quantitative trait *Y* is

$$Y_{ijk} = \mu + \alpha_i + \beta_j + (\alpha\beta)_{ij} + e_{ijk},$$

where the α_i now represent the random sibling effect for the *i*th sibship, the β_j represent the fixed genotype effect for genotype *j*, and $k=1,...,n_{ij}$. The effect of the siblings' genotypes on the quantitative trait is adjusted for the random effects of the sibship (represented by the HOUSEHOLD_ID variable in the analysis below) and sibship-by-genotype interaction.

```
%macro sibqtdt;
/************************************************************/
/* Remove sibships of size 1 from data set and create       */
/* columns of genotypes                                      */
/************************************************************/
   data sibdata;
    set offspring;
    array a{60};
    array geno{30};
    by pedigree household_id;
    if first.household_id and last.household_id then delete;
    do i=1 to 30;
     geno[i]=a[2*i-1]+a[2*i]-2;
    end;
```

```
      drop i;
    run;

  /** Loop over each marker **/
  %do m=1 %to 30;

    title "SNP &m";

    proc mixed data=sibdata;
      class geno&m household_id;
      model q1 = geno&m;
      random household_id geno&m*household_id;
    run;
  %end;
%mend;

%sibqtdt
```

Part of the results for SNP10 are shown in Output 8.11. From the "Type 3 Test of Fixed Effects" table we can conclude that SNP10 is significantly linked and associated with the trait Q1, at any significance level above 0.0001.

Output 8.11 PROC MIXED output for SNP10—sibling test of linkage and association.

```
                              SNP 10

                        The Mixed Procedure
                         Iteration History
       Iteration    Evaluations    -2 Res Log Like      Criterion
               0              1       1574.99788099
               1              2       1550.84188915     0.00000041
               2              1       1550.84168422     0.00000000
                        Convergence criteria met.

                     Covariance Parameter Estimates
                     Cov Parm              Estimate
                     household_id            2.0619
                     geno10*household_id     1.1183
                     Residual                6.3817

                            Fit Statistics
                  -2 Res Log Likelihood          1550.8
                  AIC  (smaller is better)       1556.8
                  AICC (smaller is better)       1556.9
                  BIC  (smaller is better)       1565.2

                      Type 3 Tests of Fixed Effects
                          Num      Den
                Effect     DF       DF     F Value    Pr > F
                geno10      2       28       33.26    <.0001
```

8.6 References

Allison, D. B. 1997. Transmission-disequilibrium tests for quantitative traits. *American Journal of Human Genetics* 60:676-690.

Allison, D. B., M. Heo, N. Kaplan, and E. R. Martin. 1999. Sibling-based tests of linkage and association for quantitative traits. *American Journal of Human Genetics* 64:1754-1764.

Botstein, D., R. L. White, M. Skolnick, and R. W. Davis. 1980. Construction of a genetic linkage map in man using restriction fragment length polymorphisms. *American Journal of Human Genetics* 32:314-331.

Budowle, B., and T. R. Moretti. 1999. Genotype profiles for six population groups at the 13 CODIS short tandem repeat core loci and other PCR-based loci. *Forensic Science Communications* [http://www.fbi.gov/hq/lab/fsc/backissu/july1999/budowle.htm].

Devlin, B., and N. Risch. 1995. A comparison of linkage disequilibrium measures for fine-scale mapping. *Genomics* 29:311-322.

Excoffier, L., and M. Slatkin. 1995. Maximum-likelihood estimation of molecular haplotype frequencies in a diploid population. *Molecular Biology and Evolution* 12:921-927.

Gauderman, W. J. 2003. Candidate gene association analysis for a quantitative trait, using parent–offspring trios. *Genetic Epidemiology* 25:327-338.

Guo, S. W., and E. A. Thompson. 1992. Performing the exact test of Hardy-Weinberg proportion for multiple alleles. *Biometrics* 48:361-372.

Hawley, M. E., and K. K. Kidd. 1995. HAPLO: A program using the EM algorithm to estimate the frequencies of multi-site haplotypes. *Journal of Heredity* 86:409-411.

Long, J. C., R. C. Williams, and M. Urbanek. 1995. An E-M algorithm and testing strategy for multiple-locus haplotypes. *American Journal of Human Genetics* 56:799-810.

Nielsen, D. M., M. G. Ehm, and B. S. Weir. 1999. Detecting marker-disease association by testing for Hardy-Weinberg disequilibrium at a marker locus. *American Journal of Human Genetics* 63:1531-1540.

Nielsen, D. M., and B. S. Weir. 1999. A classical setting for associations between markers and loci affecting quantitative traits. *Genetic Research* 74:271-277.

Risch, N., and K. Merikangas. 1996. The future of genetic studies of complex human diseases. *Science* 273:1516-1517.

Weir, B. S. 1979. Inferences about linkage disequilibrium. *Biometrics* 35:235-254.

Weir, B. S. 1996. *Genetic Data Analysis II*. Sunderland, MA: Sinauer Associates.

Wijsman, E. M., L. Almasy, C. I. Amos, I. Borecki, C. T. Falk, T. M. King, M. M. Martinez, D. Meyers, R. Neuman, J. M. Olson, S. Rich, M. A. Spence, D. C. Thomas, V. J. Vieland, J. S. Witte, and J. W. MacCluer. 2001. Analysis of complex genetic traits: Applications to asthma and simulated data. *Genetic Epidemiology* 21:S1-S853.

Zaykin, D. V., P. H. Westfall, S. S. Young, M. A. Karnoub, M. J. Wagner, and M. G. Ehm. 2002. Testing association of statistically inferred haplotypes with discrete and continuous traits in samples of unrelated individuals. *Human Heredity* 53:79-91.

Zhao, J. H., D. Curtis, and P. C. Sham. 2000. Model-free analysis and permutation tests for allelic associations. *Human Heredity* 50:133-139.

Chapter 9

The Extended Sib-Pair Method for Mapping QTL

Chenwu Xu and Shizhong Xu

9.1 Introduction 202
9.2 Statistical Model 203
9.3 Inferring the Proportion of Genes IBD Shared by Sibs at QTL 204
    9.3.1 Proportion of Alleles IBD at Marker Loci—Parental Genotypes Known 204
    9.3.2 Proportion of Alleles IBD at Marker Loci—Parental Genotypes Unknown 206
    9.3.3 Multipoint Estimation of Proportion of Genes IBD at QTL 207
9.4 Maximum Likelihood Estimation 208
    9.4.1 Likelihood Ratio Test 209
    9.4.2 Approximate Critical Value for Significance Test 209
9.5 Single Marker Analysis and Interval Mapping of QTL 209
9.6 Program Implementation 210
    9.6.1 Choosing the Scanning Increment 210
    9.6.2 Implementing the Multipoint Method 212
    9.6.3 Implementing the Mixed Model Methodology through PROC IML 214
    9.6.4 Implementing the Mixed Model Methodology through PROC MIXED 216
9.7 Executing the Program 216
    9.7.1 Data Preparation 217
    9.7.2 Running the Main Program 218
    9.7.3 Application Notes 219
    9.7.4 Example Demo 220
9.8 References 222

9.1 Introduction

This chapter describes a unified SAS program for single marker analysis and interval mapping of quantitative trait loci (QTL) using the extended sib-pair method under the random model methodology. The conventional sib-pair method for QTL mapping is to regress the squared phenotypic difference of the sib pair on the proportion of genes identical by descent (IBD) shared by the sibs. The extended sib-pair method is a maximum likelihood approach that uses the phenotypic values of individual sibs as the original data. The phenotypic values of different sibs within families are assumed to follow a multivariate normal distribution. The extended sib-pair method can handle multiple and variable numbers of sibs per family. The random model approach is based on the differentiated proportion of genes IBD shared by sibs at loci across the genome. The genetic covariances between sibs are modeled as a function of the proportion of genes IBD shared at the QTL. The proportion of genes IBD shared by sibs at a putative QTL position is inferred from the proportion of alleles IBD of all markers linked to the putative QTL. The variance components are estimated in the framework of mixed model methodology. Either PROC IML or PROC MIXED (SAS Institute, 1999a,b) can be used to implement the mixed model methodology for QTL mapping.

The SAS macro language is used to implement the mixed model QTL mapping procedure. The program consists of four macros. The first macro, %QTLIBD, calculates the proportion of genes IBD shared by sibs at putative QTL positions along each chromosome. This macro handles missing and partially informative markers using the multipoint method. The second macro, %QTLSCAN (including %QTLSCAN1, which is based on an optimization subroutine of PROC IML, and %QTLSCAN2, which is based on PROC MIXED), provides estimates of QTL variance components and the likelihood ratio test statistic for any putative genome positions. The third macro, %QTLCRITIC, determines the critical value used to declare statistical significance using the approximate method of Piepho (2001). The last macro, %QTLPLOT, draws SAS graphs showing the test statistic profiles of the genome.

The rapid development of molecular technology has advanced the development of dense linkage maps and further stimulated the search for genes affecting quantitative traits—that is, QTL. The study of the genetic architecture of quantitative traits using molecular markers, which includes the number and chromosomal locations of QTL, the mode of gene action, and the sizes of individual QTL, is called *QTL mapping*. There are two primary types of data used for mapping QTL: data derived from experimental line crossing experiments and data collected from undesigned field experiments or natural populations. The search for QTL has been most successful in experimental line crosses. Many statistical methods have been developed for QTL mapping using such data (Lander and Botstein, 1989; Haley and Knott, 1992; Jansen, 1993; Zeng, 1994; Xu, 1998a,b; Kao et al., 1999). These methods are classified as the fixed model approach. Numerous software packages are available for the fixed model approach to QTL mapping, such as Mapmaker/QTL, QTL Cartographer, MapQTL, MQTL, Map Manager QTX, and so on. However, the fixed model approach may not be appropriate for QTL mapping in unmanipulated outbred populations, such as in human populations. In an outbred population, there may be many different families, and the linkage phases differ from one family to another. Therefore, the search for QTL has to be made within families. However, family sizes are usually not sufficiently large to allow accurate estimation and test of QTL. In addition, both the number of loci and the number of alleles within each locus are unknown. As a consequence, the search for QTL and the estimation of QTL parameters must be based on a robust method that does not require such information. The *random model approach*, also called variance component analysis, becomes a natural choice for such a robust method.

The random model approach to QTL mapping uses the information of phenotypic covariance between genetically related individuals. The covariance is a function of the proportion of genes IBD that the two relatives share at the QTL of interest and on the overall genome. In the sib-pair analysis, the average proportion of genes IBD shared by two individuals equals 1/2, the coefficient of genetic relationship. This average IBD value is used to estimate the polygenic variance component. However, the proportion of genes IBD shared by sibs varies from one locus to another. For the same locus, the IBD proportion also varies from one sib pair to another sib pair. If the parents are not genetically related, the expectation and the variance of the locus-specific IBD are 1/2 and 1/4, respectively. The variation in the locus-dependent and sib-pair-dependent IBD coefficient is the basis for the random model methodology of QTL mapping. Because the QTL genotype is unobservable, the proportion of genes IBD shared by sibs at the QTL is also unobservable and must be inferred from the available information on linked marker loci. The estimated proportion of IBD at the QTL can take a value ranging from 0 (sharing neither gene IBD) to 1 (sharing both genes IBD). The greater the proportion of genes IBD, the more similar are the phenotypes of two sibs. In the linear model, if one also includes non-genetic effects, such as age, sex, and so on, and these effects are treated as fixed effects in the analysis, the model is called the mixed model (Henderson, 1975).

Haseman and Elston (1972) initiated the first robust QTL detection method, called sib-pair regression analysis. They simply took the regression of the squared phenotypic difference between sib pairs on the IBD values of markers. If a marker is associated with QTL, the expected regression coefficient is proportional to the (negative of) additive genetic variance explained by the linked QTL. The method was a kind of association study because the linkage relationship among markers has not been used and may be called single marker analysis. However, even if a marker is detected as significant by single marker analysis, there is no way to tell whether the significance is due to tight linkage to a small QTL or loose linkage to a large QTL. Fulker and Cardon (1994) extended the Haseman-Elston sib-pair method to interval mapping in the sense that an intermediate position bracketed by two markers can also be tested. Although statistical power has been improved with the interval mapping, the modified sib-pair method is still a least squares–based method. Therefore, information loss is still possible because the probability distribution of the phenotype has not been fully utilized. Goldgar (1990) developed a multiple-point IBD method to estimate the total amount of genetic material shared by sibs in a given chromosomal region and eventually used a maximum likelihood (ML) approach to estimate the genetic variance explained by that particular region. The ML method is a general method that can be used for any number of sibs. Although Goldgar (1990) used two flanking markers to define a chromosomal segment, the method is not interval mapping. Xu and Atchley (1995) developed an ML-based interval mapping procedure under the random model framework. Because the method uses the phenotypic values of individual sibs as the original data points (not the squared differences) and can handle multiple and variable numbers of sibs per family, we call it the extended sib-pair method. The extended sib-pair method of QTL mapping has proven to be useful in human and other outbred populations.

The purposes of this chapter are to (1) introduce the mixed model methodology of genetic mapping, (2) describe the extended sib-pair method for QTL mapping, (3) provide detailed steps of calculating the IBD matrices, and (4) introduce a SAS program for implementing the method.

9.2 Statistical Model

Let y_{ij} be the phenotypic value of the *j*th sib in the *i*th family. It can be described by the following model:

$$y_{ij} = \mu + a_{ij} + d_{ij} + g_{ij} + e_{ij},$$

where μ is the population mean, a_{ij} and d_{ij} are the additive and dominance effects of the QTL under investigation, g_{ij} is the polygenic effect, and e_{ij} is the residual error. The population mean is the fixed effect, and all remaining effects in the model are random effects with normal distributions—that is, $a_{ij} \sim N(0, \sigma_a^2)$, $d_{ij} \sim N(0, \sigma_d^2)$, $g_{ij} \sim N(0, \sigma_g^2)$, and $e_{ij} \sim N(0, \sigma_e^2)$. This model is now referred to as the random model. If we replace the population mean with $\mathbf{X}_{ij}\mathbf{b}$ to incorporate other fixed effects, the model is then called the mixed effect model, or simply the mixed model. So, the random model is simply a special case of the mixed model.

Let $\mathbf{y}_i = [y_{i1}, y_{i2}, \cdots, y_{in_i}]$ be an $n_i \times 1$ vector of the phenotypic values for n_i sibs in the ith family. The expectation and variance-covariance matrix of $\mathbf{y}_i$ are given by

$$\begin{cases} E(\mathbf{y}_i) = \mathbf{1}\mu \\ V(\mathbf{y}_i) = \mathbf{V}_i = \mathbf{\Pi}\sigma_a^2 + \mathbf{\Delta}\sigma_d^2 + \mathbf{A}\sigma_g^2 + \mathbf{I}\sigma_e^2, \end{cases} \quad (9.1)$$

where $\mathbf{1}$ is an $n_i \times 1$ vector with all entries equal to 1; $\mathbf{\Pi} = \{\pi_{uv}\}_{n_i \times n_i}$ is an $n_i \times n_i$ matrix with the element of the uth row and the vth column being the proportion of genes IBD shared by sibs u and v at the putative QTL; $\mathbf{\Delta} = \{\Delta_{uv}\}_{n_i \times n_i}$ is a dominance relationship matrix such that $\Delta_{uv} = \Pr(\pi_{uv} = 1) \; \forall \; u \neq v$ and $\Delta_{uu} = 1$; $\mathbf{A}$ is the additive relationship matrix with diagonal elements equal to unity and the other entries equal to 1/2; and $\mathbf{I}$ is the identity matrix.

9.3 Inferring the Proportion of Genes IBD Shared by Sibs at QTL

The key step to map QTL by the mixed model methodology is to infer the proportion of genes IBD, π_{uv}, and to calculate $\Delta_{uv} = \Pr(\pi_{uv} = 1) \; \forall \; u \neq v$ using molecular markers. This can be implemented by using the multipoint method (Gessler and Xu, 1999). Before inferring the proportion of genes IBD shared at QTL, we first need to calculate the proportion of alleles IBD at marker loci.

9.3.1 Proportion of Alleles IBD at Marker Loci—Parental Genotypes Known

There are four possible states of allelic sharing between a sib pair: S_{11} indicates the state of sibs sharing both alleles IBD, S_{10} indicates the paternal allele IBD, S_{01} indicates the maternal allele IBD, and S_{00} indicates no allele IBD. The proportion of alleles IBD at the locus of interest is

$$\pi_{uv} = \Pr(S_{11} \mid I_k) + \frac{1}{2}[\Pr(S_{10} \mid I_k) + \Pr(S_{01} \mid I_k)], \quad (9.2)$$

where $\Pr(S_{ij} \mid I_k)$ denotes the probability of state S_{ij} given the observed genotypes at the kth marker.

When both sib-pair and parental genotypes are known, we can use a table derived by Haseman and Elston (1972) to calculate π. They considered seven parental mating types and seven sib-pair types. Although the π can be calculated as easily as Haseman and Elston stressed, the SAS code is very long because there are so many different conditions to evaluate. The following paragraph describes a simple method to perform such a calculation of π. Based on this method, we can write a very short program to implement the calculation.

Define $A_1 A_2$ as the ordered genotype of the father and $A_3 A_4$ as the ordered genotype of the mother. By ordered genotype we mean that the first (left) allele of the genotype always comes from the paternal parent and the second (right) allele of the genotype always comes from the maternal parent. According to Mendel's law of inheritance, there are four possible progenies, each with an equal probability. The four possible progenies are $A_1 A_3, A_1 A_4, A_2 A_3,$ and $A_2 A_4$. Therefore, we have 16 possible sib-pair combinations. Each sib pair has a unique number of shared alleles. For example, the sib pair $A_1 A_3 - A_1 A_4$ share one common allele. Therefore, the IBD value for this sib pair is 1. The sib pair $A_1A_3 - A_1A_3$ share two common alleles, and thus the IBD value for this sib pair is 2. The sib pair $A_1A_3 - A_2A_4$ share no common alleles at all, and thus the IBD value for this sib pair is 0. The 16 sib pairs and their shared IBD values are listed in Table 9.1.

Table 9.1 The 16 possible sib pairs and their shared IBD values.

		Sib two			
		A_1A_3	A_1A_4	A_2A_3	A_2A_4
	A_1A_3	2	1	1	0
Sib one	A_1A_4	1	2	0	1
	A_2A_3	1	0	2	1
	A_2A_4	0	1	1	2

If we do not observe the genotypes of the locus, each sib pair has a probability of 1/16 to take any one of the 16 possible pairs. Therefore, the expected IBD value shared by a random sib pair of a full-sib family is

$$\frac{1}{16}(2+2+2+2+1+1+1+1+1+1+1+1) = 1,$$

leading to an expected or average proportion of allele IBD of 1/2. The denominator, 2, represents a diploid—i.e., each individual carries two alleles at a single locus.

We now further describe the algorithm for calculating the proportion of alleles IBD given the marker genotypes of both parents and the sibs. Remember that the order of the four possible progenies in any full-sib family is A_1A_3, A_1A_4, A_2A_3, and A_2A_4. Let us now consider a sib pair generated from the mating type $ab \times cd$. The four possible progenies are $ac, ad, bc,$ and bd, where ac is equivalent to A_1A_3, ad is equivalent to A_1A_4, and so on. This is the situation where we can fully observe the allelic inheritance of the progeny. Consider another mating type, $ab \times ab$; the four possible progenies are $aa, ab, ba,$ and bb. In this case, aa is equivalent to A_1A_3 and bb is equivalent to A_2A_4. However, ab and ba are indistinguishable. If we observe ab, we still do not know whether it is A_1A_4 or A_2A_3, and thus we assign an equal probability to each possible genotype—i.e., $1/2$ for A_1A_4 and $1/2$ for A_2A_3. We now assign the individual ab to the four possible ordered genotypes A_1A_3, A_1A_4, A_2A_3, and A_2A_4 by four probabilities: 0, 1/2, 1/2, and 0. Similarly, we assign the individual aa to the four ordered genotypes by probabilities 1, 0, 0, and 0. Consider another mating type, $ab \times cc$, where the four possible progenies are $ac, ac, bc,$ and bc. Compared with the ordered genotypes A_1A_3, A_1A_4, A_2A_3, and A_2A_4, we can see that if we observe an ac progeny, we do not know whether it is A_1A_3 or A_1A_4, and thus we assign an equal probability to each one of the two genotypes, $1/2$ for A_1A_3 and $1/2$ for A_1A_4. Therefore, the four probabilities for ac are $1/2, 1/2, 0$, and 0. Similarly, if we observe a bc progeny, we assign $1/2$ to A_2A_3 and $1/2$ to A_2A_4, and thus the four probabilities for bc are $0, 0, 1/2$, and $1/2$. In general, let q_1, q_2, q_3, and q_4 be the four possible probabilities for sib one, where $\sum_{i=1}^{4} q_i = 1$, and let g_1, g_2, g_3 and g_4 be the four possible probabilities for sib two, where $\sum_{i=1}^{4} g_i = 1$. After finding the four probabilities for both sibs, we can construct the following 4×4 joint probability table (Table 9.2).

Table 9.2 The 4×4 joint probability table of sib pairs given observed genotypes.

		Sib two			
		g_1	g_2	g_3	g_4
	q_1	q_1g_1	q_1g_2	q_1g_3	q_1g_4
Sib one	q_2	q_2g_1	q_2g_2	q_2g_3	q_2g_4
	q_3	q_3g_1	q_3g_2	q_3g_3	q_3g_4
	q_4	q_4g_1	q_4g_2	q_4g_3	q_4g_4

We now combine Table 9.1 and Table 9.2 to calculate the probability of state S_{ij} given the observed marker genotype:

$$\begin{aligned}
\Pr(S_{11} \mid I_k) &= q_1g_1 + q_2g_2 + q_3g_3 + q_4g_4 \\
\Pr(S_{10} \mid I_k) &= q_1g_2 + q_2g_1 + q_3g_4 + q_4g_3 \\
\Pr(S_{01} \mid I_k) &= q_1g_3 + q_2g_4 + q_3g_1 + q_4g_2 \\
\Pr(S_{00} \mid I_k) &= q_1g_4 + q_2g_3 + q_3g_2 + q_4g_1.
\end{aligned} \quad (9.3)$$

Using Equation 9.2, we can derive the expected proportion of alleles IBD shared by sib pairs at marker locus given the observed marker genotypes:

$$\begin{aligned}
\pi_{uv} &= q_1g_1 + q_2g_2 + q_3g_3 + q_4g_4 + \frac{1}{2}(q_1g_2 + q_2g_1 + q_3g_4 + q_4g_3 + q_1g_3 + q_2g_4 + q_3g_1 + q_4g_2) \\
&= \sum_{i=1}^{4} q_i g_i + \frac{1}{2}(q_1 + q_4)(g_2 + g_3) + \frac{1}{2}(q_2 + q_3)(g_1 + g_4).
\end{aligned}$$

9.3.2 Proportion of Alleles IBD at Marker Loci—Parental Genotypes Unknown

When parental genotypes are unknown, Haseman and Elston (1972) and Elston (1990) suggested the following general estimator of π for any pair of relatives:

$$\pi = \frac{f_{11}\Pr(I_k \mid S_{11}) + f_{10}\Pr(I_k \mid S_{10}) + f_{01}\Pr(I_k \mid S_{01})}{f_{11}\Pr(I_k \mid S_{11}) + f_{10}\Pr(I_k \mid S_{10}) + f_{01}\Pr(I_k \mid S_{01}) + f_{00}\Pr(I_k \mid S_{00})},$$

where $f_{11}, f_{10}, f_{01},$ and f_{00} are the prior probabilities that the pair of relatives share both genes IBD, paternal gene IBD, maternal gene IBD, and no alleles IBD, respectively. For full sibs, $f_{11}, f_{10}, f_{01},$ and f_{00} are all 1/4. $\Pr(I_k \mid S_{ij})$ is the probability of the observed marker genotypes given that the sib pairs are at state S_{ij} of IBD sharing. The probabilities of various pairs of marker genotypes of sibs are given in

Table 9.3. From Bayes's theorem, we can find the posterior probability of state S_{ij} given the observed marker genotypes for full sibs

$$\Pr(S_{ij} \mid I_k) = \frac{\Pr(I_k \mid S_{ij})}{\sum \Pr(l_k \mid S_{ij})},$$

where the summation is for all states of S_{ij}.

Table 9.3 Probabilities of various pairs of marker genotypes of sibs conditional on the number of genes IBD. (Source: Elston, 1990.)

Type	Marker pairs	$\Pr(I_k \mid S_{11})$	$\Pr(I_k \mid S_{10})$	$\Pr(I_k \mid S_{01})$	$\Pr(I_k \mid S_{00})$
I	$M_j M_j, M_j M_j$	f_j^2	$f_j^3/2$	$f_j^3/2$	f_j^4
II	$M_j M_j, M_k M_k$	0	0	0	$2f_j^2 f_k^2$
III	$M_j M_j, M_j M_k$	0	$f_j^2 f_k$	$f_j^2 f_k$	$4f_j^3 f_k$
IV	$M_j M_j, M_k M_l$	0	0	0	$4f_j^2 f_k f_l$
V	$M_j M_k, M_j M_k$	$2f_j f_k$	$f_j f_k(f_j + f_k)/2$	$f_j f_k(f_j + f_k)/2$	$4f_j^2 f_k^2$
VI	$M_j M_k, M_j M_l$	0	$f_j f_k f_l$	$f_j f_k f_l$	$8f_j^2 f_k f_l$
VII	$M_j M_k, M_l M_m$	0	0	0	$8f_j f_k f_l f_m$

Note that j, k, l and m index different alleles and f_j, f_k, f_l, and f_m are the frequencies of the corresponding alleles.

9.3.3 Multipoint Estimation of Proportion of Genes IBD at QTL

Similar to Equation 9.2, we now estimate the proportion of genes IBD shared by sibs u and v at putative QTL by

$$\pi_{uv} = \Pr(S_{11} \mid I_M) + \frac{1}{2}[\Pr(S_{10} \mid I_M) + \Pr(S_{01} \mid I_M)],$$

where I_M denotes genotypes of all M markers on the same chromosome. From Bayes's theorem,

$$\Pr(S_{ij} \mid I_M) = \frac{\Pr(S_{ij}) \Pr(I_M \mid S_{ij})}{\sum \Pr(S_{ij}) \Pr(I_M \mid S_{ij})} = \frac{\Pr(I_M \mid S_{ij})}{\sum \Pr(I_M \mid S_{ij})}$$

for a constant $\Pr(S_{ij})$. Note that here $\Pr(S_{11} \mid I_M)$ is equivalent to $\Delta_{uv} = \Pr(\pi_{uv} = 1) \; \forall \; u \neq v$. In matrix notation, we have

$$\Pr(I_M \mid S_{ij}) = \mathbf{1}^T \mathbf{D}_1 \mathbf{T}_{12} \mathbf{D}_2 \cdots \mathbf{T}_{kq} \mathbf{D}_{(..)} \mathbf{T}_{q(k+1)} \cdots \mathbf{D}_{M-1} \mathbf{T}_{(M-1)M} \mathbf{D}_M \mathbf{1},$$

where $\mathbf{1} = [1, 1, 1, 1]$ and $\mathbf{D}_k$ is a 4×4 matrix for the kth marker with the form

$$\mathbf{D}_k = \begin{bmatrix} \Pr(S_{11} | I_k) & 0 & 0 & 0 \\ 0 & \Pr(S_{10} | I_k) & 0 & 0 \\ 0 & 0 & \Pr(S_{01} | I_k) & 0 \\ 0 & 0 & 0 & \Pr(S_{00} | I_k) \end{bmatrix}.$$

$\Pr(S_{ij} | I_k)$ is the probability of state S_{ij} at the kth marker as described above. $\mathbf{D}_{(..)}$ is analogous to matrix $\mathbf{D}_k$ except that it is for the putative QTL. Because the QTL genotype is unobservable, we must consider four possible $\mathbf{D}_{(..)}$, one for each genotype. For example, $\mathbf{D}_{(11)}$ is a matrix with all elements equal to zero except that the first row and first column element takes unity. $\mathbf{T}_{..}$ is a transition matrix linking neighboring markers or a marker and the putative QTL at the position under investigation, defined as

$$\mathbf{T}_{(M-1)M} = \begin{bmatrix} c^2 & c(1-c) & (1-c)c & (1-c)^2 \\ c(1-c) & c^2 & (1-c)^2 & (1-c)c \\ (1-c)c & (1-c)^2 & c^2 & c(1-c) \\ (1-c)^2 & (1-c)c & c(1-c) & c^2 \end{bmatrix},$$

where $c = r^2 + (1-r)^2$ and r is the recombination fraction between positions $M-1$ and M.

9.4 Maximum Likelihood Estimation

Under the assumption of the normal distribution of the data $\mathbf{y}_i$, we have the following joint density function of observing a particular vector of data:

$$f(\mathbf{y}_i) = \frac{1}{(2\pi)^{n_i/2} |\mathbf{V}_i|^{1/2}} \exp\{-\frac{1}{2}(\mathbf{y}_i - \mathbf{1}\mu)^T \mathbf{V}_i^{-1}(\mathbf{y}_i - \mathbf{1}\mu)\}.$$

The overall log-likelihood for n independent families is

$$L = \sum_{i=1}^{n} \log[f(\mathbf{y}_i)]. \tag{9.4}$$

This likelihood function relates to the position of the QTL through π. The unknown parameters are $\boldsymbol{\theta} = \{\mu, \sigma_a^2, \sigma_d^2, \sigma_g^2, \sigma_e^2\}^T$. The maximum likelihood estimates (MLE) of Equation 9.4 can be implemented via the Newton-Raphson or the simplex algorithm (Nelder and Mead, 1965).

9.4.1 Likelihood Ratio Test

Define the log-likelihood value evaluated at the MLE of parameters $\boldsymbol{\theta}$ as

$$L(\hat{\boldsymbol{\theta}}) = \sum_{i=1}^{n} \log[f(\mathbf{y}_i \mid \hat{\boldsymbol{\theta}})].$$

The null hypothesis is $H_0 : \sigma_a^2 = \sigma_d^2 = 0$; i.e., there is no QTL segregating at the tested position. The maximum likelihood under the null hypothesis is denoted by L_0. The likelihood ratio test statistic,

$$\mathrm{LR} = -2(L_0 - L),$$

is used to test the significance of QTL at the putative position.

9.4.2 Approximate Critical Value for Significance Test

We adopt a quick method proposed by Piepho (2001) to compute an approximate critical value for QTL detection. The method can control the genome-wise Type I error rate and use the variation and correlation of the likelihood ratio (LR) test statistics across the genome locations. The upper bound of the genome-wise Type I error rate is estimated by

$$\gamma = m \Pr(\chi_p^2 > C) + (\sum_{i=1}^{m} v_l) C^{(1/2)(p-1)} e^{(-C/2)} 2^{(-p/2)} / \Gamma(p/2), \tag{9.5}$$

where γ is the assigned genome-wise Type I error rate; C is the critical value for the LR test statistic; p is the number of genetic effects for the putative QTL; m is the number of chromosomes; and v_l is the value of v for the lth chromosome, which is defined as

$$v_l = |\sqrt{T(\rho_0)} - \sqrt{T(\rho_1)}| + |\sqrt{T(\rho_1)} - \sqrt{T(\rho_2)}| + \cdots + |\sqrt{T(\rho_{L-1})} - \sqrt{T(\rho_L)}|, \tag{9.6}$$

where $T(\rho)$ is the LR test statistic at the putative QTL position ρ in centiMorgans (cM), and $\rho_1, \cdots, \rho_L$ are the successive turning points of $\sqrt{T(\rho)}$. Equation 9.6 can be calculated simply by taking the absolute differences between successive square roots of $T(\rho)$ on the fine grid, such as between 1 cM and 2 cM, and summing these across the chromosome. The unknown critical value C in Equation 9.5 is solved numerically using the bisection procedure.

9.5 Single Marker Analysis and Interval Mapping of QTL

The extended sib-pair methods of QTL mapping may be implemented with either single marker analysis or interval mapping. In single marker analysis, we simply report the test statistics and the estimated QTL variances at the marker locations. The practitioner then decides which and how many markers are associated with the quantitative trait. However, the exact positions of QTL cannot be located by single marker analysis. In contrast, the interval mapping procedure can evaluate the QTL effect in virtually any location of the genome. If the position happens to overlap with a marker, we will have a more accurate estimate of the QTL variance. If the position happens to be somewhere between two markers, we can infer the IBD of that position using flanking marker information or information from all markers on the same chromosome (the so-called multipoint method). The test statistic becomes a continuous function of the genome location, called the *test statistic profile*. From the test statistic profile, we can visually identify significant QTL. These two strategies of QTL mapping are incorporated into our program by assigning an argument in the macro %QTLIBD, which is described next.

9.6 Program Implementation

We used the SAS macro language to implement the extended sib-pair method above. The program consists of four macros, including %QTLIBD, %QTLSCAN, %QTLCRITIC, and %QTLPLOT. The first macro, %QTLIBD, is used to calculate the proportion of genes IBD shared by sibs at putative QTL position along a chromosome. This macro handles missing and partially informative markers using the multipoint method (Gessler and Xu, 1999). The second macro, %QTLSCAN, provides estimates of QTL variance components and calculation of test statistic for any putative position of the genome. The maximum likelihood method and likelihood ratio test are used in the analysis. The third macro, %QTLCRITIC, determines the critical value used to declare statistical significance using the approximate method of Piepho (2001). This method is simple and fast because it requires no permutation resampling of the data. The last macro, %QTLPLOT, plots the result to visually identify the QTL position on the corresponding chromosome.

There are two ways to implement the mixed model methodology of QTL mapping. One way is to exclusively use PROC IML by calling one of the non-linear programming subroutines, such as NLPNMS, the non-linear programming by the Nelder-Mead simplex method (1965). The macro implemented by calling the optimization subroutine of the IML procedure is named %QTLSCAN1. The other way is to invoke PROC MIXED for each putative position. To scan the entire genome, we need the SAS macro to repeatedly call PROC MIXED. The macro implemented by PROC MIXED is named %QTLSCAN2.

We found that the PROC IML subroutine calling approach is many times faster than the PROC MIXED approach, but the latter is more general so it can handle more complicated models.

9.6.1 Choosing the Scanning Increment

A chromosome is a continuous entity, but the genome scanning approach can evaluate only a finite number of points along the chromosome. Therefore, we need to define the scanning increment along the chromosome. The scanning increment is usually assigned by 1 cM or 2 cM in most QTL mapping software packages. There is a problem with the fixed value of the scanning increment. If an interval bracketed by markers is not an integer, a marker position may not be evaluated. The ideal situation would be that all markers be evaluated and their test statistics included in the test statistic profile. We wrote a subroutine to choose a variable increment value with an upper limit, denoted by INT. With this upper limit, the actual increment for all intervals will be less than or equal to INT. In addition, each marker position is guaranteed to be evaluated, whether it takes an even number, odd number, or non-integer position. If we choose an INT to be sufficiently large, only marker positions are evaluated, which is equivalent to single marker analysis. We assign an argument INT in the %QTLIBD macro to control the scanning increment. It is implemented by the following statements:

```
do i=1 to nrow(chmk);
   point=0;
   do j=1 to chmk[i,2]-1;
      nintvl=(ccm[i,j+1]-ccm[i,j])/&int;
      if nintvl<=1 then nintvl=1;
      else nintvl=int(nintvl+0.99);
      intvl1=(ccm[i,j+1]-ccm[i,j])/nintvl;
      if j=chmk[i,2]-1 then ttt=nintvl;
      else ttt=nintvl-1;
      do k=0 to ttt;
```

```
            point=point+1;
            kk=ccm[i,j]+intvl1*k;
            ppp=ppp//(i||point||kk);
         end;
      end;
   end;
```

The macro variable INT is an argument of the macro %QTLIBD. We can control the scanning increment by this upper limit. For example, assume that there are two chromosomes. The first chromosome has three markers and their positions are 0, 5, and 8 cM, respectively. The second chromosome has two markers and their positions are 2 and 7 cM, respectively. We now define the following two matrices:

$$\mathbf{chmk} = \begin{bmatrix} 1 & 3 \\ 2 & 2 \end{bmatrix} \text{ and } \mathbf{ccm} = \begin{bmatrix} 0 & 5 & 8 \\ 2 & 7 & 0 \end{bmatrix}.$$

In matrix **chmk**, the first column stores chromosome IDs and the second column stores the number of markers for the corresponding chromosome. In matrix **ccm**, the first row is the positions of markers on the first chromosome; the second row is the marker positions on the second chromosome, and so on. Note that the number of columns for matrix **ccm** equals the number of markers on the chromosome with the maximum number of markers. The statements above will generate a matrix named **ppp**, which stores the chromosome IDs (the first column), the point IDs to be evaluated within each chromosome (the second column), and the actual positions in cM corresponding to the evaluated points (the third column). For example, if INT=1, we have

$$\mathbf{ppp} = \begin{bmatrix} 1 & 1 & 1 & 1 & 1 & 1 & 1 & 1 & 1 & 2 & 2 & 2 & 2 & 2 \\ 1 & 2 & 3 & 4 & 5 & 6 & 7 & 8 & 9 & 1 & 2 & 3 & 4 & 5 & 6 \\ 0 & 1 & 2 & 3 & 4 & 5 & 6 & 7 & 8 & 2 & 3 & 4 & 5 & 6 & 7 \end{bmatrix}^T.$$

If we choose INT=2, we get

$$\mathbf{ppp} = \begin{bmatrix} 1 & 1 & 1 & 1 & 1 & 1 & 2 & 2 & 2 & 2 \\ 1 & 2 & 3 & 4 & 5 & 6 & 1 & 2 & 3 & 4 \\ 0 & 1.67 & 3.33 & 5 & 6.5 & 8 & 2 & 3.67 & 5.33 & 7 \end{bmatrix}^T.$$

We can see that although the scanning increments are not the same in different flanking marker intervals, every marker point is included in the scanning points. Furthermore, if we choose a sufficiently large value for INT—say, 10 in this example—we get

$$\mathbf{ppp} = \begin{bmatrix} 1 & 1 & 1 & 2 & 2 \\ 1 & 2 & 3 & 1 & 2 \\ 0 & 5 & 8 & 2 & 7 \end{bmatrix}^T.$$

In this situation, we will scan only the marker points along the entire genome and the program will perform the single marker analysis. Generally, this assigned large value for argument INT should not be less than the maximum marker interval along the entire genome if you want to perform single marker analysis.

9.6.2 Implementing the Multipoint Method

The following code defines a module named MPM($\mathbf{m}$, $\mathbf{p}_0$) to perform the multipoint method for calculating the proportion of genes IBD shared by sibs at each assigned QTL position:

```
start mpm(m,p0);
   n=nrow(m);
   p1=p0;
   p2=p0;
   p1[1,]=p0[1,];
   p2[n,]=j(1,4,1);
   do i=2 to n;
      cm=m[i]-m[i-1];
      r=(1-exp(-2*cm/100))/2;      /* Haldane's map function */
      c=r**2+(1-r)**2;
      t=j(2,2,0);
      t[1,1]=c;
      t[1,2]=1-c;
      t[2,1]=1-c;
      t[2,2]=c;
      tt=t@t;                       /* tt is transition matrix */
      d=diag(p0[i,]);
      p1[i,]=p1[i-1,]*tt*d;
      j=n+2-i;
      cm=m[j]-m[j-1];
      r=(1-exp(-2*cm/100))/2;      /* Haldane's map function */
      c=r**2+(1-r)**2;
      t=j(2,2,0);
      t[1,1]=c;
      t[1,2]=1-c;
      t[2,1]=1-c;
      t[2,2]=c;
      tt=t@t;                       /* tt is transition matrix */
      d=diag(p0[j,]);
      p2[j-1,]=p2[j,]*d*tt;
   end;
   p=p1#p2#p0;
   p=p/(p[,+]@j(1,4,1));
   return (p);
finish mpm;
```

In this module, $\mathbf{m}$ is a vector storing all the putative positions (i.e., the third column of matrix $\mathbf{ppp}$), $\mathbf{p}_0$ is an input matrix storing the probabilities of IBD states of all putative positions of the current chromosome before the multipoint method is invoked, and $\mathbf{p}$ is an output matrix storing the probabilities of IBD states for all assigned putative positions after the multipoint method is invoked.

We now use an example to demonstrate the multipoint method above. Consider three markers located at 0, 5, and 8 cM, respectively, on a chromosome. The marker genotypes of the father, the mother, and their two sibs are shown in Table 9.4. Consider the first marker with alleles arranged in the particular order shown in Table 9.4. The four possible genotypes that a sib can take are 25, 21, 15, and 11. Both sibs take the third one, and thus $\mathbf{q} = [0 \ 0 \ 1 \ 0]$ and $\mathbf{g} = [0 \ 0 \ 1 \ 0]$. For the second marker, the four genotypes become 44, 45, 54, and 55. Sib 1 takes the second or third genotype with an equal probability, and sib 2 takes the fourth genotype. Therefore, $\mathbf{q} = [0 \ 0.5 \ 0.5 \ 0]$ and $\mathbf{g} = [0 \ 0 \ 0 \ 1]$ for the second marker.

Table 9.4 Marker genotypes of the full-sib family used in the example.

Member	Marker 1	Marker 2	Marker 3
Father	21	45	52
Mother	51	45	34
Sib 1	15	54	23
Sib 2	15	55	24

Note that integers 1 to 5 represent different alleles at the marker locus.

Using the same approach, we get the corresponding vectors for the third markers as $\mathbf{q} = [0\ 0\ 1\ 0]$ and $\mathbf{g} = [0\ 0\ 0\ 1]$. These vectors are used to convert the genotypic probabilities into the IBD sharing probabilities using Equation 9.3, which are

$$\Pr(S_{11} | M_1) = 1,\ \Pr(S_{10} | M_1) = \Pr(S_{01} | M_1) = \Pr(S_{00} | M_1) = 0$$
$$\Pr(S_{11} | M_2) = \Pr(S_{00} | M_2) = 0,\ \Pr(S_{10} | M_2) = \Pr(S_{01} | M_2) = 0.5$$
$$\Pr(S_{11} | M_3) = \Pr(S_{01} | M_3) = \Pr(S_{00} | M_3) = 0,\ \Pr(S_{10} | M_3) = 1.$$

We now have

$$\mathbf{m} = \begin{bmatrix} 0 \\ 5 \\ 8 \end{bmatrix} \text{ and } \mathbf{p}_0 = \begin{bmatrix} 1 & 0 & 0 & 0 \\ 0 & 0.5 & 0.5 & 0 \\ 0 & 1 & 0 & 0 \end{bmatrix}.$$

Calling module MPM($\mathbf{m}, \mathbf{p}_0$), we get

$$\mathbf{p} = \begin{bmatrix} 1 & 0 & 0 & 0 \\ 0 & 0.996 & 0.004 & 0 \\ 0 & 1 & 0 & 0 \end{bmatrix}.$$

This assumes that we evaluate only the three marker positions. If we want to evaluate positions for every 1 cM, we have

$$\mathbf{m} = \begin{bmatrix} 0 & 1 & 2 & 3 & 4 & 5 & 6 & 7 & 8 \end{bmatrix}^T$$

and

$$\mathbf{p}_0 = \begin{bmatrix} 1 & 0.25 & 0.25 & 0.25 & 0.25 & 0 & 0.25 & 0.25 & 0 \\ 0 & 0.25 & 0.25 & 0.25 & 0.25 & 0.5 & 0.25 & 0.25 & 1 \\ 0 & 0.25 & 0.25 & 0.25 & 0.25 & 0.5 & 0.25 & 0.25 & 0 \\ 0 & 0.25 & 0.25 & 0.25 & 0.25 & 0 & 0.25 & 0.25 & 0 \end{bmatrix}^T.$$

Rows in matrix $\mathbf{p}_0$ index different scanning positions. Note that the elements of rows 2 through 5 and rows 7 through 8 are all 1/4. This is because these positions are putative positions without observation on the genotype. By calling the MPM module, we get

$$\mathbf{p} = \begin{bmatrix} 1 & 0.798 & 0.598 & 0.399 & 0.2 & 0 & 0.002 & 0.002 & 0 \\ 0 & 0.199 & 0.398 & 0.596 & 0.796 & 0.996 & 0.995 & 0.996 & 1 \\ 0 & 0.002 & 0.003 & 0.003 & 0.003 & 0.004 & 0.002 & 0 & 0 \\ 0 & 0 & 0.001 & 0.001 & 0.001 & 0 & 0.002 & 0.002 & 0 \end{bmatrix}^T.$$

Similarly, if we select 2 cM as the upper limit of the increment, we have

$$\mathbf{m} = \begin{bmatrix} 0 & 1.67 & 3.33 & 5 & 6.5 & 8 \end{bmatrix}^T \text{ and } \mathbf{p}_0 = \begin{bmatrix} 1 & 0.25 & 0.25 & 0 & 0.25 & 0 \\ 0 & 0.25 & 0.25 & 0.5 & 0.25 & 1 \\ 0 & 0.25 & 0.25 & 0.5 & 0.25 & 0 \\ 0 & 0.25 & 0.25 & 0 & 0.25 & 0 \end{bmatrix}^T.$$

By calling the MPM module, we get

$$\mathbf{p} = \begin{bmatrix} 1 & 0.664 & 0.334 & 0 & 0.002 & 0 \\ 0 & 0.332 & 0.662 & 0.996 & 0.996 & 1 \\ 0 & 0.003 & 0.003 & 0.004 & 0.001 & 0 \\ 0 & 0 & 0.001 & 0 & 0.002 & 0 \end{bmatrix}^T.$$

9.6.3 Implementing the Mixed Model Methodology through PROC IML

PROC IML in SAS offers a set of optimization subroutines for maximizing a continuous nonlinear function, such as NLPNRA, NLPNMS, etc. The following statements define the function and call the NLPNMS subroutine to solve for the ML solution:

```
start fun3(b) global(nf,index,ind,y);              /* start module */
   l=0;
   do i=1 to nf;
      yi=y[index[i]+1:index[i+1]];                 /* yi is vector of phenotypes */
      c=i(ind[i]);                                 /* of the ith sibship */
      do u=1 to ind[i]-1;
         do v=u+1 to ind[i];
            c[u,v]=0.5;
            c[v,u]=0.5;                            /* c is equivalent to A in */
         end;                                      /* Equation 9.1 */
      end;
      v=c*exp(b[2])+i(ind[i])*exp(b[3]);           /* exp(b[2])=Vg */
      d=det(v);                                    /* exp(b[3])=Ve */
      vi=inv(v);
      l=l-0.5*log(d)-0.5*(yi-b[1])`*vi*(yi-b[1]);  /* L is joint */
   end;                                            /* likelihood function */
   return (l);
finish fun3;                                       /* finish module */

b0=j(3,1,0);
optn={1,0};
call nlpnms(rc,br,"fun3",b0,optn);
l0=fun3(br`);
```

The FUN3 module specifies the objective function L under the null hypothesis $H_0: \sigma_a^2 = \sigma_d^2 = 0$. The starting values of the parameter vector are represented by vector $\mathbf{b}_0 = j(3,1,0) = [0,0,0]$. The first element of $\mathbf{b}_0$ corresponds to the population mean μ. The second and third elements of $\mathbf{b}_0$ correspond to the natural logarithms of σ_g^2 and σ_e^2, respectively. The first element of vector **optn** specifies whether the problem is a minimization (0) or maximization (1) problem. The second element specifies the amount of printed output.

Similarly, the FUN4 module that follows specifies the objective function L under the full model, where the number of parameters is 5, including μ, σ_a^2, σ_d^2, σ_g^2, and σ_e^2. The starting point is represented by $\mathbf{b}_0 = j(5,1,0) = [0,0,0,0,0]$. Again, the last four elements of vector $\mathbf{b}_0$ are the natural logarithms of the four variance components.

```
start fun4(b) global(nf,y,p11,pi,index,index1,ind); /*start module*/
    l=0;
    do i=1 to nf;
        yi=y[index[i]+1:index[i+1]];
        p1=p11[index1[i]+1:index1[i+1]];
        p2=pi[index1[i]+1:index1[i+1]];
        m=i(ind[i]);
        n=m;
        c=m;
        z=0;
        do u=1 to ind[i]-1;
            do v=u+1 to ind[i];
                z=z+1;
                m[u,v]=p2[z];
                m[v,u]=m[u,v];
                n[u,v]=p1[z];                         /* m,n and c are */
                n[v,u]=n[u,v];              /* equivalent to Pi,Delta and A */
                c[u,v]=0.5;                 /* in Equation 9.1, respectively */
                c[v,u]=c[u,v];
            end;
        end;
        v=m*exp(b[2])+n*exp(b[3])+c*exp(b[4])+i(ind[i])*exp(b[5]);
        d=det(v);
        vi=inv(v);
        l=l-0.5*log(d)-0.5*(yi-b[1])`*vi*(yi-b[1]);
    end;                                    /* L is joint likelihood function */
    return (l);
finish fun4;                                /* finish module */

b0=j(5,1,0);
optn={1,0};
call nlpnms(rc,br,"fun4",b0,optn);
var=exp(br[2:5]);
l=fun4(br`);
```

The preceding two modules involve a simple method of reparameterization. Although the parameters can be directly optimized subject to boundary constraints, we found that it is easier to use the simple method of reparameterization, which uses the simplex algorithm to search the unknowns in the appropriate parameter space. Because variance components cannot be negative, we actually search the elements of vector **b** in the entire real domain but take the exponentials to convert into non-negative variance components.

In both modules FUN3 and FUN4, the global variables include NF, **y**, $\mathbf{p}_{11}$, $\mathbf{p}_i$, **ind**, **index**, and **index1**. Variable NF is the number of families. Vectors **y**, $\mathbf{p}_{11}$, and $\mathbf{p}_i$ are column vectors of phenotypic values, probability of state S_{11} at QTL, and the proportion of genes IBD at QTL, respectively. Vector **ind** is an NF by 1 vector with the *i*th entry equal to the number of sibs in the *i*th family. The **index** vector is an NF+1 by 1 vector defined as **index** =0//cusum(**ind**), where cusum(**ind**) is cumulative sums of vector **ind** and '//' represents vertical concatenation of matrices. Similarly, vector **index1** is also 0 concatenated vertically with cumulative sums of vector **ind1**. Vector **ind1** is also an NF by 1 vector with the *i*th entry equal to the number of possible sib-pair combinations in the *i*th family.

9.6.4 Implementing the Mixed Model Methodology through PROC MIXED

Alternatively, we can use PROC MIXED to implement the mixed model methodology of QTL mapping. The following SAS code demonstrates this alternative approach:

```
ods output fitstatistics=fs;
proc mixed data=trait;
    class id f sib;
    model y=;
    repeated id/subject=intercept type=lin(2) ldata=tem0;
run;
```

This is the method for estimating the polygenic variance under the null hypothesis $H_0 : \sigma_a^2 = \sigma_d^2 = 0$. The ODS statement in the first line is used to direct the output table FITSTATISTICS generated by PROC MIXED to a data set named FS. The data set taken by PROC MIXED includes the phenotypic data TRAIT and the coefficient matrices data set TEM0 associated with the TYPE=LIN(2) option in the REPEATED statement. The TEM0 data set must contain the variables PARM, ROW, COL, and VALUE. The PARM variable denotes which of the coefficient matrices is currently being constructed, and the ROW, COL, and VALUE variables specify the matrix values. Unspecified elements of these matrices are set to zero by default.

The following statements are used to implement the estimation of variance components and the calculation of likelihood value by PROC MIXED under the full model:

```
ods output covparms=cp;
ods output fitstatistics=fs;
proc mixed data=trait;
    class id f sib;
    model y=;
    repeated id/subject=intercept type=lin(4) ldata=tem;
run;
```

The ODS statements in the first two lines are used to direct the output tables COVPARMS and FITSTATISTICS to data sets named CP and FS, respectively. The coefficient matrices data set TEM is associated with the TYPE=LIN(4) option in the REPEATED statement. Now the number of parameters in the TEM data set is four, rather than two as in the reduced model.

The default method in PROC MIXED is REML (restricted maximum likelihood method). If you want to perform other methods such as ML and MIVQUE0, you need to add the option METHOD=ML or METHOD=MIVQUE0 to PROC MIXED.

9.7 Executing the Program

Data required for the extended sib-pair method of QTL mapping include (1) marker map, (2) marker genotypes, and (3) phenotypic values of all individuals in the mapping population. These data can be saved separately in three external files, the map file, the marker genotype file, and the trait file. Although the format for these files may be arbitrary, we recommend that users edit and save these files in a Microsoft Excel spreadsheet with a comma-delimited format. Thus, users need to prepare these three files based on the required input format.

9.7.1 Data Preparation

The map file, MAP.CSV, stores the marker map information; e.g., in our simulated example, the map file is shown as follows:

```
1   m1    0
1   m2   10
1   m3   20
1   m4   30
1   m5   40
1   m6   50
1   m7   60
1   m8   70
1   m9   80
1   m10  90
1   m11 100
```

In this file, columns 1 through 3 are the chromosome ID (numerical variable), the marker name (character variable), and the marker position (numerical variable) measured in cM within each chromosome. The chromosome ID ranges from 1 to NCHR, the number of chromosomes.

The marker genotype file, MARK.CSV, stores the genotypes for all individuals at all markers in the order specified in the map file. Four families from our simulated example are copied here:

```
1 1 0 0 4 1 5 1 3 5 4 5 3 5 4 1 4 4 4 3 4 2 1 4 5 1
1 2 0 0 3 5 4 2 4 2 2 3 1 1 4 5 5 5 4 2 5 2 2 3 4 5
1 3 1 2 1 3 1 4 5 4 5 2 3 1 1 4 4 5 3 4 2 5 4 2 1 4
1 4 1 2 1 3 5 4 3 4 4 2 5 1 1 4 4 5 4 2 4 2 1 3 5 5
1 5 1 2 4 3 5 2 3 2 4 2 3 1 1 4 4 5 4 4 4 5 1 2 5 4
1 6 1 2 1 3 1 4 5 4 5 2 5 1 1 5 4 5 3 2 2 2 4 3 5 5
2 1 0 0 3 3 3 3 1 1 4 4 5 5 4 1 2 4 4 2 3 3 1 3 3 2
2 2 0 0 5 3 1 3 5 3 3 4 3 2 1 1 3 2 5 2 5 3 5 5 1 4
2 3 1 2 3 3 3 3 1 3 4 4 5 2 1 1 4 2 2 5 3 5 3 5 2 1
2 4 1 2 3 3 3 3 1 5 4 4 5 3 4 1 2 3 4 5 3 5 1 5 3 1
2 5 1 2 3 5 3 1 1 5 4 3 5 3 4 1 4 3 2 5 3 5 3 5 2 4
2 6 1 2 3 3 3 3 1 3 4 4 5 2 4 1 2 3 4 5 3 5 1 5 2 1
3 1 0 0 3 3 2 3 4 1 1 2 1 2 3 4 3 4 2 1 4 2 1 1 2 4
3 2 0 0 3 4 3 2 4 3 5 3 2 5 4 4 1 2 2 2 3 2 4 2 5 5
3 3 1 2 3 4 3 2 1 3 2 3 1 5 3 4 3 2 2 2 4 2 1 2 4 5
3 4 1 2 3 4 2 2 4 3 1 3 1 5 3 4 3 2 2 2 4 2 1 2 4 5
3 5 1 2 3 3 3 3 1 4 2 5 2 5 4 4 4 2 1 2 2 2 1 2 4 5
3 6 1 2 3 3 3 2 1 3 1 3 1 5 3 4 3 2 2 2 4 2 1 2 2 5
4 1 0 0 4 5 4 4 1 2 1 2 4 5 4 3 4 1 1 4 1 4 2 5 3 1
4 2 0 0 5 2 2 4 4 3 5 2 3 1 1 5 5 2 3 2 1 4 5 3 5 3
4 3 1 2 4 5 4 2 1 4 1 2 5 1 3 5 1 2 4 2 4 4 5 3 1 3
4 4 1 2 5 2 4 4 1 3 1 2 4 1 4 1 4 5 1 3 1 1 2 5 3 5
4 5 1 2 4 2 4 4 1 3 1 2 4 1 4 5 4 5 1 3 1 1 2 5 3 5
4 6 1 2 4 2 4 4 2 3 2 2 5 1 4 5 1 2 4 2 4 4 5 3 1 3
```

In the marker genotype file, columns 1 through 4 store the family ID, the individual ID within family, the father ID, and the mother ID for each member, respectively. Note that these four ID variables are all numeric variables. The family ID should be numbered from 1 to NF, the number of families, and the individual ID should be numbered from 1 to the number of individuals within family. Columns 5 and 6 contain the allelic forms of the two alleles of the first marker carried by all individuals including parents and their sibs. These allelic forms are also formatted as numeric variables. Columns 7 and 8 contain the allelic forms of the two alleles of the second marker carried by all individuals, and so on. Note that the first two members in each family are the genotypic information of father and mother, respectively. In addition, if the parental genotypes or the sibs' genotypes are unknown, the allelic forms of all markers should be denoted as 0.

In the trait file, columns 1 through 4 are the individual ID, family ID, sib ID within each family, and actual phenotypic value of the trait. The individual ID is arranged from 1 to NIND, the number of individuals, and the trait file, TRAIT.CSV, stores the phenotypic values of all sibs for all NF full-sib families. The first 20 observations from a simulated example are presented here:

```
 1   1   1    3.812992
 2   1   2   -1.10829
 3   1   3   -2.33176
 4   1   4    2.51982
 5   2   1   -0.81486
 6   2   2    5.570843
 7   2   3   -3.44317
 8   2   4   -1.80353
 9   3   1    0.967894
10   3   2   -2.94383
11   3   3   -0.94575
12   3   4   -2.34805
13   4   1   -3.71269
14   4   2    2.277921
15   4   3    5.148251
16   4   4   -5.24088
17   5   1    2.629425
18   5   2    3.092444
19   5   3    5.313678
20   5   4   -2.28822
```

9.7.2 Running the Main Program

We grouped the four SAS macros, %QTLIBD, %QTLSCAN, %QTLCRITIC, and %QTLPLOT, in one file named QTLSCAN.SAS. Before calling them, you need to open the QTLSCAN.SAS file in SAS and submit these macros.

The main program named MP.SAS contains two blocks of statements. The first block specifies the file references for the external data. The first three files are user-defined input data files, and the last file is the user-defined result output file. The second block calls the macros to implement the QTL mapping.

Suppose that the three data files shown above are all stored in the 'c:\sasbook' folder and the result file is also stored in the same folder; the main program is listed here.

```
filename map 'c:\sasbook\map.csv';        ❶
filename mark 'c:\sasbook\mark.csv';
filename trait 'c:\sasbook\trait.csv';
filename result 'c:\sasbook\result.csv';

%qtlibd(int=1);    ❷
%qtlscan1;         ❸
%qtlcritic;        ❹
%qtlplot;          ❺
```

❶ Note that if you decide to save the external files in another folder or by another filename, you need to change the file path in the FILENAME statements to match the file references of your own files.

❷ Macro %QTLIBD calculates the proportion of genes IBD at putative QTL position by the multipoint method (Gessler and Xu, 1999). The argument used in this macro is INT, the scanning increment.

❸ %QTLSCAN1 macro reads the trait file and a SAS data set created by the %QTLIBD macro. The mapping result will be saved in an external file. The folder or directory used for saving the result file and the filename depend on the FILENAME statement mentioned before. The result file RESULT.CSV contains the following variables from 0 to 20 cM on the chromosome as shown here:

```
1   1    0   1.296208  0.371315  0.973539  7.204615  5.143183
1   2    1   1.225842  0.501811  0.978778  7.138644  5.321803
1   3    2   1.133466  0.639823  1.001807  7.069582  5.481386
1   4    3   1.022954  0.776568  1.043596  7.000369  5.614197
1   5    4   0.902389  0.903968  1.101021  6.936001  5.713112
1   6    5   0.77795   1.01356   1.170711  6.880705  5.77281
1   7    6   0.653358  1.103174  1.252236  6.834312  5.790828
1   8    7   0.539409  1.163943  1.336323  6.803179  5.768101
1   9    8   0.439921  1.198     1.423316  6.78412   5.708818
1   10   9   0.342122  1.214144  1.517167  6.772652  5.619601
1   11  10   0.26297   1.207978  1.602716  6.773758  5.508246
1   12  11   0.209749  1.417622  1.555445  6.664085  6.42072
1   13  12   0.132734  1.644936  1.523835  6.543561  7.424322
1   14  13   0.043466  1.871195  1.507635  6.421898  8.49302
1   15  14   1.52E-05  2.041892  1.47355   6.326137  9.588637
1   16  15   3.44E-05  2.157028  1.425376  6.257939  10.66463
1   17  16   4.93E-05  2.239795  1.393949  6.205164  11.68211
1   18  17   7.85E-05  2.28962   1.377712  6.171288  12.60966
1   19  18   0.000101  2.309214  1.376997  6.153326  13.42484
1   20  19   0.000139  2.297894  1.391887  6.151266  14.11477
1   21  20   0.000219  2.260091  1.418384  6.165024  14.67571
```

Columns 1 through 8 are the chromosome ID, point, scanning position, additive genetic variance of QTL, dominance genetic variance of QTL, polygenic variance, residual variance, and likelihood ratio test statistic, respectively. Similarly, if you want to map QTL using PROC MIXED, you need to submit the statement

```
%qtlscan2;
```

❹ The macro %QTLCRITIC finds the approximate threshold (C) at the genome-wise Type I error rate using the quick method of Piepho (2001).

❺ The macro %QTLPLOT plots the QTL test statistic profile for each chromosome.

9.7.3 Application Notes

Users are required to provide the three external files: the map file, the marker genotype file, and the trait file. After running the program in SAS, the result file will automatically be created in the assigned folder and the QTL mapping profile will be charted in the GRAPH window (Output 9.1). From the profile, we can visually identify whether significant QTL have been detected and where they are located. Note that the result file is also saved in a Microsoft Excel spreadsheet in a comma-delimited format. The following list summarizes the important steps in running the program:

1. Users need to simultaneously open the QTLMAP.SAS file and the MP.SAS file in the SAS Enhanced Editor. QTLMAP.SAS is a SAS macro program that has four macros, including %QTLIBD, %QTLSCAN, %QTLCRITIC, and %QTLPLOT. The MP.SAS file is the main program to call the macros. You run QTLMAP.SAS first and then run MP.SAS after, and wait for the result. In other words, you should run the macro program first before you call the macros in the main program.

Output 9.1 The QTL mapping profile for the simulated data.

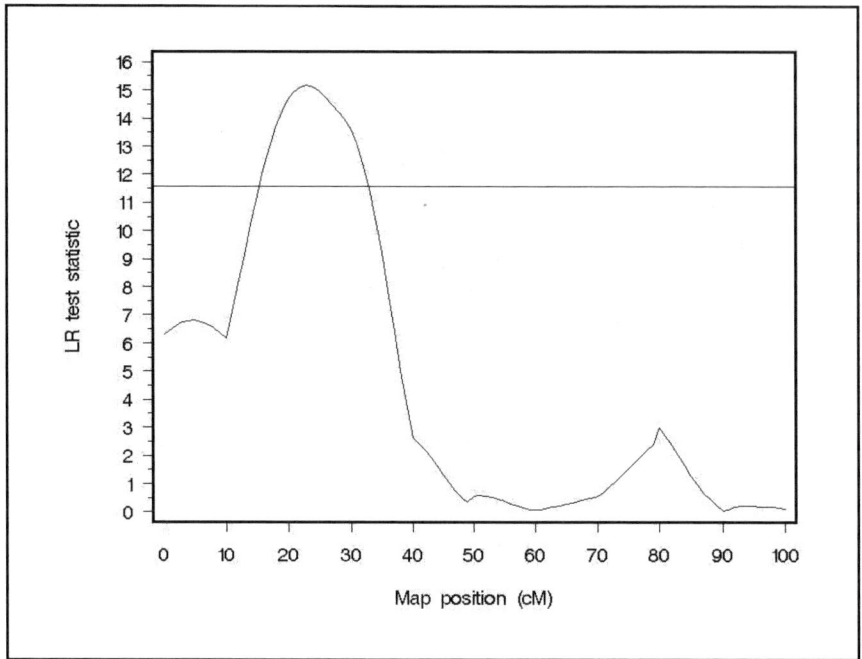

2. Users need to check the first four lines (making corrections as needed) of the main program before running the main program. These four lines are the SAS FILENAME statements that link the SAS file references to the external files. These external files include the map file, marker genotype file, trait file, and result file. The first three files are all input files provided by the users; the last file is the output file used to store the mapping result. Users should ensure that these external files and their paths are correctly saved on their computer.

3. All missing alleles should be replaced by 0, including any missing alleles of sibs.

9.7.4 Example Demo

We provide two examples for demonstrating our program. The first example is a data set simulated for 250 independent full-sib families, each with four sibs (the number of sibs per family can vary in real data analysis). There is one chromosome with 11 evenly spaced markers. The length of this chromosome is 100 cM. A single QTL with 20% heritability is located at position 25 cM on the chromosome. The three external input files of a random sample have been shown above, along with the external result file. The QTL mapping profile produced by calling %QTLSCAN1 macro has been shown in Output 9.1.

The second example is a real data set provided by Autism Genetic Resource Exchange (Geschwind et al., 2001). AGRE is the world's first collaborative gene bank for autism disease, sponsored by Cure Autism Now. This data set involves 113 full-sib families, each with at least two sibs. The total number of sibs in the data set is 254. A set of 281 microsatellite markers, each with 226 parent and 254 sib genotypes, were used to map genes controlling head circumference (cm), a typical quantitative trait. The program detected three QTL, respectively, on chromosomes 6, 9, and 22. The three QTL mapping profiles are shown in Outputs 9.2–9.4, respectively, and the mapping results are listed in Table 9.5. Note that the estimates of variance components may be biased because of the confounding between the residual variance and the polygenic variance due to small sample size.

Output 9.2 QTL mapping profile for head circumference (cm) on chromosome 6.

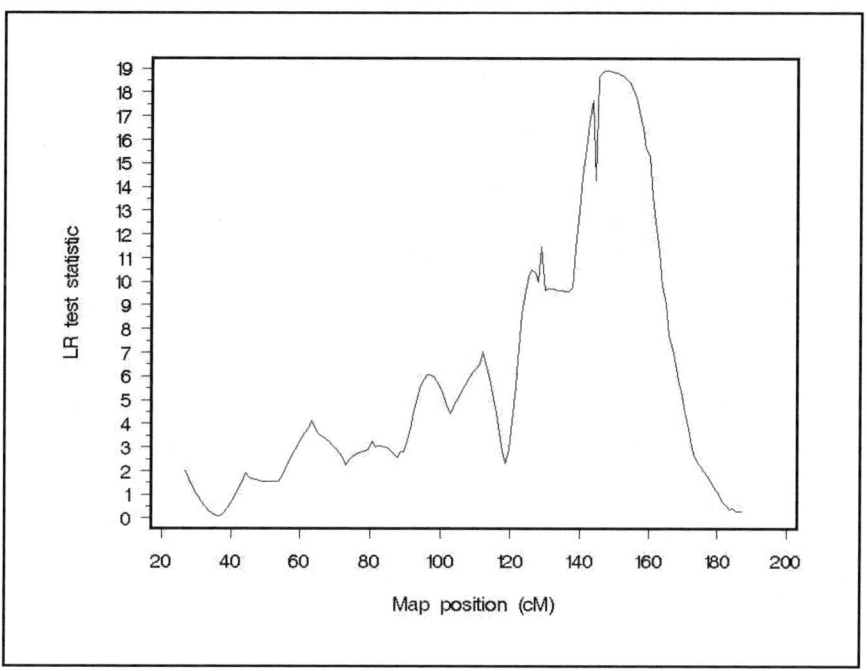

Output 9.3 QTL mapping profile for head circumference (cm) on chromosome 9.

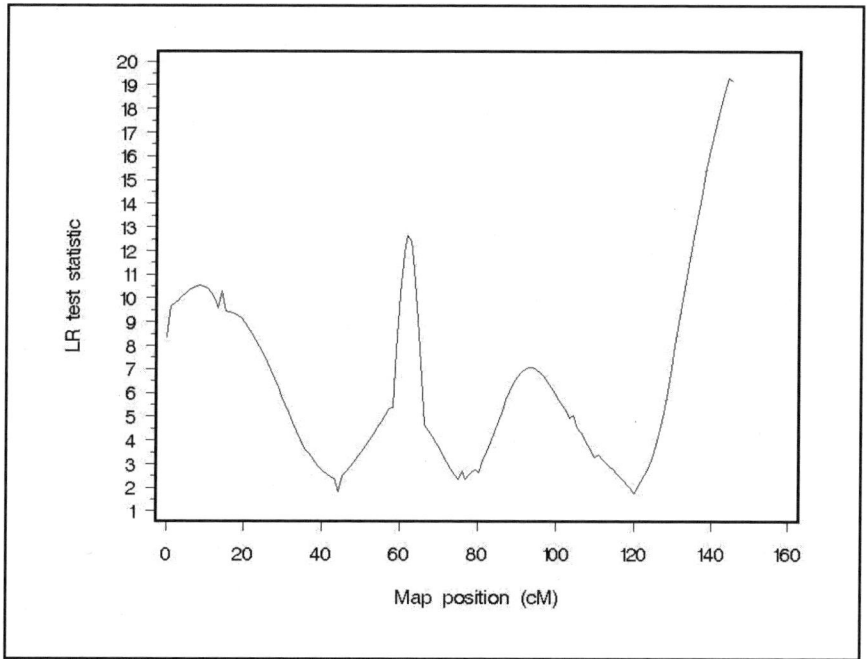

Output 9.4 The QTL mapping profile for head circumference (cm) on chromosome 22.

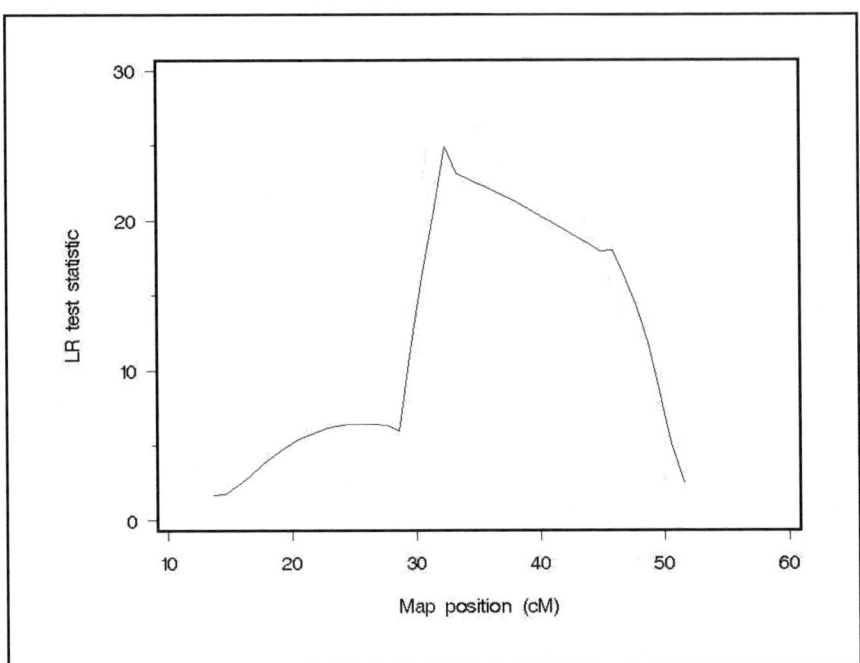

Table 9.5 Mapping results for head circumference (cm) from the AGRE data.

Chromosome	Position (cM)	Likelihood ratio test statistic	Additive genetic variance	Dominance genetic variance	Polygenic variance	Residual variance
6	147.4	18.90	0.00	3.69	2.27	0.00
9	144.2	19.31	0.15	4.18	1.57	0.00
22	32.4	24.97	4.32	0.00	0.39	1.23

The program source code and simulated data can be downloaded from the companion Web site for this book at support.sas.com/companionsites. If you want to access AGRE's genotypic and phenotypic data, you are required to contact AGRE for a user name and password. Detailed information for accessing the data can be found on the AGRE Web site at http://www.agre.org. Of course, if you want to demonstrate the real data used in this presentation, you need to contact us for information about the family and sib identifications and the marker names. These programs were written in SAS Release 8.2 and run in both the Windows and UNIX operating environments. To scan a genome of size 100 cM in 1 cM increments with 250 full-sib families, each with four sibs, the program takes about 10 minutes on a Pentium 4 PC by calling %QTLSCAN1.

9.8 References

Elston, R. C. 1990. A general linkage method for the detection of major genes. In *Advances in Statistical Methods for Genetic Improvement of Livestock*, ed. D. Gianola and K. Hammond, 495-506. New York: Springer-Verlag.

Fulker, D. W., and L. R. Cardon. 1994. A sib-pair approach to interval mapping of quantitative trait loci. *Am. J. Hum. Genet.* 54:1092-1103.

Geschwind, D. H., J. Sowinski, C. Lord, P. Iversen, J. Shestack, P. Jones, L. Ducat, and S. J. Spence. 2001. The autism genetic resource exchange: A resource for the study of autism and related neuropsychiatric conditions. *Am. J. Hum. Genet.* 69:463-466.

Gessler, D. D. G., and S. Xu. 1999. Multipoint genetic mapping of quantitative trait loci with dominant markers in outbred populations. *Genetica* 105:281-291.

Goldgar, D. E. 1990. Multipoint analysis of human quantitative genetic variation. *Am. J. Hum. Genet.* 47:957-967.

Haley, C. S., and S. A. Knott. 1992. A simple regression method for mapping quantitative trait loci in line crosses using flanking markers. *Heredity* 69:315-324.

Haseman, J. K., and R. C. Elston. 1972. The investigation of linkage between a quantitative trait and a marker locus. *Behav. Genet.* 2:3-19.

Henderson, C. R. 1975. Best linear unbiased estimation and prediction under a selection model. *Biometrics* 31:423-447.

Jansen, R. C. 1993. Interval mapping of multiple quantitative trait loci. *Genetics* 135:205-211.

Kao, C. H., Z. B. Zeng, and R. D. Teasdale. 1999. Multiple interval mapping for quantitative trait loci. *Genetics* 152:1203-1216.

Lander, E. S., and S. D. Botstein. 1989. Mapping Mendelian factors underlying quantitative traits using RFLP linkage maps. *Genetics* 121:185-199.

Nelder, J. A., and R. Mead. 1965. A simplex method for function minimization. *Comput. J.* 7:308-313.

Piepho, H. P. 2001. A quick method for computing approximate thresholds for quantitative trait loci detection. *Genetics* 157:425-432.

SAS Institute Inc. 1999. *SAS/IML User's Guide, Version 8*. Cary, NC: SAS Institute Inc.

SAS Institute Inc. 1999. *SAS/STAT User's Guide, Version 8*. Cary, NC: SAS Institute Inc.

Xu, S. 1998a. Iteratively reweighted least squares mapping of quantitative trait loci. *Behav. Genet.* 28: 341-355.

Xu, S. 1998b. Further investigation on regression method of mapping quantitative trait loci. *Heredity* 80:364-373.

Xu, S., and W. R. Atchley. 1995. A random model approach to interval mapping of quantitative trait loci. *Genetics* 141:1189-1197.

Zeng, Z. B. 1994. Precision mapping of quantitative trait loci. *Genetics* 136:1457-14

224

Chapter 10

Bayesian Mapping Methodology

Guilherme J. M. Rosa and Robert J. Tempelman

10.1 Introduction 225
10.2 Two-Point Linkage Analysis 226
    10.2.1 Inference Regarding Recombination Rates 227
    10.2.2 Example 228
    10.2.3 Testing for Linkage 230
10.3 Three-Point Analysis 231
10.4 Genetic Map Construction 238
10.5 QTL Analysis 240
    10.5.1 Example 242
    10.5.2 Results 246
10.6 Final Comments 249
10.7 References 249

10.1 Introduction

The advent of molecular markers has created opportunities for a better understanding of genetic inheritance, important for human genetic disease control (Ott, 1999), and for developing novel strategies for genetic improvement in agriculture (Soller, 1978). Molecular markers are used, for example, to construct framework maps of the genome of livestock species, which represent important steps in the ongoing effort to sequence the entire genome of various organisms. In addition, the location and effects of quantitative trait loci (QTL) can be inferred by combining information from marker genotypes and phenotypic scores of individuals in a population in linkage disequilibrium, such as in experiments with line crosses, e.g., using backcross (BC) or F_2 progenies (Weller, 1986; Knott and Haley, 1992). QTL information, in turn, may be used for marker assisted selection (MAS) schemes in animal or plant breeding (Fernando and Grossman, 1989; Lande and Thompson, 1990).

Statistical analyses in genetics are generally conducted in a classical framework, such as by using maximum likelihood techniques. However, a rising interest in Bayesian statistics in genetics has been observed in recent years (Gianola and Fernando, 1986; Shoemaker et al., 1999). The Bayesian approach, in contrast to classical procedures, allows the complete exploration of the posterior distributions, properly accounting for all uncertainty regarding nuisance parameters in the model. In addition, the adoption of the Bayesian paradigm simplifies the interpretation of the results (Shoemaker et al., 1999).

The analytical implementation of Bayesian inference using high-dimensional models, however, is cumbersome and generally unfeasible (Gelman et al., 2003). The application of Bayesian approaches in genetics and animal breeding, therefore, was not widespread until the advent of Markov chain Monte Carlo (MCMC) methods (Gianola, 2000). Early applications of Gibbs sampling in animal breeding, for example, were presented by Wang et al. (1993) and Sorensen et al. (1994). Highly complex model formulations can be currently analyzed using MCMC algorithms, which have been employed in many areas of genetics, including animal breeding and selection (Gianola, 2000; Blasco, 2001; Gianola et al., 2002), linkage analysis (George et al., 1999; Rosa et al., 2002), and QTL studies (Satagopan et al., 1996; Uimari and Hoeschele, 1997; Sillanpaa and Arjas, 1999; Bink et al., 2000).

The majority of the available software for genetic analysis employs classical approaches, such as regression or maximum likelihood. SAS/Genetics software, for example, offers a set of procedures for studying allelic (PROC ALLELE) and haplotype (PROC HAPLOTYPE) frequencies, testing for Hardy-Weinberg equilibrium, and other applications (see Chapter 8, "Gene Frequencies and Linkage Disequilibrium"). For Bayesian inference in genetics, on the other hand, there is a relatively little software designed for specific analyses and situations.

SAS software was originally developed to perform statistical analyses within the classical, frequentist context. In recent years, however, SAS has been developing some Bayesian-oriented procedures. The PRIOR statement in PROC MIXED, for example, enables the user to carry out a sampling-based Bayesian analysis for a restricted class of variance component models.

Nevertheless, for more complex models such as those encountered in some applications in genetic analysis, a fully Bayesian analysis can be conducted only by developing specific codes. Fortunately, SAS is a very flexible interface/language, with a powerful and reliable (pseudo-)random number generation system, which allows the implementation of virtually any model specification under a Bayesian setting.

In this chapter, we discuss some examples of Bayesian applications in genetic analyses, and show how the SAS language and SAS/IML can be used to implement MCMC techniques to approximate the posterior analyses. Through simple examples related to genetic map construction and QTL analysis, the implementation of the Gibbs sampler and the Metropolis-Hastings (MH) algorithm is illustrated.

Specific codes using the (pseudo-)random number functions are used to generate samples from the posterior distributions, and SAS/INSIGHT, or the UNIVARIATE, CAPABILITY, or KDE procedure, is used to analyze the posterior samples and to approximate features of the posterior distributions of interest.

10.2 Two-Point Linkage Analysis

Genetic linkage maps play a prominent role in many areas of genetics, such as QTL analysis and map-based cloning of genes (Jansen et al., 2001). While the order and distances between genetic markers will ultimately be obtained from sequence data, currently one must rely on genetic linkage or radiation hybrid data in most species (Broman and Weber, 1998). In the next few sections, we review some Bayesian and MCMC methods for linkage analysis and genetic map construction, and discuss how they can be implemented using SAS tools.

10.2.1 Inference Regarding Recombination Rates

Consider a backcross (BC) involving two completely homozygous lines for each of two loci (A and B), as illustrated here (showing all four possible BC offspring):

$$L_1 \quad \begin{matrix} a_1|a_1 \\ b_1|b_1 \end{matrix} \quad \times \quad \begin{matrix} a_2|a_2 \\ b_2|b_2 \end{matrix} \quad L_2$$

$$F_1 \quad \begin{matrix} a_1|a_2 \\ b_1|b_2 \end{matrix} \quad \times \quad \begin{matrix} a_1|a_1 \\ b_1|b_1 \end{matrix} \quad L_1$$

$$BC \quad \begin{matrix} a_1|a_1 \\ b_1|b_1 \end{matrix} \quad \begin{matrix} a_2|a_1 \\ b_2|b_1 \end{matrix} \quad \begin{matrix} a_1|a_1 \\ b_2|b_1 \end{matrix} \quad \begin{matrix} a_2|a_1 \\ b_1|b_1 \end{matrix}$$

Probabilities: $(1-\theta)/2 \quad (1-\theta)/2 \quad \theta/2 \quad \theta/2$

The probability of an F_1 individual producing a recombinant gamete (a_1b_2 or a_2b_1) is θ, the recombination rate. Suppose a random sample of n independent BC progenies is genotyped for both loci, to infer the recombination rate θ. The number of recombinant individuals (r) in the progeny is a realization of a binomial process with parameters n and θ, $r \sim Bin(n,\theta)$. The sampling model is then

$$\Pr(r|n,\theta) = \binom{n}{r} \theta^r (1-\theta)^{n-r}, \quad r = 0, 1, \ldots, n \tag{10.1}$$

The maximum likelihood estimator of θ is $\hat{\theta} = r/n$, with an asymptotic standard error given by $\sqrt{\hat{\theta}(1-\hat{\theta})/n}$.

For testing the null hypothesis that $\theta = 0.5$ (no linkage between the two loci) against the hypothesis that $\theta < 0.5$ (meaning that the two loci are linked to each other), a log likelihood ratio test is often used:

$$\text{LRT} = \log \frac{L(\hat{\theta}|r)}{L(\theta=0.5|r)} = r\log(2\hat{\theta}) + (n-r)\log[2(1-\hat{\theta})],$$

where 2 × LRT has an asymptotic chi-square distribution with 1 degree of freedom, under the null hypothesis. Alternatively, the log base 10 is generally used in genetics for computing the likelihood ratio test, which is called the LOD score:

$$\text{LOD} = \log_{10} \frac{L(\hat{\theta}|r)}{L(\theta=0.5|r)} = r\log_{10}(2\hat{\theta}) + (n-r)\log_{10}[2(1-\hat{\theta})] = (\log 10)^{-1} \text{LRT}.$$

In a Bayesian context, the sampling model (Equation 10.1) is multiplied by a prior distribution for θ, and inferences are drawn from the posterior distribution $p(\theta \mid r)$, given by (Gelman et al., 2003)

$$p(\theta \mid r) \propto p(\theta) p(r \mid \theta).$$

A conjugate prior for θ is the beta distribution. Its density, defined by two parameters (α and β), is given by

$$p(\theta) \propto \theta^{\alpha-1}(1-\theta)^{\beta-1}.$$

Using such a prior leads to the following posterior density:

$$p(\theta \mid r) \propto \theta^{r+\alpha-1}(1-\theta)^{n-r+\beta-1},$$

which is the kernel of a beta distribution with parameters $(r+\alpha)$ and $(n-r+\beta)$.

Inferences regarding θ are drawn from the posterior distribution. For example, you may be interested in the posterior mean of θ, as well as the posterior variance, credibility sets, etc.

All these features describing the posterior distribution of θ can be obtained analytically in this simple, one-parameter model. Here, instead, we use SAS to sample from the posterior distribution of θ to illustrate how Monte Carlo approximations can be obtained. For more complex models, an analytical treatment of the posterior distribution is often not possible, and a numerical approximation is necessary.

If k independent samples of θ are obtained, its density may be approximated by using, for example, a kernel density estimator given by

$$\hat{p}[\theta] \approx \frac{1}{kh} \sum_{i=1}^{k} K\left(\frac{\theta - \theta^{(i)}}{h}\right),$$

where $K(.)$ is a kernel function, and h is a fixed constant defining the smoothness of the estimated curve (Scott, 1992). Thus, the posterior mean and variance of θ, for example, are approximated by

$$E[\theta \mid r] \approx \overline{\theta} = \frac{1}{k} \sum_{i=1}^{k} \theta^{(i)} \quad \text{and} \quad \operatorname{var}[\theta \mid r] \approx \frac{1}{k} \sum_{i=1}^{k} (\theta^{(i)} - \overline{\theta})^2.$$

Also, an approximate $100(1-\alpha)\%$ credibility set is given by the percentiles $P_{\alpha/2}$ and $P_{1-\alpha/2}$. All these computations can be performed using SAS/INSIGHT software, for example.

10.2.2 Example

Suppose $n = 120$ and $r = 32$. The following SAS code can be used to generate samples from the posterior distribution of θ. To sample from a beta distribution with parameters a and b in SAS, generate $x1$ and $x2$ from gamma distributions with common shape and scale parameters equal to a and b, and take $x1/(x1 + x2)$.

```
proc iml;
  k = 3000;  * Monte Carlo sample size;
  n = 120;   * Total number of progeny (sample size);
  r = 32;    * Number of recombinant progeny;
```

```
 a1 = 1; b1 = 1; * hyper-parameters (Beta prior);
  theta = j(k,1,0); * Vector to store the MC samples;
   * Parameters of the posterior (Beta) distribution;
     a2 = r + a1; b2 = n - r + b1;

do i=1 to k;
 gamma1=2*RANGAM(0,a2);
 gamma2=2*RANGAM(0,b2);
 theta[i] = gamma1/(gamma1+gamma2);
end;

create sample from theta [colname={rate}];
 append from theta;
close sample;

quit;

title1 'Posterior Distribution of the Recombination Rate';
 title2 'Histogram and Kernel Density Estimation';
   symbol c=blue interpol = join value= none width = 1;
     proc capability data=sample;
       histogram rate / kernel (color=red);
run;
```

The analysis of the posterior samples of θ using PROC CAPABILITY is given in Output 10.1. The histogram and the kernel density estimate of the posterior distribution of the recombination rate θ are presented in Output 10.2. The Monte Carlo approximation for the posterior mean and variance of θ are 0.2691 and 0.0015, respectively. A central 90% credibility set is given by [0.2070; 0.3365].

Output 10.1 Monte Carlo approximation for the posterior inference on the recombination rate.

```
                    The CAPABILITY Procedure
                       Variable:   RATE

                            Moments
   N                          3000    Sum Weights              3000
   Mean                    0.26918398  Sum Observations      807.55194
   Std Deviation           0.03926389  Variance              0.00154165
   Skewness                0.21943533  Kurtosis              0.0336313
   Uncorrected SS          222.003463  Corrected SS          4.62341828
   Coeff Variation         14.5862667  Std Error Mean        0.00071686

                      Basic Statistical Measures
         Location                         Variability
     Mean      0.269184       Std Deviation          0.03926
     Median    0.268288       Variance               0.00154
     Mode          .          Range                  0.26548
                              Interquartile Range    0.05200

                  Quantile       Estimate
                  100% Max      0.430162413
                  99%           0.367509314
                  95%           0.336526438
                  90%           0.320804675
                  75% Q3        0.294496347
                  50% Median    0.268288208
                  25% Q1        0.242492131
                  10%           0.219072541
                  5%            0.207042528
                  1%            0.183424017
```

Numerical approximations, as indicated before, are not needed for this model, as these posterior features could be obtained analytically. The example here is given just to illustrate how Monte Carlo methods can be used when an approximation is inevitable.

From the posterior distribution in Output 10.2 we can also calculate the probability that θ lies in a certain parameter subspace. This probability can be approximated using the Monte Carlo samples as well. For example, if we want to estimate the probability that θ is less than 0.20, we approximate it by the proportion of samples below 0.20 in the posterior simulations (in our example this probability would be 0.028).

Output 10.2 Histogram and kernel density estimate of the posterior distribution of the recombination rate.

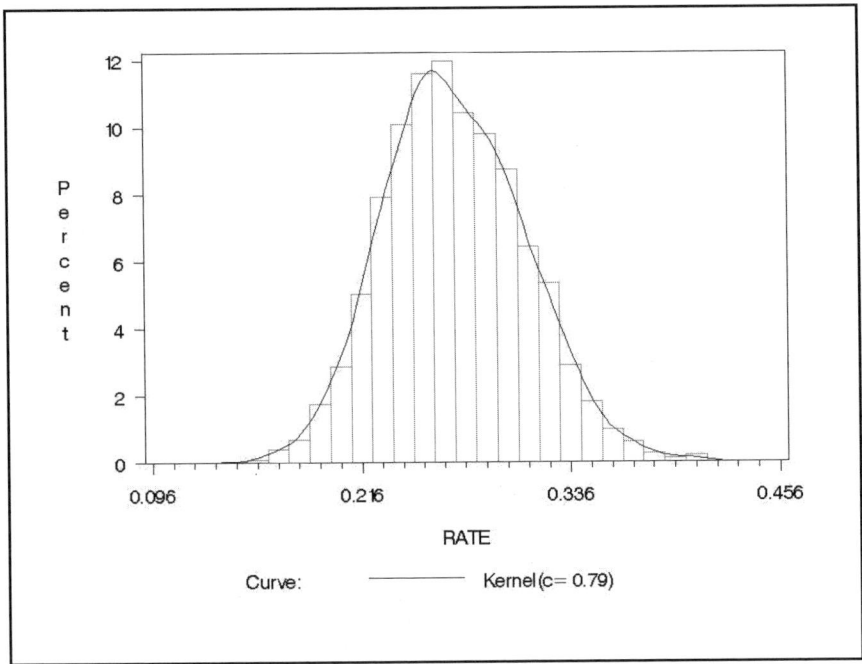

10.2.3 Testing for Linkage

The first step in a map construction is to test for linkage between all possible pairs of markers. Markers at different linkage groups should present recombination rates of 0.5, whereas markers within the same linkage group should present recombination rates less than 0.5. In practice, markers are split into linkage groups according to the results of the statistical tests. Ideally, every marker should present at least one significant linkage to other markers within a linkage group.

This task can be translated into a series of pairwise (markers i and j) testing of $H_0 : \theta_{ij} = 0.5$ vs. $H_a : \theta_{ij} < 0.5$. In a frequentist approach, likelihood ratios may be used for such tests. In a Bayesian context, model comparison can be done using Bayes factors, which is given by the ratio of posterior to prior odds, namely,

$$BF = \frac{p(\mathbf{y}\,|\,\text{Model 1})}{p(\mathbf{y}\,|\,\text{Model 2})},$$

where

$$p(\mathbf{y} | \text{Model } m) = \int p(\mathbf{y} | \boldsymbol{\theta}_m, \text{Model } m) p(\boldsymbol{\theta}_m | \text{Model } m) d\boldsymbol{\theta}_m$$

is the marginal likelihood, $\boldsymbol{\theta}_m$ is the vector of parameters of Model m, and $p(\boldsymbol{\theta}_m | \text{Model } m)$ is the prior density ($m = 1, 2$). For a review of Bayesian hypothesis testing and model selection see, for example, Raftery (1996). Newton and Raftery (1994) discuss strategies to approximate the marginal likelihoods from the MCMC samples for each competing model.

Once linkage groups are formed, we want to order the markers within each cluster and to estimate recombination rates between them. In the example in Section 10.3, the order and the recombination rates between markers are inferred by considering data from a phase-known triple backcross mating.

10.3 Three-Point Analysis

We consider in this section a situation with three linked marker loci. We show how Bayesian methodology can be used to simultaneously estimate loci order and recombination rates, allowing for any degree of interference.

The interference is defined as $I = 1 - \gamma$, where γ is the coefficient of coincidence. The coefficient γ is a measure of deviation from independence of recombinations in different intervals between genes on a chromosome. As an example, consider three loci with order ABC. Under independence, a double recombination occurs with probability $\theta_{AB} \times \theta_{BC}$, where θ_{ij} is the recombination rate between loci i and j. The deviation from independence is then given by (Muller, 1916)

$$\gamma = \frac{g_{11}}{\theta_{AB}\theta_{BC}} = \frac{1}{2}\frac{(\theta_{AB} + \theta_{BC} - \theta_{AC})}{\theta_{AB}\theta_{BC}},$$

where g_{11} is the frequency of double recombinations. For a comprehensive discussion on interference, refer, for example, to Ott (1999).

Consider the data presented in Ott (1999) relative to 20 offspring from a phase-known triple backcross mating. Assume that 17 offspring are non-recombinant. Among the three recombinant offspring, let $n_{AB} = 1$ be recombinant for AB, $n_{BC} = 2$ offspring recombinant for BC, and $n_{AC} = 3$ offspring recombinant for AC.

For a given order of the loci, let k_{11} be the number of doubly recombinant (in intervals 1 and 2) individuals; let k_{10} and k_{01} be the number of individuals with recombination only in intervals 1 and 2, respectively; and let $k_{00} = n - k_{10} - k_{01} - k_{11}$ be the number of non-recombinant individuals. Then, the number of recombinations in intervals 1 and 2, and between flanking markers, is given by $R_1 = k_{10} + k_{11}$, $R_2 = k_{01} + k_{11}$, and $R_3 = k_{10} + k_{01}$, respectively. Table 10.1 shows the results of the analysis for the three possible locus orders.

Table 10.1 Number of offspring interpreted in each possible order (adapted from Ott, 1999).

	Order ABC (CBA)			Order BAC (CAB)			Order BCA (ACB)		
	R_2	N_2	Total	R_2	N_2	Total	R_2	N_2	Total
R_1	0	1	1	1	0	1	2	0	2
N_1	2	17	19	2	17	19	1	17	18
Total	2	18	20	3	17	20	3	17	20

Note: R_i = recombination in interval i; N_i = non-recombination in interval i.

Ott (1999) discusses a maximum likelihood approach for ordering the loci. The likelihood is proportional to a multinomial process, given by

$$L(g_{11}, g_{10}, g_{01}, g_{00} \mid k_{11}, k_{10}, k_{01}, k_{00}) \propto (g_{11})^{k_{11}} (g_{10})^{k_{10}} (g_{01})^{k_{01}} (g_{00})^{k_{00}},$$

where g_{ij} are the probabilities of each class of individuals. For each locus order, estimates of g_{ij} and γ are obtained subject to the constraint $0 \leq \gamma \leq 1$, and the LOD score is calculated. The maximum likelihood estimate of the locus order (and of g_{ij} and γ) is taken to be the one presenting the highest LOD score.

Here, we show how a Bayesian MCMC approach can be used to make inferences regarding locus order, recombination fractions, and the coefficient of coincidence. Beta or Dirichlet processes can be used as priors for recombination fractions or probabilities g_{ij}, but conditioning on the restriction $0 \leq \gamma \leq 1$. To implement the MCMC scheme we use here a Metropolis-Hastings algorithm comprising four steps:

Start from any locus order (denoted by $\lambda^{(t)}$) and probabilities $\mathbf{g}^{(t)} = (g_{00}, g_{01}, g_{10}, g_{11})'$.

1. Sample a new order λ^* (candidate) from an equal probability (uniform) distribution, i.e., $\Pr(\lambda^* = k) = 1/3$, k = ABC, BCA, BAC.

2. Draw candidates θ_1^* and θ_2^* from proposal densities $\pi(.)$ (such as beta or uniform distributions, for example), conditional to the candidate order λ^*.

3. Draw also θ_3^*, but restricted to the values $(\theta_1^* + \theta_2^* - 2\theta_1^*\theta_2^*) \leq \theta_3^* \leq \min(\theta_1^* + \theta_2^*, 1/2)$, as imposed by the constraint $0 \leq \gamma \leq 1$ on the parameter space. Calculate $\mathbf{g}^* = (g_{00}^*, g_{01}^*, g_{10}^*, g_{11}^*)'$ as

$$\begin{cases} g_{11}^* = (\theta_1^* + \theta_2^* - \theta_3^*)/2 \\ g_{10}^* = (\theta_1^* - \theta_2^* + \theta_3^*)/2 \\ g_{01}^* = (-\theta_1^* + \theta_2^* + \theta_3^*)/2 \\ g_{00}^* = 1 - g_{01}^* - g_{10}^* - g_{11}^* \end{cases}$$

4. Decision:

$$(\lambda^{(t+1)}, \boldsymbol{\theta}^{(t+1)}) = \begin{cases} (\lambda^*, \boldsymbol{\theta}^*) & \text{with probability } \alpha^{(t)} \\ (\lambda^{(t)}, \boldsymbol{\theta}^{(t)}) & \text{with probability } (1-\alpha^{(t)}) \end{cases},$$

where

$$\alpha^{(t)} = \min\left[1, \frac{p(\lambda^*, \mathbf{g}^* | \mathbf{y})}{p(\lambda^{(t)}, \mathbf{g}^{(t)} | \mathbf{y})} \frac{\pi(\mathbf{g}^{(t)})}{\pi(\mathbf{g}^*)}\right].$$

From the MCMC output, a posterior probability is calculated for each order as $\Pr(\lambda_i) = n_i / n$, where n_i is the number of samples with order λ_i, and n is the total number of MCMC samples. In addition, the marginal (and joint) posterior distribution for any (set of) parameter(s) of interest, such as θ_{AB}, θ_{BC}, θ_{AC}, g_{00}, g_{10}, g_{01}, g_{11}, or γ, can be approximated using histograms or a kernel density estimator procedure.

The SAS/IML code used for implementing this analysis is presented here:

```
proc iml;

* k1, k2, k3 and k4 obtained considering the order A-B-C;
  k1=0; k2=1; k3=2; k4=17;

* Starting points;
  order=1;
   ab=0.05; bc=0.11; ac=0.15;
    c1=(ab+bc-ac)/2;
     c2=(ab-bc+ac)/2;
      c3=(bc+ac-ab)/2;
       c4=1-c1-c2-c3;
        c=(ab+bc-ac)/(2*ab*bc);
         p_old=k1*log(c1)+k2*log(c2)+k3*log(c3)+k4*log(c4);
iter=5000;
theta=j(iter,5,0);

do i=1 to iter;
 do j=1 to 300;

r1=0.5*uniform(0); r2=0.5*uniform(0);
 r3min=r1+r2-2*r1*r2;
  r3max=min(r1+r2,.5);
   r3=r3min+(r3max-r3min)*uniform(0);

c1=(r1+r2-r3)/2;
 c2=(r1-r2+r3)/2;
  c3=(r2+r3-r1)/2;
   c4=1-c1-c2-c3;

cand=1+int(3*uniform(0));
 if cand=1 then goto order1;
 if cand=2 then goto order2;
 if cand=3 then goto order3;

* Order A-B-C;
order1:
p_new=k1*log(c1)+k2*log(c2)+k3*log(c3)+k4*log(c4);
 if p_new-p_old > log(uniform(0)) then do;
 ab=r1; bc=r2; ac=r3;
 c=(r1+r2-r3)/(2*r1*r2);
```

```
    order=cand;
    p_old=p_new;
  end;
  goto another;

* Order B-C-A;
order2:
  p_new=k3*log(c1)+k1*log(c2)+k2*log(c3)+k4*log(c4);
  if p_new-p_old > log(uniform(0)) then do;
    bc=r1; ac=r2; ab=r3;
    c=(r1+r2-r3)/(2*r1*r2);
    order=cand;
    p_old=p_new;
  end;
  goto another;

* Order B-A-C;
order3:
  p_new=k2*log(c1)+k1*log(c2)+k3*log(c3)+k4*log(c4);
  if p_new-p_old > log(uniform(0)) then do;
    ab=r1; ac=r2; bc=r3;
    c=(r1+r2-r3)/(2*r1*r2);
    order=cand;
    p_old=p_new;
  end;
  goto another;

another:

 end;

 theta[i,1]=order;
 theta[i,2]=ab; theta[i,3]=bc;
 theta[i,4]=ac; theta[i,5]=c;

end;

theta=T(1:iter)||theta;

create MH from theta [colname={iter order ab bc ac c}];
 append from theta;
close MH;

quit;

proc print data=MH(obs=20);
run;

proc freq data=MH;
  table order;
run;

title1 'Recombination Rate Between Loci A and B';
 title2 'Histogram and Kernel Density Estimation';
   symbol c=blue interpol = join value= none width = 1;
    proc capability data=MH;
      var ab bc ab c;
      histogram ab / kernel (color=red);
run;
```

Some results of the MCMC output are shown next. First, the PROC FREQ table gives the estimated probabilities for each different locus order (Output 10.3). It is shown that the order ABC is by far the most probable order given the data.

Output 10.3 Estimated probabilities for each locus order.

```
                         The FREQ Procedure

                                      Cumulative     Cumulative
        ORDER    Frequency   Percent    Frequency     Percent

          1        4635       92.70       4635        92.70
          2          22        0.44       4657        93.14
          3         343        6.86       5000       100.00
```

Next, we discuss the analysis of the posterior samples of the recombination rate between A and B (θ_{AB}). Similar analyses apply for θ_{AC} and θ_{BC} as well.

The estimated marginal posterior density of θ_{AB} (integrated over all different orders and values of coefficient of coincidence) is given in Outputs 10.4 and 10.5. Some features of this distribution may be described by the posterior mean and standard deviation, in this case equal to 0.089 and 0.060, respectively. The mode (Mo = 0.052) or the median (Md = 0.079) are alternative estimates of central tendency.

Output 10.4 Monte Carlo approximation for the posterior inference regarding the recombination rate between loci A and B.

```
                     The CAPABILITY Procedure
                          Variable:  AB

                              Moments
     N                          5000    Sum Weights               5000
     Mean                   0.08923678   Sum Observations     446.183903
     Std Deviation          0.05969863   Variance             0.00356393
     Skewness               1.21383451   Kurtosis             2.11483514
     Uncorrected SS         57.6320829   Corrected SS         17.8160678
     Coeff Variation        66.8991294   Std Error Mean       0.00084427

                     Basic Statistical Measures
              Location                        Variability
     Mean       0.089237    Std Deviation               0.05970
     Median     0.077090    Variance                    0.00356
     Mode          .        Range                       0.47298
                            Interquartile Range         0.07615

                     Tests for Location: Mu0=0
          Test           -Statistic-         -----p Value------
          Student's t    t    105.6975       Pr > |t|    <.0001
          Sign           M       2500        Pr >= |M|   <.0001
          Signed Rank    S    6251250        Pr >= |S|   <.0001
```

(continued on next page)

Output 10.4 *(continued)*

```
            Quantiles (Definition 5)
      Quantile                Estimate
      100% Max              0.47478101268
      99%                   0.27140495217
      95%                   0.20481792276
      90%                   0.17143574726
      75% Q3                0.12005261081
      50% Median            0.07709031439
      25% Q1                0.04390344212
      10%                   0.02494911443
      5%                    0.01697715734
      1%                    0.00737568410
```

Output 10.5 Histogram and kernel density estimate of the posterior distribution of the recombination rate between loci A and B.

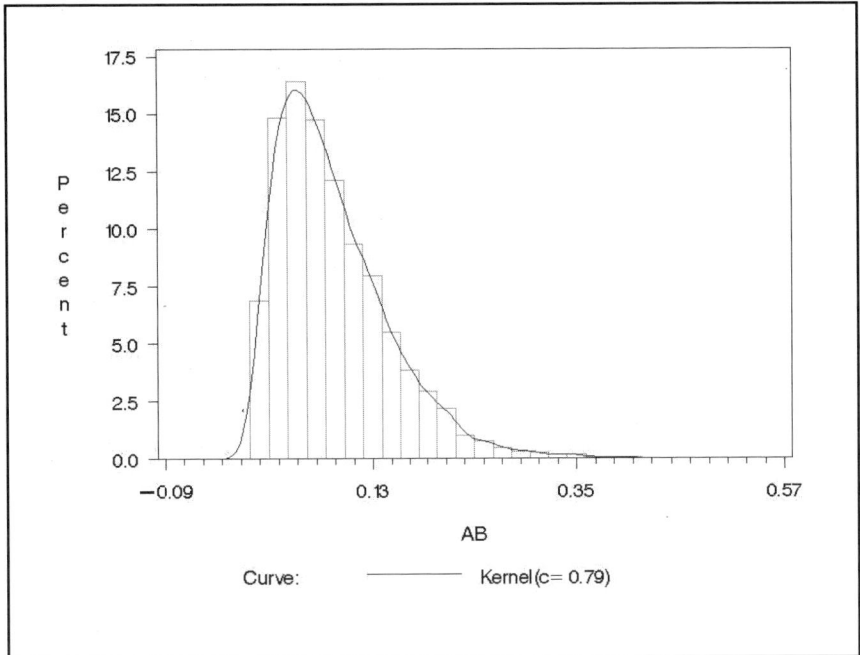

As discussed in the previous example (two-point linkage analysis), probability statements can be drawn directly from the posterior distributions. For example, it is suggested here that the probability that θ_{AB} is larger than 0.2048 is 0.05.

It is well known that inferences regarding the coefficient of coincidence γ are generally imprecise when dealing with relatively small data sets (Ott, 1999). The histogram in Output 10.7, based on PROC CAPABILITY results in Output 10.6, denotes low information from the data to modify the prior (uniform distribution) considered for this parameter. Under these circumstances, the choice of prior will basically dictate the shape of the posterior distribution of γ. The user may try different prior alternatives to study their impact on posterior inferences in this example. The user may also want to analyze bigger data sets, where the likelihood dominates the prior.

Output 10.6 Monte Carlo approximation for the posterior inference regarding the coefficient of coincidence.

```
                        The CAPABILITY Procedure
                             Variable:  C

                                  Moments
N                            5000        Sum Weights               5000
Mean                   0.51631973        Sum Observations    2581.59867
Std Deviation          0.28801392        Variance            0.08295202
Skewness              -0.0747963         Kurtosis             -1.200658
Uncorrected SS         1747.60748        Corrected SS          414.67714
Coeff Variation        55.7820865        Std Error Mean        0.00407313

                        Basic Statistical Measures
             Location                        Variability
         Mean     0.516320         Std Deviation             0.28801
         Median   0.529025         Variance                  0.08295
         Mode     .                Range                     0.99926
                                   Interquartile Range       0.49599

                        Tests for Location: Mu0=0
         Test              -Statistic-       -----p Value------
         Student's t    t   126.7623         Pr > |t|    <.0001
         Sign           M      2500          Pr >= |M|   <.0001
         Signed Rank    S   6251250          Pr >= |S|   <.0001

                         Quantiles (Definition 5)
                        Quantile            Estimate
                        100% Max         0.999706326518
                        99%              0.992123711152
                        95%              0.952693107982
                        90%              0.907349455360
                        75% Q3           0.767085850596
                        50% Median       0.529024618691
                        25% Q1           0.271098457869
                        10%              0.106649183485
                        5%               0.057843797867
                        1%               0.011723233392
```

Output 10.7 Histogram relative to the posterior distribution of the coefficient of coincidence.

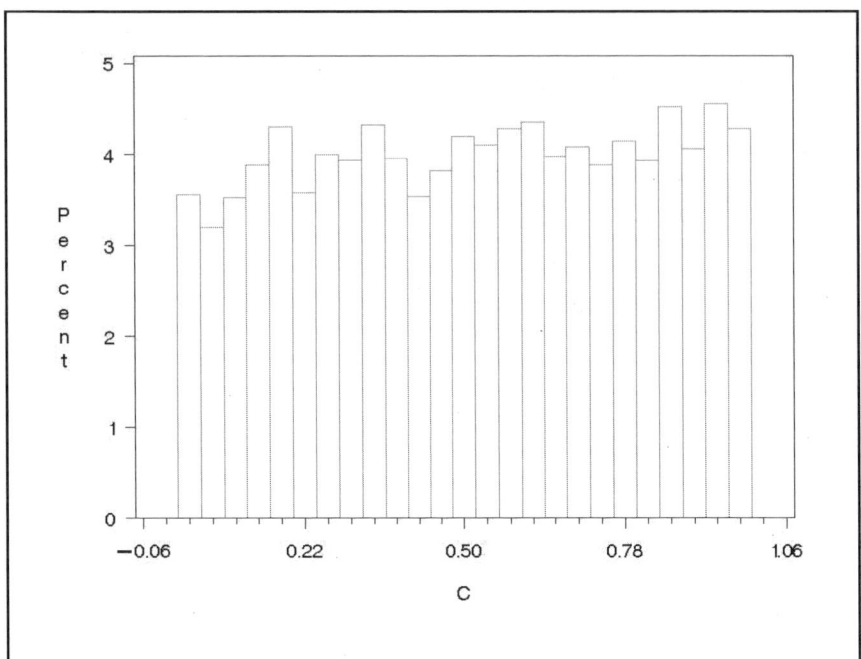

10.4 Genetic Map Construction

A Bayesian approach for genetic map construction was presented by George et al. (1999) and by Rosa et al. (2002). Consider, for example, a double haploid or backcross design, and denote the genotype of q markers for the individual i as $\mathbf{g}_i = [g_{i1}, g_{i2}, \ldots, g_{iq}]'$, where $g_{ij} = 0$ if the individual i ($i = 1, 2, \ldots, n$) is recessive homozygous for the locus j ($j = 1, 2, \ldots, q$), and 1 otherwise. Assuming non-interference (Haldane map function; i.e., I = 0) and a sample of n independent individuals, the likelihood of λ and $\boldsymbol{\theta}$ is given by

$$L(\lambda, \boldsymbol{\theta} | \mathbf{G}) = p(\mathbf{G} | \lambda, \boldsymbol{\theta}) \propto \prod_{i=1}^{n} \prod_{j=1}^{q-1} \theta_j^{k_{ij}} (1 - \theta_j)^{1 - k_{ij}} ,$$

where $k_{ij} = |g_{ij} - g_{i,j+1}|$, λ is the order of the genetic marker loci, and θ_j is the recombination rate between the loci j and $j + 1$.

A prior distribution for λ and $\boldsymbol{\theta}$ may be expressed as $p(\lambda, \boldsymbol{\theta}) = p(\boldsymbol{\theta} | \lambda) p(\lambda)$, where $p(\lambda)$ is a probability distribution over the different orders for the q markers, and

$$p(\boldsymbol{\theta} | \lambda) = \prod_{j=1}^{q-1} p(\theta_j | \lambda) ,$$

where $\theta_j | \lambda$ has a conjugate beta distribution.

The posterior density of λ and $\boldsymbol{\theta}$ is given by $p(\lambda, \boldsymbol{\theta} | \mathbf{G}) \propto p(\mathbf{G} | \lambda, \boldsymbol{\theta}) p(\boldsymbol{\theta} | \lambda) p(\lambda)$. As this density is analytically intractable for inference (George et al., 1999; Rosa et al., 2002), numerical or MCMC methods, such as the Gibbs sampling and Metropolis-Hastings algorithms (Gilks et al., 1996; Robert and

Casella, 1999), must be employed to approximate the marginal distributions of interest. The conditional posterior distributions deriving from it are needed for MCMC implementation.

The fully conditional posterior distribution of each recombination rate θ_j is shown to be a beta distribution. The updating for the gene order λ involves moves between a set of models, as the recombination rates have different meanings for distinct ordering. George et al. (1999) presented a reversible jump algorithm, for which recombination rates are converted into map distances, and reverted to new recombination rates after shifting a randomly selected marker around a pivot marker. Alternatively, Rosa et al. (2002) discussed a Metropolis-Hastings scheme for the MCMC updating of λ and $\boldsymbol{\theta}$, simultaneously. A new gene ordering is proposed according to a candidate generator density $q(.)$, and new recombination rates are simulated for this new order. The Markov chain moves from the current state to the candidate state with certain acceptance probability, as discussed in Section 10.2.

The complete SAS/IML code (file MAP.SAS) used for this analysis can be found on the companion Web site for this book at support.sas.com/companionsites. Here, we describe some alternatives for sampling candidate orders for the Metropolis-Hastings algorithm. For example, to sample a complete new order of m markers, independently from the current order, simply use the following code:

```
do j=1 to m;
onew[j]=uniform(0);
end;
onew=rank(onew);
if onew[>:<] > onew[<:>] then onew=onew[m:1];
```

The last line of this code specifies that the marker numbered as 1 is always to the left of the marker numbered as m. This is to ensure the same order is always written in the same direction. So, for example, an order such as 51243 is switched to 34215.

When the number of markers increases, the total number of possible different orders (given by $m!/2$) is extremely large. In these cases, the algorithm above is inefficient, as it will be sampling very unlikely (or even impossible) orders. As a consequence, the Markov chain will remain stuck in a specific order for some iterations, and it will jump when a new candidate order is accepted.

Alternatively, candidate orders may be proposed by considering smaller changes in the current order (Rosa et al., 2002). For example, a new order may be generated by switching two adjacent loci, as shown here:

```
u=1+int(m*uniform(0));
onew=o;
if u = m then do;
  onew[1]=o[m]; onew[m]=o[1];
end;
else do;
  onew[u]=o[u+1]; onew[u+1]=o[u];
end;
if onew[>:<] > onew[<:>] then onew=onew[m:1];
```

Another alternative would be switching any two loci (adjacent or not), as shown in the following code, or flipping a segment of the linkage group.

```
onew=o;
u1=1+int(m*uniform(0));
do;
again:
 u2=1+int(m*uniform(0));
 if u2=u1 then goto again;
end;
onew[u1]=o[u2];
 onew[u2]=o[u1];
   if onew[>:<] > onew[<:>] then onew=onew[5:1];
```

These strategies for generating candidate orders improve the mixing properties of the MCMC, so the chain presents more frequent (but smaller) moves. As for any MCMC implementation, however, it is important to monitor the convergence and the dependence between successive iterations. Care should be taken to set adequate burn-in period and thinning intervals for sub-sampling (Robert and Casella, 1999; Sorensen and Gianola, 2002).

The file MAP.SAS reads a data set with 5 markers and 60 BC progeny. The results are expressed as probabilities for each different marker ordering (Output 10.8). In addition, recombination rates or chromosomal position of each marker can be easily obtained, marginally or conditionally to a specific order.

Output 10.8 Estimated probabilities for each locus order.

```
                         The FREQ Procedure

                                          Cumulative     Cumulative
     order    Frequency      Percent       Frequency       Percent

     12345         8202        82.02            8202         82.02
     12354           11         0.11            8213         82.13
     12435          577         5.77            8790         87.90
     12453            6         0.06            8796         87.96
     12534           49         0.49            8845         88.45
     12543          342         3.42            9187         91.87
     21345          692         6.92            9879         98.79
     21354            1         0.01            9880         98.80
     21435           62         0.62            9942         99.42
     21534            7         0.07            9949         99.49
     21543           51         0.51           10000        100.00
```

In this example, the most plausible marker order is the 12345, with a probability of 0.82. The orders 12435 and 21345 present probabilities of about 0.06 each. All the remaining orders together have a probability of about 0.05.

10.5 QTL Analysis

Molecular markers are used by a large number of researchers to study genetic inheritance, as an attempt to identify genes associated with human diseases (Ott, 1999) or genes with major effects on traits of agronomical importance (Soller, 1978; Lynch and Walsh, 1998).

It is not our purpose to give a comprehensive discussion of Bayesian approaches for QTL analysis and gene mapping. Instead, we review some of the concepts and strategies for the implementation of such analyses using an MCMC Bayesian methodology. We first discuss the analysis of a single QTL model for an F_2 or BC experiment, and then illustrate how SAS software can be used to implement Gibbs sampling and Metropolis-Hastings algorithms within this context.

A simple linear additive model to describe the phenotype y_i for the ith individual ($i = 1, 2,\ldots,n$) in an F_2 population may be expressed as (Satagopan et al., 1996; Broman, 2001; Doerge, 2002)

$$y_i = \mu + q_i\alpha + 1 - |q_i|\delta + \varepsilon_i,$$

where α and δ represent the additive and the dominance effects, respectively, and ε_i is a random residual with mean 0 and variance σ^2. Here, q_i denotes the QTL genotype for the individual i, where q_i assumes the values -1, 0, or 1 if the individual i is recessive homozygous, heterozygous, or dominant homozygous, respectively. In this case, the parameters μ, α, and δ determine the expected response, given the QTL genotype q_i.

After an error distribution is specified, such as normal, this leads to the density $p(y_i | q_i, \boldsymbol{\theta})$, where $\boldsymbol{\theta} = [\mu, \alpha, \delta, \sigma^2]'$. Given the location ($\tau$) of the putative QTL and the marker genotypes ($\boldsymbol{G}$), the probability distribution of the QTL genotypes (q_i) can be modeled in terms of the recombination between the QTL locus and the markers. In F_2 experiments, for example, the probability distribution $p(q_i | \tau, \boldsymbol{G})$ is a generalized Bernoulli process.

Considering n independent observations, the likelihood of τ and $\boldsymbol{\theta}$ is

$$L(\tau, \boldsymbol{\theta} | \boldsymbol{y}, \boldsymbol{G}) = \prod_{i=1}^{n} \sum_{q_i} p(y_i | q_i, \boldsymbol{\theta}) p(q_i | \tau, G).$$

In a Bayesian context, the likelihood is combined with a prior $p(\tau, \boldsymbol{\theta})$, and marginal posterior inferences are obtained via MCMC integration of the joint posterior density

$$p(\tau, \boldsymbol{\theta}, \mathbf{q} | \boldsymbol{y}, \boldsymbol{G}) \propto p(\boldsymbol{y} | \mathbf{q}, \boldsymbol{\theta}) p(\mathbf{q} | \tau, \boldsymbol{G}) p(\tau, \boldsymbol{\theta}).$$

The joint prior $p(\tau, \boldsymbol{\theta})$ is generally assumed as

$$p(\tau, \boldsymbol{\theta}) = p(\mu) p(\alpha) p(\delta) p(\sigma^2) p(\tau),$$

where $p(\mu)$, $p(\alpha)$, and $p(\delta)$ are Gaussian; $p(\sigma^2)$ is an inverted gamma density; and $p(\tau)$ is uniform.

The Gibbs sampling works by drawing iteratively from the fully conditional distributions of μ, α, δ, σ^2, $\boldsymbol{q}$, and τ, which are described below. It has been shown (Satagopan et al., 1996) that the conditional distributions of μ, α, and δ are Gaussian. The RANNOR (or alternatively, the NORMAL) function of SAS returns a variate generated from a normal distribution with mean 0 and variance 1. A normal variate X with mean m and variance var can be generated as follows: x = m + SQRT(var) * RANNOR(seed).

Updating σ^2 is done by drawing a random sample from an inverted gamma distribution. This is easily done using the RANGAM function of SAS. To generate a variate from a gamma distribution with scale and shape parameters a and b, use b*RANGAM(seed, a). The inverted gamma variate is obtained simply from the inverse of the gamma variate.

The distribution of τ does not have a closed form, but a Metropolis-Hastings step can be used for sampling from it (Satagopan et al., 1996; Sorensen and Gianola, 2002).

10.5.1 Example

The data set discussed here was first presented by Ferreira et al. (1995a) and Ferreira et al. (1995b), and reanalyzed by Satagopan et al. (1996) within a Bayesian context. It refers to a double haploid (DH) progeny of *Brassica napus* used to detect QTLs for flowering time. There were 104 individuals with flowering data (phenotype) and genotypes of 10 markers (including 9% of missing data) from linkage group 9. The data file is organized with each row representing one individual and columns representing the phenotypic observations (first column) and marker genotypes (additional 10 columns). In the analysis that follows, the genetic marker map (as published by Ferreira et al., 1995b) is considered known, with markers 1 to 10 having positions (cM) 0.0, 8.8, 20.6, 27.4, 34.2, 42.9, 53.6, 64.1, 69.2, and 83.9.

The complete SAS/IML code (BRASSICA.SAS) for performing the analysis of this data set is posted on the companion Web site for this book at support.sas.com/companionsites. Here, we simply discuss the main sections of the program in order to facilitate its adaptation to other situations and scenarios by the user.

The data file BRASSICA.TXT is read into SAS/IML, and partitioned into two components: a vector **y** of phenotypes, and a 104 × 10 matrix (**m**) with the marker genotypes. The parameters (μ, α, σ^2, and τ) of the model are coded as MU, A, VAR, and DIST.

The fully conditional posterior distribution of μ is normal with mean $\sum_{i=1}^{n}(y_i - q_i\alpha)$ and variance σ^2/n.

Similarly, the fully conditional distribution of α is normal with mean $\frac{1}{n}\mathbf{q}'(\mathbf{y}-\mu)$ and variance σ^2/n.
Random variates from these conditional distributions are generated using this program:

```
* Sampling from the conditional distribution of mu;
mu=(y-q*a)[+]/num;
 muvar=var/num;
  mu=mu+sqrt(muvar)*normal(0);

* Sampling from the conditional distribution of a;
a=T(q)*(y-mu)/num;
 avar=var/num;
  a=a+sqrt(avar)*normal(0);
```

The fully conditional distribution of σ^2 is shown to be an inverted gamma with parameters $a1 = \text{alpha} + n/2$ and $a2 = (\text{beta} + ss/2)^{-1}$, where

$$ss = \sum_{i=1}^{n}(y - \mu - q_i\alpha)^2.$$

A variate from this distribution is obtained by inverting a random draw from a gamma distribution, as follows:

```
* Sampling from the conditional distribution of var;
ss=T(y-mu-q*a)*(y-mu-q*a);
 a1=alpha+num/2;
  a2=1/(beta+ss/2);
   ar=a2*rangam(0,a1);
    var=1/var;
```

Given the position τ of the QTL, the genotypes **q** are drawn from independent Bernoulli distributions with probabilities modeled in terms of recombination rates between the QTL and the flanking markers and a penetrance function (Satagopan et al., 1996; Sorensen and Gianola, 2002). For example, if the QTL is flanked by markers with genotypes –1 and 1 for a certain individual, the probabilities that the QTL is –1 or 1 are proportional, respectively, to

$$(1-t)u\exp\left\{-\frac{1}{2\sigma^2}(y_i-(\mu-\alpha))^2\right\}$$

and

$$t(1-u)\exp\left\{-\frac{1}{2\sigma^2}(y_i-(\mu+\alpha))^2\right\}.$$

In the code that follows, the four different genotypes of flanking markers (–1, –1), (–1,1), (1, –1), and (1,1) are coded as 1, –3, 3, and –1, respectively. Each individual is classified into one of the possible situations and the appropriate probabilities are calculated.

```
* Generation of QTL genotypes;

flank=w[,int]-2*w[,int+1];
prob=j(num,2,0);
do j=1 to num;

if flank[j] = 3 then do;
 prob[j,1]=(1-u)*t*exp(-0.5*(y[j]-mu+a)**2/var);
 prob[j,2]=(1-t)*u*exp(-0.5*(y[j]-mu-a)**2/var);
end;

if flank[j] = -1 then do;
 prob[j,1]=t*u*exp(-0.5*(y[j]-mu+a)**2/var);
 prob[j,2]=(1-t)*(1-u)*exp(-0.5*(y[j]-mu-a)**2/var);
end;

if flank[j] = -3 then do;
 prob[j,1]=(1-t)*u*exp(-0.5*(y[j]-mu+a)**2/var);
 prob[j,2]=(1-u)*t*exp(-0.5*(y[j]-mu-a)**2/var);
end;

if flank[j] = 1 then do;
 prob[j,1]=(1-t)*(1-u)*exp(-0.5*(y[j]-mu+a)**2/var);
 prob[j,2]=(t*u)*exp(-0.5*(y[j]-mu-a)**2/var);
end;

end;

prob=prob[,1]/(prob[,1]+prob[,2]);

do j=1 to num;
 if prob[j]>uniform(0) then q[j]=-1;
 else q[j]=1;
end;
```

For the Metropolis-Hastings step used to simulate from the conditional distribution of the QTL position, we use the following code:

```
* Generation of Distance;

ll=max(0, dist-incre);
 ul=min(.839, dist+incre);
   range=ul-ll;
     k=range*uniform(0)+ll;

dd=d//k;
 dd=rank(dd); pos=k; k=dd[10];

dd=0//d;

r=(1-exp(-2*(pos-dd[k])))/2;
s=(1-exp(-2*(dd[k+1]-pos)))/2;

dd=((1-q)||(1+q))/2;

flank=w[,int]-2*w[,int+1];

prob=j(num,2,0);
do j=1 to num;

if flank[j] = 3 then do;
 prob[j,1]=(1-u)*t/(t*(1-u)+u*(1-t));
 prob[j,2]=(1-t)*u/(t*(1-u)+u*(1-t));
end;

if flank[j] = -1 then do;
 prob[j,1]=(t*u)/(t*u+(1-t)*(1-u));
 prob[j,2]=(1-t)*(1-u)/(t*u+(1-t)*(1-u));
end;

if flank[j] = -3 then do;
 prob[j,1]=(1-t)*u/(t*(1-u)+u*(1-t));
 prob[j,2]=(1-u)*t/(t*(1-u)+u*(1-t));
end;

if flank[j] = 1 then do;
 prob[j,1]=(1-t)*(1-u)/(t*u+(1-t)*(1-u));
 prob[j,2]=(t*u)/(t*u+(1-t)*(1-u));
end;

end;

pq_old=log(prob); pq_old=dd#pq_old;
 pq_old=pq_old[+]+log(min(.839,pos+incre)-max(0, pos-incre));

flank=w[,k]-2*w[,k+1];

prob=j(num,2,0);
do j=1 to num;

if flank[j] = 3 then do;
 prob[j,1]=(1-s)*r/(r*(1-s)+s*(1-r));
 prob[j,2]=(1-r)*s/(r*(1-s)+s*(1-r));
end;

if flank[j] = -1 then do;
 prob[j,1]=(r*s)/(r*s+(1-r)*(1-s));
 prob[j,2]=(1-r)*(1-s)/(r*s+(1-r)*(1-s));
end;

if flank[j] = -3 then do;
 prob[j,1]=(1-r)*s/(r*(1-s)+s*(1-r));
 prob[j,2]=(1-s)*r/(r*(1-s)+s*(1-r));
end;
```

```
    if flank[j] = 1 then do;
     prob[j,1]=(1-r)*(1-s)/(r*s+(1-r)*(1-s));
     prob[j,2]=(r*s)/(r*s+(1-r)*(1-s));
    end;

   end;

   pq_new=log(prob); pq_new=dd#pq_new;
   pq_new=pq_new[+]+log(range);
   if pq_new-pq_old > log(uniform(0)) then do;
    dist=pos; int=k;
    t=r; u=s;
   end;
```

Some missing genotypes are presented in the data set. Those must be sampled on each iteration of the Gibbs sampling as well. Bernoulli distributions with probabilities modeled in terms of recombination rates between missing genotype locus and its flanking markers are used for missing data imputation.

```
   * Missing marker data imputation;

   dd=0//d;
    dd=dd[1:int]//dist//dd[int+1:10];
     rr=j(10,1,0);
   do j=1 to 10;
    rr[j]=(1-exp(-2*(dd[j+1]-dd[j])))/2;
   end;

   mq=m[,1:int]||q||m[,int+1:10];
    wq=w[,1:int]||q||w[,int+1:10];

   flank=wq[,2];
   do ind=1 to num;
   if mq[ind,1]=-99 then do;
    if flank[ind]=-1 then p=1-rr[1];
   else p=rr[1];
   if p > uniform(0) then wq[ind,1]=-1;
   else wq[ind,1]=1;
   end;
   end;

   flank=wq[,10];
   do ind=1 to num;
   if mq[ind,11]=-99 then do;
    if flank[ind]=-1 then p=1-rr[10];
   else p=rr[10];
   if p > uniform(0) then wq[ind,11]=-1;
   else wq[ind,11]=1;
   end;
   end;

   do mid=1 to 9;
   flank=wq[,mid]-2*wq[,mid+2];
   do ind=1 to num;
   if mq[ind,mid+1]=-99 then do;

   if flank[ind]=3 then p=rr[mid]*(1-rr[mid+1])/(rr[mid]*(1-rr[mid+1])
              +(1-rr[mid])*rr[mid+1]);
   if flank[ind]=-1 then p=rr[mid]*rr[mid+1]/(rr[mid]*rr[mid+1]
              +(1-rr[mid])*(1-rr[mid+1]));
   if flank[ind]=-3 then p=(1-rr[mid])*rr[mid+1]/(rr[mid]*(1-rr[mid+1])
              +(1-rr[mid])*rr[mid+1]);
```

```
if flank[ind]=1 then p=(1-rr[mid])*(1-rr[mid+1])/(rr[mid]*rr[mid+1]
               +(1-rr[mid])*(1-rr[mid+1]));

if p>uniform(0) then wq[ind,mid+1]=-1;
else wq[ind,mid+1]=1;

end;
```

10.5.2 Results

The Markov chain length was set to run 400,000 cycles with samples saved every 200 iterations, as suggested by Satagopan et al. (1996). To illustrate the results, we discuss here just the marginal posterior distributions of the variance and the position of the QTL. The marginal posterior density of σ^2, as well as some features of it, is given in Outputs 10.9 and 10.10. It is seen that the marginal posterior density of σ^2 is slightly skewed to the right, with posterior mean, median, and mode equal to 0.0625, 0.0614, and 0.0588, respectively. A central 95% credibility set is given by the limits [0.0461; 0.0845].

Output 10.9 Histogram and kernel density estimate of the posterior distribution of the residual variance.

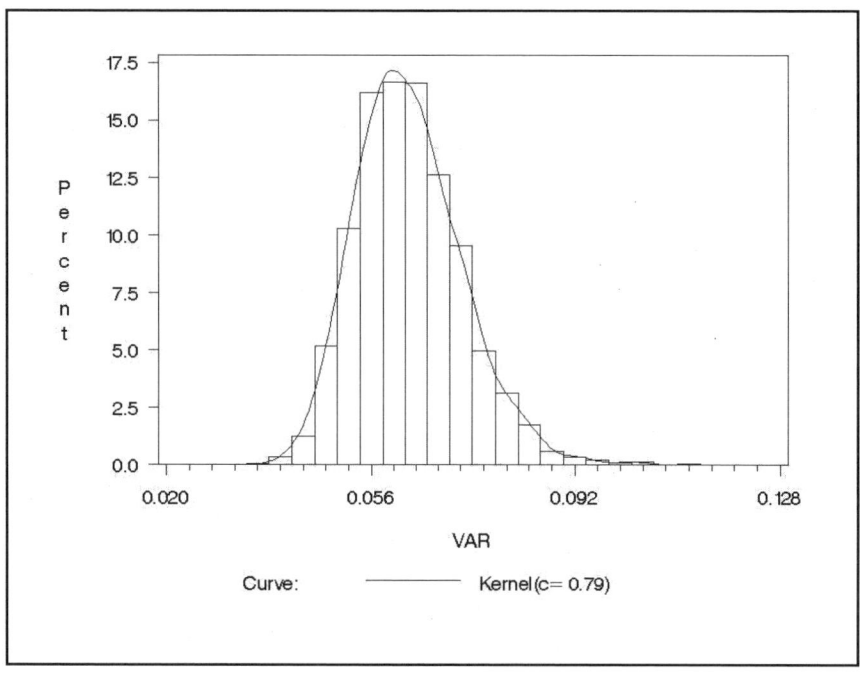

Output 10.10 Monte Carlo approximation for the posterior inference on the residual variance.

```
                         The CAPABILITY Procedure
           Variable:  VAR

                                   Moments
           N                          10000    Sum Weights               10000
           Mean                    0.06260194  Sum Observations      626.019394
           Std Deviation           0.00965539  Variance              0.00009323
           Skewness                0.62093108  Kurtosis              0.91735419
           Uncorrected SS          40.1222006  Corrected SS          0.93217241
           Coeff Variation         15.4234685  Std Error Mean        0.00009655

                           Basic Statistical Measures
                 Location                      Variability
           Mean      0.062602     Std Deviation             0.00966
           Median    0.061709     Variance                  0.0000932
           Mode         .         Range                     0.09628
                                  Interquartile Range       0.01272

           Quantiles (Definition 5)
           Quantile        Estimate
           100% Max       0.1324797186
           99%            0.0891275456
           95%            0.0799416237
           90%            0.0750554140
           75% Q3         0.0684825978
           50% Median     0.0617090729
           25% Q1         0.0557649226
           10%            0.0509790111
           5%             0.0485007595
           1%             0.0440746503
           0% Min         0.0361989435
```

For the position of the QTL, the marginal posterior is bimodal, with a higher mode between markers 9 and 10 and a smaller mode between markers 5 and 6 (Output 10.11), as discussed by Satagopan et al. (1996). Additional results are given in Output 10.12. It can be seen, for example, that the probability that the QTL is located between markers 9 (69.2 cM) and 10 (83.9 cM) is greater than 0.90.

Output 10.11 Histogram and kernel density estimate of the posterior distribution of the position (cM) of the QTL.

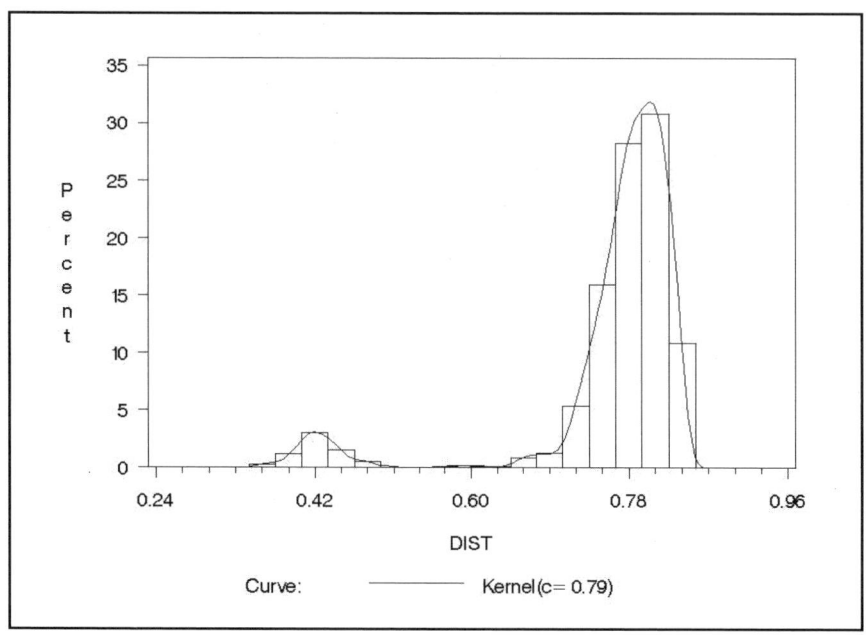

Output 10.12 Monte Carlo approximation for the posterior inference regarding the position (cM) of the QTL.

```
                         The CAPABILITY Procedure
           Variable:  DIST

                                Moments
           N                         10000    Sum Weights              10000
           Mean                  0.76477333    Sum Observations      7647.73327
           Std Deviation         0.08906631    Variance              0.00793281
           Skewness              -3.0501908    Kurtosis              9.11739375
           Uncorrected SS        5928.10256    Corrected SS          79.3201367
           Coeff Variation       11.6461052    Std Error Mean        0.00089066

                          Basic Statistical Measures
                Location                       Variability
           Mean      0.764773    Std Deviation               0.08907
           Median    0.786712    Variance                    0.00793
           Mode         .        Range                       0.50695
                                 Interquartile Range         0.05075

           Quantiles (Definition 5)
           Quantile           Estimate
           100% Max         0.838993240
           99%              0.837141042
           95%              0.831383554
           90%              0.825095866
           75% Q3           0.809595281
           50% Median       0.786712109
           25% Q1           0.758846632
           10%              0.722765225
           5%               0.464949442
           1%               0.402010001
           0% Min           0.332045705
```

10.6 Final Comments

The use of SAS tools to perform MCMC computations for Bayesian implementation of linkage and QTL analyses was illustrated with some simple examples. More general models have been presented in the literature to address, for example, different experimental designs and outbred populations (Hoeschele et al., 1997; Sillanpaa and Arjas, 1999; Bink et al., 2000), to estimate the number of QTLs (Satagopan and Yandell, 1996; Waagepetersen and Sorensen, 2001), and to study more complex genetic effects and interactions between QTLs (Yi and Xu, 2002).

SAS programs and SAS/IML can be used to perform such analyses in a reliable, efficient way. The user of Bayesian MCMC methodologies, however, should be aware of the issues relative to the definition of prior distributions (such as the use of improper priors or the influence of subjective priors on posterior distributions), MCMC convergence monitoring, and the problem of serial correlation between samples. For a comprehensive discussion of Bayesian methods and MCMC, refer to Gelman et al. (2003) and Robert and Casella (1999). A good treatment of MCMC applications in quantitative genetics is presented by Sorensen and Gianola (2002).

10.7 References

Bink, M. C. A. M., L. L. G. Janss, and R. L. Quaas. 2000. Markov chain Monte Carlo for mapping a quantitative trait locus in outbred populations. *Genet. Res.* 75:231-241.

Blasco, A. 2001. The Bayesian controversy in animal breeding. *J. Anim. Sci.* 79(8):2023-2046.

Broman, K. W. 2001. Review of statistical methods for QTL mapping in experimental crosses. *Lab Animal* 30:44-52.

Broman, K. W., and J. L. Weber. 1998. Estimation of pairwise relationships in the presence of genotyping errors. *Am. J. Hum. Genet.* 63:1563-1564.

Doerge, R. W. 2002. Mapping and analysis of quantitative trait loci in experimental populations. *Nature Reviews Genetics* 3:43-52.

Fernando, R. L., and M. Grossman. 1989. Marker assisted selection using best linear unbiased prediction. *Genet. Sel. Evol.* 21:467-477.

Ferreira, M. E., P. H. Williams, and T. C. Osborn. 1995. RFLP mapping of *Brassica napus* using double haploid lines. *Theor. Appl. Genet.* 89:615-621.

Ferreira, M. E., J. Satagopan, B. S. Yandell, P. H. Williams, and T. C. Osborn. 1995. Mapping loci controlling vernalization requirement and flowering time in *Brassica napus*. *Theor. Appl. Genet.* 90:727-732.

Gelman, A., J. B. Carlin, H. S. Stern, and D. B. Rubin. 2003. *Bayesian Data Analysis*. 2d ed. New York: Chapman & Hall/CRC.

George, A. W., K. L. Mengersen, and G. P. Davis. 1999. A Bayesian approach to ordering gene markers. *Biometrics* 55:419-429.

Gianola, D. 2000. Statistics in animal breeding. *J. Am. Stat. Assoc.* 95:296-299.

Gianola, D., and R. L. Fernando. 1986. Bayesian methods in animal breeding theory. *J. Anim. Sci.* 63: 217-244.

Gianola, D., R. Rekaya, G. J. M. Rosa, and A. Sanches. 2002. Advances in Bayesian methods for quantitative genetic analysis. *Seventh World Congress on Genetics Applied to Livestock Production*, Montpellier, France, August 19–23, 2002. CD-ROM.

Gilks, W. R., S. Richardson, and D. J. Spiegelhalter. 1996. *Markov Chain Monte Carlo in Practice*. London: Chapman & Hall.

Hoeschele, I., P. Uimari, F. E. Grignola, Q. Zhang, and K. M. Cage. 1997. Advances in statistical methods to map quantitative trait loci in outbred populations. *Genetics* 147:1445-1457.

Jansen, J., A. G. de Jong, and J. W. van Ooijen. 2001. Constructing dense genetic linkage maps. *Theor. Appl. Genet.* 102:1113-1122.

Knott, S. A., and C. S. Haley. 1992. Aspects of maximum-likelihood methods for the mapping of quantitative trait loci in line crosses. *Genet. Res.* 60:139-151.

Lande, R., and R. Thompson. 1990. Efficiency of marker-assisted selection in the improvement of quantitative traits. *Genetics* 124(3):743-756.

Lynch, M., and B. Walsh. 1998. *Genetics and Analysis of Quantitative Traits.* Sunderland, MA: Sinauer Associates.

Muller, J. 1916. The mechanism of crossing over. *Amer. Naturalist* 50:193-207.

Newton, M. A., and A. E. Raftery. 1994. Approximate Bayesian inference by the weighted likelihood bootstrap (with discussion). *Journal of the Royal Statistical Society B* 56:3-48.

Ott, J. 1999. *Analysis of Human Genetic Linkage.* Baltimore: Johns Hopkins University Press.

Raftery, A. E. 1996. Hypothesis testing and model selection. In *Markov Chain Monte Carlo in Practice*, ed. W. R. Gilks, S. Richardson, and D. J. Spiegelhalter. London: Chapman & Hall.

Robert, C. P., and G. Casella. 1999. *Monte Carlo Statistical Methods.* New York: Springer-Verlag.

Rosa, G. J. M., B. S. Yandell, and D. Gianola. 2002. A Bayesian approach for constructing genetic maps when genotypes are miscoded. *Genet. Sel. Evol.* 34(3):353-369.

Satagopan, J. M., and B. S. Yandell. 1996. Estimating the number of quantitative trait loci via Bayesian model determination. Special Contributed Paper Session on Genetic Analysis of Quantitative Traits and Complex Diseases, Biometrics Section, Joint Statistical Meetings, Chicago, IL.

Satagopan, J. M., B. S. Yandell, M. A. Newton, and T. C. Osborn. 1996. A Bayesian approach to detect quantitative trait loci using Markov chain Monte Carlo. *Genetics* 144:805-816.

Scott, D. W. 1992. *Multivariate Density Estimation: Theory, Practice, and Visualization.* New York: John Wiley & Sons.

Shoemaker, J. S., I. S. Painter, and B. S. Weir. 1999. Bayesian statistics in genetics: A guide for the uninitiated. *Trends Genet.* 15:354-358.

Sillanpaa, M. J., and E. Arjas. 1999. Bayesian mapping of multiple quantitative trait loci from incomplete outbred offspring data. *Genetics* 151:1605-1619.

Soller, M. 1978. Use of loci associated with quantitative effects in dairy-cattle improvement. *Animal Production* 27:133-139.

Sorensen, D., and D. Gianola. 2002. *Likelihood, Bayesian, and MCMC Methods in Quantitative Genetics.* New York: Springer-Verlag.

Sorensen, D. A., C. S. Wang, J. Jensen, and D. Gianola. 1994. Bayesian-analysis of genetic change due to selection using Gibbs sampling. *Genet. Sel. Evol.* 26:333-360.

Uimari, P., and I. Hoeschele. 1997. Mapping-linked quantitative trait loci using Bayesian analysis and Markov chain Monte Carlo algorithms. *Genetics* 146:735-743.

Waagepetersen, R., and D. Sorensen. 2001. A tutorial on reversible jump MCMC with a view toward applications in QTL-mapping. *Int. Stat. Rev.* 69(1):49-61.

Wang, C. S., J. J. Rutledge, and D. Gianola. 1993. Marginal inferences about variance-components in a mixed linear model using Gibbs sampling. *Genet. Sel. Evol.* 25(1):41-62.

Weller, J. I. 1986. Maximum-likelihood techniques for the mapping and analysis of quantitative trait loci with the aid of genetic markers. *Biometrics* 42:627-640.

Yi, N. J., and S. Z. Xu. 2002. Mapping quantitative trait loci with epistatic effects. *Genet. Res.* 79:185-198.

Chapter 11

Gene Expression Profiling Using Mixed Models

Greg Gibson and Russ Wolfinger

11.1 Introduction 251
11.2 Theory of Mixed Model Analysis of Microarray Data 252
11.3 Mixed Model Analysis of Two-Color Microarray Data Using
  PROC MIXED 254
  11.3.1 Data Preparation 254
  11.3.2 Computation of Relative Fluorescence Intensities 255
  11.3.3 Computation of Gene-Specific Significance Models 257
  11.3.4 Explore and Interpret the Results 258
11.4 ANOVA-Based Analysis of Affymetrix Data Using
  SAS Microarray Solution 260
  11.4.1 Download Data 260
  11.4.2 Examine Data 262
  11.4.3 Submit the Analysis 263
  11.4.4 Specifications for Mixed Model Analysis 264
  11.4.5 Submit the Analysis 266
11.5 Discussion 276
11.6 References 277

11.1 Introduction

Gene expression profiling is the monitoring of differences in the level of expression of thousands of individual genes across a series of treatments. The technology for gene expression profiling emerged in the 1990s with the development of two microarray platforms, commonly known as cDNA microarrays and oligonucleotide gene chips (Schena et al., 1996; Chee et al., 1996; see *The Chipping Forecast I* and *II* (1999; 2002) for two recent series of reviews). In essence, anywhere from several hundred to tens of thousands of DNA probes are spotted at high density onto a glass slide or silicon cassette, and a mixture of labeled target messenger RNA derivatives are hybridized to this array. The intensity of fluorescence associated with each spot is assumed to be proportional to the amount of transcript for the particular gene to which each probe corresponds. Statistical methods are then applied to analyze intensity levels of each gene across treatments and to evaluate the significance of differences between treatments.

There are essentially three major statistical challenges in the first phase of analysis of any gene expression profiling data set. The first is normalization of the data to remove systemic biases that may affect all of the genes simultaneously, such as differences in the amount of mRNA that is labeled in a particular replicate of a treatment (Yang and Speed, 2002; Quackenbush, 2002). The second is assessment of the contribution of biological and experimental sources of error to variation in the expression of each individual gene. We advocate performing these two steps sequentially (Wolfinger et al., 2001) as described below, but note that they can also be combined into a single model (Kerr et al., 2000; Churchill, 2002). The third challenge is organizing significantly differentially expressed genes into interpretable units, such as by hierarchical clustering (Eisen et al., 1998; Slonim, 2002). Most of these clustering operations can be performed in JMP or SAS, and will not be considered further in this chapter.

Microarray analysis is now used for many more purposes than simply identifying a list of genes that are differentially expressed over a series of conditions. Many of these applications involve methods discussed in other chapters in this volume, such as QTL mapping. Expression QTL, or eQTL, are quantitative trait loci for transcript abundance (Schadt et al., 2003; Yvert et al., 2003). They have been identified by simultaneously profiling gene expression and genotyping recombinant progeny of phenotypically divergent parents. Co-localization of eQTL and QTL peaks may become a potent tool for identification of the actual genes that are responsible for genetic variation, and similarly correlation of gene expression and phenotypic variation across a set of outbred lines has potential for finding the genes that matter. A variety of data mining approaches are also being applied to problems such as finding regulatory motifs associated with co-expressed genes; defining regulatory networks in time series data; and exploring information content in combined transcript, protein, metabolic, and text data.

While an in-depth discussion and examples of advanced topics like eQTL are beyond the scope of this chapter, we will focus on a foundational technique for microarray data analysis: the mixed model analysis of variance. The next section discusses some critical issues with this kind of approach. We then focus on specific methods for both two-color and Affymetrix data, and conclude with some summarizing observations.

11.2 Theory of Mixed Model Analysis of Microarray Data

Data normalization is typically performed on log base 2 transformed raw intensity measures, perhaps after some type of background subtraction according to the software that is used to capture the data. Oftentimes background subtraction simply adds noise to data, but where washing of the array leaves a portion of the microarray covered with a high background signal, it is advisable to subtract it. Log transformation both transforms the data toward normality (raw gene expression measurements tend to consist of a large number of genes at low intensity, and a few at very high intensity, with a distribution that is right skewed), and use of the base 2 scale ensures symmetry such that, for example, twofold increases or decreases are represented by the same difference (+/−1) on the measurement scale.

Several approaches can be taken to normalization. The most direct approach is simply to center the data such that the mean value of each log base 2 intensity measure is zero. Constructing residuals from a simple linear mixed model with array as a random effect and dye as a fixed effect is an easy way to implement this. Note the specific model you choose can depend upon the technology: Affymetrix GeneChip technology employs a single biotin label, so each array is represented by a single measurement, whereas microarrays usually employ competitive hybridization with two different fluorescent dyes (typically Cy3 and Cy5). You may also include the treatment terms in this linear model, which has the virtue of assessing any systemic effects of a particular treatment or interaction at the level of the whole transcriptome. For example, males may be found to give consistently higher expression values than females in a particular experiment. Dye effects are very often found to be significant in this type of analysis. Another kind of normalization seeks to ensure that the variance of individual hybridizations is constant, by dividing the centered log intensities by the standard deviation for the particular array and dye (Durbin et al., 2002). Speed and coworkers have advocated use of a local transformation, such as a loess transformation on each array, to reduce this source of bias, as well as adjustment for technical effects such as pin differences (Yang et al., 2002). In any case, the output of this step of the analysis is a "relative fluorescence intensity" (RFI) measure that expresses the observed transcript abundance relative to some baseline such as the sample mean for each replicate of each treatment.

Gene-specific models are then fit on these RFI values to assess the significance and magnitude of treatment effects for each gene, one gene at a time (Wolfinger et al., 2001). Mixed models are very useful here because nearly all microarray experiments involve fixed and random factors. Typically the treatment effects (for example, sex, genotype, drug, tissue type, time point, environment), as well as the dye when two colors are employed, are fixed effects. On the other hand, biological and technical effects (for example, array, spot, individual animal or plant) are assumed to be random and drawn from a normal distribution. It is essential to fit the array and dye effects in the gene-specific mixed models, despite the fact that they are also fit in the global normalization model. The array term controls for the correlation between the two measures from the same probe on microarrays, or between multiple probes for a single gene on a gene chip. Failure to fit this effect can strongly bias the analyses. The dye term controls for the fact that Cy3 and Cy5 do not incorporate into target RNA or cDNA with the same efficiency for each gene: hence the magnitude of dye effects varies greatly among genes, and can be particularly large for genes expressed at low levels. Plots of transformed data and residuals against predicted values for each array can provide hints of systemic biases in the data and visual assessment of model fit.

As always, the selected mixed model must be appropriate for the experimental objectives. For example, complete models including dye and array interaction terms with each of the treatment effects may not be feasible because of lack of sufficient degrees of freedom to estimate them or because of lack of biological interest. It will generally be informative to include some pairwise and higher interactions among treatment effects. However, since microarray experiments are expensive, it is not always practical to include sufficient replicates to justify all contrasts, and in this case the main effects and one or two interactions of particular interest may be modeled.

The model must also accommodate the experimental design. For example, both split-plot and loop designs are commonly employed. *Split-plot designs* are those in which direct contrasts are nested within a higher treatment. For example, in an experiment involving both sexes of two genotypes, all arrays might contrast genotypes within a sex (Jin et al., 2001). With mixed models it is nevertheless possible to fit a sex effect as well as to assess the interaction terms. The virtues of this approach are that it maximizes power for the direct contrasts and that it is quite flexible to imbalances in the design. *Loop designs* are randomized incomplete block designs that involve contrasts of different treatment types such that every comparison is represented by a direct contrast on a microarray (Kerr and Churchill, 2001; Churchill, 2002). So long as the design is balanced, they have the virtue that statistical power is equivalent for each treatment. Random effects should be selected carefully to represent the appropriate covariance structure of the data. For example, if the same genes are spotted more than once in an array, then effects for both array and spot-within-array can be included. Mixed models are capable of handling a very wide range of experimental designs including missing arrays, although it must be appreciated that statistical power varies across comparisons.

Potentially the most important theoretical issue is how to assess the significance of differences in gene expression. By fitting gene-specific models, treatment effects are determined independently for each gene: that is, variance components associated with each gene are used to assess significance. It can be argued that variance components such as that from the residual error are often mostly experimental noise and should not be gene-specific, and hence that a more reasonable approach is to fit a common error model, where the error term is determined by permutation (Kerr et al., 2000). However, as the number of treatments increases, and the number of replicates of each treatment rises, the contribution of experimental error may be more gene-specific and gene-specific error models are more justified. In practice, the gene-specific approach probably increases power in relation to small changes in expression, but may lose power for those genes where a few arrays are compromised by bad spots. Compromise empirical Bayes strategies are also possible (Friedman et al., 2000; Efron and Tibshirani, 2002).

A final prevalent issue relating to gene expression profiling is adjustment for the large number of multiple comparisons. Bonferroni corrections will usually be overconservative since gene expression is not independent (regulatory mechanisms introduce covariation) and interaction effects may also be correlated. A considerable body of literature on false discovery rates and setting of statistical thresholds is emerging (e.g., Tusher et al., 2001; Storey and Tibshirani, 2003). In our view, in practice, thresholds should be taken as a guide in conjunction with biological information and the aims of the experiment. If the goal is to make broad comparisons of the genomic response to a treatment, a high false-positive rate may be preferred over failure to detect differences, whereas if the aim is to single out a small number of genes for further analysis, a low false-positive rate is required.

11.3 Mixed Model Analysis of Two-Color Microarray Data Using PROC MIXED

We now present a step-by-step approach to analyzing two-color array data using mixed models. Refer to Chu et al. (2002) for a similar discussion of Affymetrix GeneChip data, and to Section 11.4.

11.3.1 Data Preparation

1a) Background subtraction
Most scanning software gives you a variety of options for subtracting a background signal from each spot intensity measure, as reviewed by Yang et al. (2001). These include subtraction of the mean or median pixel values within a square surrounding each spot, or of a local background covering several spots. More advanced procedures subtract the background from the true signal intensity peaks in a histogram of pixel intensities for each spot. In many cases, though, we have found that background subtraction simply adds noise, so unless there is a compelling reason to use it, such as clear regional background effects on the array, this step can be eliminated.

1b) Data scaling
The (background-subtracted) intensity values from the two channels from each array can be plotted against one another to check the quality of the data—namely that data for all genes whose transcription levels have remained essentially unchanged should fall on a straight diagonal line. The range of the values should be nearly the same for all channels, and neither channel should be saturated. Saturation is implied if one or both of the channels have a maximum value that is the same for more than 20 or so spots. In this case, the plot of the two sets of intensity values for an array will show a plateau. The magnitude of the maximum value can be changed by multiplication by a scaling factor, but no transformation can remove the plateau. This phenomenon arises because there is an upper limit to the intensity value that any pixel can take in an image file ($2^{16} = 65,536$), so if the fluorescence is too high, resolution of differences at the high end of the scale will be lost. The only way to deal with saturation is to rescan the slide at a lower laser

setting; however, this can decrease resolution at the low end, so an optimal compromise is often the best that can be achieved.

1c) Local transformation

Some authors also advocate local scaling of the data from each array to remove biases in the efficiency of labeling by the two dyes that appear to be dependent on the concentration of the transcripts (Yang et al., 2002). Thus, when the ratio of the intensity measurements for the two dyes for each spot is plotted against the product of the measures, a banana-shaped curve is often observed instead of a horizontal scatter. Loess or other smoothing spline transformations can be employed to remove this bias. With an ANOVA approach, some of the bias is accounted for by fitting the dye effect in the gene-specific models, and plots of the residuals from these models can reveal additional biases.

1d) Log base 2 transformation

Log transformation ensures that the data are approximately normally distributed within each gene, which improves statistical performance of the analysis of variance. The method is fairly robust to departures from normality, but non-transformed expression data are typically highly skewed since the median expression level is much lower than the mean. The log base 2 scale is convenient because each unit after transformation corresponds to a twofold difference. That is, 10 is half the expression level of 11; 14 is 8-fold greater than 11 and 16-fold greater than 10; and so on.

11.3.2 Computation of Relative Fluorescence Intensities

2a) Create a SAS data set

After these *preliminary* steps, the data should be prepared, typically in the form of a tab-delimited text file, with one row for each measurement as follows:

```
CLONE   ARRAY   DYE   TREAT1   TREAT2    RAW     LOG2I
C001    A01     CY3   MAL      GENO1      250    7.966
C001    A01     CY5   FEM      GENO1     2000   10.966
C001    A02     CY3   FEM      GENO2    10000   13.288
C001    A02     CY5   MAL      GENO2     4000   11.966
 .       .       .     .        .          .       .
 .       .       .     .        .          .       .
C999    A24     CY5   FEM      GENO2    64000   15.966
```

The CLONE column designates the clone (spot position) in the array, and the ARRAY column designates the number of the array itself. The DYE column designates which dye was used. The two TREAT columns categorize each different treatment, such as sex or genotype. The RAW column is the raw fluorescence intensity measure (after background subtraction), and LOG2I is the log base 2 transformed intensity measure.

A separate file is typically kept that relates the CLONE identifier to the GENE name, as the gene identity is not relevant during the statistical analysis. However, sometimes the same gene may be represented by two or more clones or by replicate spots, and so you may want to add an additional column indicating the replicate.

Note that as many treatments can be added to the list as desired, and these can take on any number of different categories. Some will be binary (for example, male or female sex; with or without drug), and some will have multiple levels (for example, a series of genotypes or time points). Labels can be alphabetical, numeric, or alphanumeric. Other columns might also be included, such as PIN number, slide BATCH, or X and Y coordinate values for constructing "pseudo-image" plots. Missing values are entered as a "." The number of rows in the final file should be the same as [the number of clones × the number of arrays × the number of dyes] plus 1 (for the header).

2b) Import the file into SAS

Suppose that the file containing the data has been imported into the WORK directory of SAS using the Import Wizard, and that you gave it the name MADATA. The comment "NOTE: WORK.MADATA was successfully created." should appear at the bottom of the Log pane. To confirm that the data were successfully imported, type the following into the Editor pane and submit:

```
proc print data=work.madata(obs=20);
run;
```

The first 20 rows of data should appear exactly as they do in the text file, but with an extra column at the beginning labeled "Obs" (for observation). SAS may think there are missing observations if there are empty rows beneath the last row of data in the text file. To avoid this problem, simply ensure that there are no spaces after the last entry in the text file, by deleting any empty space if necessary.

2c) Fit the *normalization* mixed model ANOVA

The next step is to compute the relative fluorescence intensities for each clone relative to the sample mean, namely centralization of the data. A short SAS program must be written in the Editor pane as follows:

```
proc mixed data=work.madata;      ❶
   class array dye;                ❷
   model value = dye / outp=rfi;   ❸
   random array array*dye;         ❹
run;
```

❶ The first line simply tells SAS which *procedure* (PROC) to run, and identifies the data file, which you previously imported into the (temporary) WORK directory.

❷ The class variables ARRAY and DYE will be included in the analysis. Pin and Batch might also be included if they are of interest, or you may choose to use a model with the treatment and even interaction effects. In practice, inclusion of treatment terms generally has little effect on the residuals so long as the array and dye terms are fit appropriately.

❸ The MODEL statement requests to fit the raw fluorescence values as a function of dye. The OUTP=RFI option tells SAS to create a new data set called RFI (for relative fluorescence intensity). It will contain the data needed for the next stage, in particular, the residuals (original minus predicted values) from this model.

❹ The random effects variables in the analysis are array and the ARRAY × DYE interaction.

2d) Save the relative fluorescence intensities

If desired, the RFIs can be saved permanently by submitting the following SAS code:

```
data sasuser.rfi;
   set work.rfi;
run;
```

This creates a permanent copy of the data in your SASUSER directory. You can also export it to another location as a text file using the Export Data command. Since the output will be too large for Excel, save it as a .txt file.

11.3.3 Computation of Gene-Specific Significance Models

3a) Sort data by clones
In order to perform mixed model ANOVAs on the expression profiles for individual genes, the expression data must be sorted so that each clone forms a batch of data. If this was not done initially, it can be done in SAS using the SORT procedure as follows:

```
proc sort data=sasuser.rfi;
   by clone treat1 treat2 dye;
run;
```

This procedure overwrites the permanent SASUSER.RFI file that was created in the previous step. In this example, the data are sorted by clone, then by treatments, then by dye levels. For most purposes, the only sort that really matters is the clone sort, as this ensures that all of the values for each clone are grouped into a contiguous block.

3b) Perform clone-specific ANOVAs
PROC MIXED is extremely versatile in terms of the statistical models you can specify. It fits each of the parameters in the model simultaneously using restricted maximum likelihood estimation. To assess the magnitude and significance of the effects of two treatments and their interaction on the relative fluorescence intensity of each clone, enter the following program into the Editor pane of SAS and submit it:

```
ods exclude all;  ❶
ods noresults;
proc mixed data=sasuser.rfi;
   by clone;  ❷
   class array dye treat1 treat2;
   model resid = treat1 treat2 treat1*treat2 dye / outp=sasuser.rfir;  ❸
   random array;  ❹
   lsmeans dye treat1 treat2 treat1*treat2 / diff;  ❺
   ods output covparms=sasuser.covparms tests3=sasuser.tests3  ❻
      lsmeans=sasuser.lsms diffs=sasuser.diffs;
run;
ods exclude none;
ods results;
```

❶ The first two ODS statements ensure that the results are not listed in the SAS output window. This avoids filling up this window unnecessarily.

❷ The BY statement tells SAS to compute the ANOVAs separately for each clone. PROC MIXED will fit a distinct model to each BY group, and so will often take several minutes to finish. You can check your SAS log during the run to see which clone is being processed. If for some reason you want to stop execution, click the **Break** (exclamation point) button.

❸ All of the variables in the MODEL statement are assumed to be fixed effects. This is appropriate in cases where the treatment effects are imposed by the biology (for example, sex or drug treatment) or where they were chosen for a specific reason and can be replicated at will (for example, extreme genotypes).

❹ ARRAY is considered a random effect and is equivalent to a spot effect if there are no replicate spots in the array. If there are replicate spots, you should include another random effect like SPOT*ARRAY to account for the different levels of correlation. The random effect assumption is justified by considering that spot effects will typically fluctuate according to an accumulation of many small random perturbations and are therefore assumed to arise from a normal probability distribution by appeal to the central limit theorem.

The ARRAY term must be included to control for variation among arrays. Fitting this term essentially substitutes for the rationale of using ratios—it takes into account the fact that pairs of observations in an array are correlated. You can run ANOVAs without this term, and oftentimes more significant effects will appear as a result; however, these results are usually misleading because of the ill-fitted model.

❺ The LSMEANS statement calculates the mean value for each treatment category, adjusted for the other terms in the model. Depending on the experimental design and degree to which the data are balanced, these may or may not be different from the actual means.

❻ The ODS OUTPUT statement outputs the results to four SAS data sets. The SASUSER.COVPARMS data set contains the variance component estimates for each clone. The SASUSER.TESTS3 data set contains the SAS Type 3 F-tests and p-values of each of the model terms. The SASUSER.LSMS data set contains the least squares means and the SASUSER.DIFFS data set contains the differences of the least square means for each of the effects along with associated t-tests and p-values.

To construct *custom* hypothesis tests, use ESTIMATE and CONTRAST statements. See the information about PROC MIXED in SAS Help and Documentation for details on how to construct these.

11.3.4 Explore and Interpret the Results

The output data sets from Section 11.3.3 have unique observations for each clone and model parameter estimate, and can therefore have thousands of records. Statistical and graphical summaries are imperative for efficient and effective investigation of dozens of potential research questions. To this end, we recommend the use of JMP software because of its feature-rich combination of common statistical methods and dynamic displays. Simply open the desired SAS data set in JMP and begin exploring. Keep in mind that JMP stores all results in memory, so some operations may be slow for large data sets, depending upon your hardware.

One good place to begin is with the SASUSER.RFIR data set. It contains both the normalized raw data and residuals for each of the gene-specific models, and these are useful for quality control and for checking model assumptions. Because the models have heterogeneous variance components, it's usually best to standardize the residuals to allow intercomparability. This can be accomplished by adding one or more PROC MIXED MODEL statement options available in SAS 9.1. You can then construct displays like Q-Q plots and residual-versus-predicted scatter plots in JMP in order to look for unusual data patterns or outliers.

For treatment difference summarization, a very useful graphical display is the volcano plot, shown in Figure 11.1. This plot is so named because each of the points explodes from a base of nonsignificant effects into a somewhat symmetric scatter of spots covering the full range of significance and magnitude of effects. The axes of the plot are fold change between two treatments (usually the difference in least

Figure 11.1 Typical volcano plots showing significance against fold difference in gene expression.

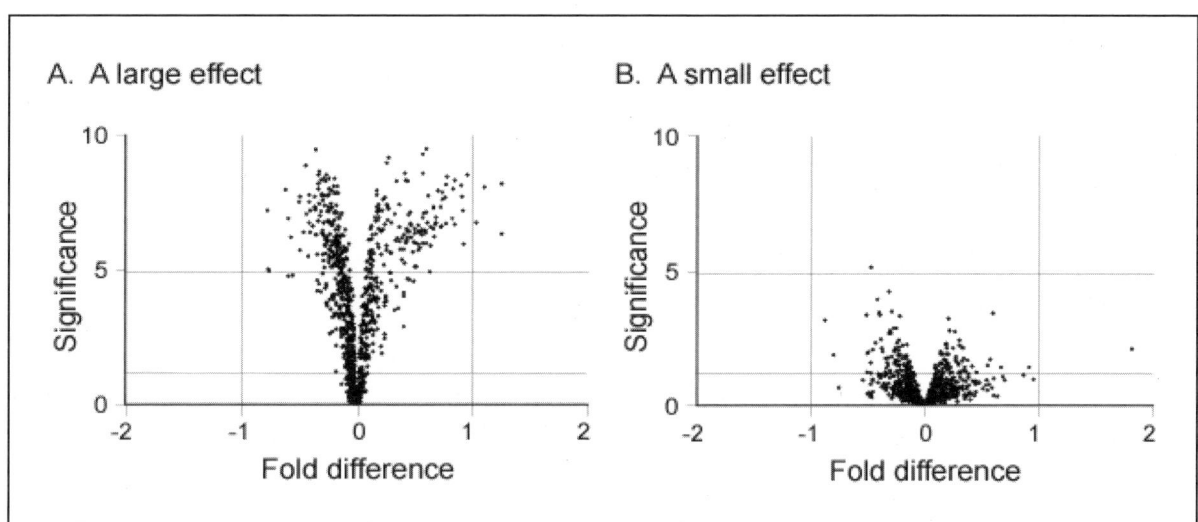

squares means on the log base 2 scale) on the abscissa, and significance on the ordinate as the negative logarithm of the *p*-value (equivalently, the logarithm of $1/p$). Figure 11.1 provides an immediate graphical comparison of the relationship between magnitude and significance, the symmetry and directionality of effects, and comparison of two or more treatments. In these plots, *fold difference* typically refers to the difference in least square mean estimates of the log base 2 scale intensity measures of the two treatments being contrasted, and *significance* is the negative logarithm of the *p*-value associated with the difference. The two vertical lines represent arbitrary twofold cutoffs, but these can also be computed as statistical thresholds if a common error model for all genes is employed. The upper horizontal hairline represents a typical multiple comparison experiment-wise threshold, in this case for $p = 10^{-5}$, from ANOVA of each gene, while the lower horizontal hairline represents the gene-wise 0.05 threshold. Figure 11.1A presents a typical plot for a highly significant biological effect, showing asymmetry toward upregulation in one of the treatments (to the right), and a large fraction of genes above the experiment-wise cutoff. Figure 11.1B shows a typical plot of a nonsignificant effect or a technical term such as the dye effect. Many genes are above the gene-wise threshold, and several genes are near or above a two-fold change without being significant overall.

Plotting gene effects this way highlights four classes of relationship between magnitude and significance of effect. Nonsignificant genes with small changes and little difference are at the base of the volcano, while genes with large and highly significant effects are toward the top left or right, indicating which direction the effect is in. Hover-text capability in JMP for identification of gene names or annotation provides a nice way to quickly visualize which families of genes are being co-regulated in the same manner. To the bottom left and right are genes that have apparently large effects, but if the variance among measurements is high, these will not be significant and hence might be false positives when decisions are made strictly on the basis of fold change. By contrast, genes in the top-center portion of a volcano plot are false negatives that would be missed by fold-change criteria, but which reproducibly show a consistent difference between treatments. Even a small change such as 1.3-fold may have dramatic biological consequences, and one of the prime virtues of statistical analysis is that it draws attention to such genes. Comparison of the relative explosiveness of volcano plots for different treatments provides a quick view of the contribution of each pairwise comparison to the transcriptional variance in a large and complex experiment.

Practically, you can construct volcano plots on the least squares means differences by first executing the following:

```
data sasuser.diffs;
   set sasuser.diffs;
   NegLog10p = -log10(Probt);
run;
```

Then you open this data set in JMP. Click **Graph → Overlay Plot**, select NEGLOG10p as the Y variable and ESTIMATE as the X variable, and then select all of the treatment variable names, including those beginning with an underscore, as the BY variables. This creates a set of linked volcano plots for each least squares mean difference. To enhance these displays, you can merge gene annotation information with this data set and use various fields as labels or colors.

11.4 ANOVA-Based Analysis of Affymetrix Data Using SAS Microarray Solution

This section steps you through a basic set of analyses of an experiment conducted using Affymetrix gene chips using SAS Microarray Solution 1.0.1. For the sake of time and simplicity, we will use only the MAS 5.0 summary statistics here; however, much more precise inferences can usually be obtained using the probe-level data obtained from Affy CEL files (Chu et al., 2002). Such probe-level analyses are similar in form to the ones described here, but they are more complex because they directly account for probe-specific effects.

11.4.1 Download Data

Download the Asthma-Atopy microarray data set GDS266.SOFT.TXT from the publicly available Gene Expression Omnibus (GEO): www.ncbi.nlm.nih.gov/geo/gds/gds_browse.cgi, under the title **Asthma and Atopy (HG-U133A)**. Edit the downloaded SOFT text file to remove the header information. The first row of the new file should contain the column names ID_REF, IDENTIFIER, GSM3921, GSM3923, and so on. Save this new text file with the name GDS266.soft.txt. The arrays are divided into five categories as described in Table 11.1.

Table 11.1 Summary of data available in the Asthma-Atopy experiment.

Asthma	Atopy	Coding of Variable trt	Number of Arrays
None	No	asthma0_atopy0	4
None	Yes	asthma0_atopy1	5
Mild	No	asthma1_atopy0	2
Mild	Yes	asthma1_atopy1	13
Severe	Yes	asthma2_atopy1	5

Note that the combination of asthma=severe and atopy=no is missing from these data, so direct inferences about certain interactions will be impossible. We will use a one-way ANOVA coding for the treatment combinations and construct interesting comparisons directly.

The data from the 29 arrays can be loaded into SAS Microarray Solution using its generic input engine. The result is one tall SAS data set with all of the expression measurements stacked into one variable, along with other variables for genes, arrays, and treatment. Alternatively, you can construct the tall data set directly in SAS using the following code. Replace the path with the one on your machine.

```
%let path = C:\Documents and Settings\<yourID>\My Documents\BISC\Asthma\;

proc import out=aa datafile="&path.GDS266.soft.txt" dbms=tab replace;
   getnames=yes;
   datarow=2;
run;

libname a "&path";

/*---log2 transform MAS 5.0 statistics---*/
%let narray = 29;
data a.aa;
   set aa;
   array gsm{*} gsm:;
   do array = 1 to &narray;
      gsm[array] = log2(gsm[array]);
   end;
   drop array;
run;

/*---create tall data set by hand-coding treatments for one-way ANOVA,
     note asthma2_atopy0 combination is not present in the data---*/
data a.aatall;
   set a.aa;
   array g{*} gsm:;
   do array = 1 to &narray;
      log2mas5 = g[array];
      if (array <= 13)      then trt = "asthma1_atopy1";
      else if (array <= 15) then trt = "asthma1_atopy0";
      else if (array <= 20) then trt = "asthma2_atopy1";
      else if (array <= 25) then trt = "asthma0_atopy1";
      else if (array <= 29) then trt = "asthma0_atopy0";
      else trt = "Unknown";
      output;
   end;
   drop gsm:;
run;

/*---note that some of the control spots are replicated 2-4 times on the chip,
so take means of these to create a single obs per gene on each array---*/
proc sort data=a.aatall;
   by array id_ref;

proc means data=a.aatall noprint;
   by array id_ref;
   var log2mas5;
   id identifier trt;
   output out=aatallm(drop=_type_ _freq_) mean=log2mas5;
run;

/*---normalize by subtracting medians---*/
proc stdize data=aatallm out=a.aatallstd method=median;
   by array;
   var log2mas5;
run;

/*---change id ref for later merge with annotation---*/
data a.aatallstd;
   set a.aatallstd;
   rename id_ref = Probe_Set_ID;
   drop identifier;
run;
```

262   Part 2: Molecular Genetics

```
/*---load trimmed u133 annotation file from netaffx.com---*/
proc import out=a.u133anno
      datafile= "&path.HG-U133A_annot_trim.csv"
         dbms=csv replace;
      getnames=yes;
      datarow=2;
run;
```

Note the MAS 5.0 statistics have been log base 2 transformed to enable interpretations in terms of log base 2 fold change, and the data from each chip have been pre-standardized to have median 0. Also, an annotation file from www.netaffx.com appropriate for this U133A gene chip is also loaded into a SAS data set in order to provide descriptive information about each gene.

11.4.2 Examine Data

In SAS Microarray Solution, click **Analysis → Analytical Process Selector**, then click **ArrayGroupCorrelation**. This initial analysis is designed to perform a check on data quality and consistency, as well as to enable some ad hoc exploration. The idea is to plot the data from each chip against those from other chips in the same treatment group, letting each point in the plots represent a gene. Input for the process is shown as two screen shots (Display 11.1A, B; note the scroll bar on the right is moved down in 11.1B).

Display 11.1 Example input shown across two screens (A and B).

A)

B)

11.4.3 Submit the Analysis

The underlying SAS macro will execute, creating a new SAS data set and a JMP script. The JMP script automatically generates a set of scatterplot matrices from the Multivariate platform in JMP, the first of which is shown in Display 11.2 (for the asthma0_atopy0 group).

Display 11.2 Scatterplot matrix for the asthma example.

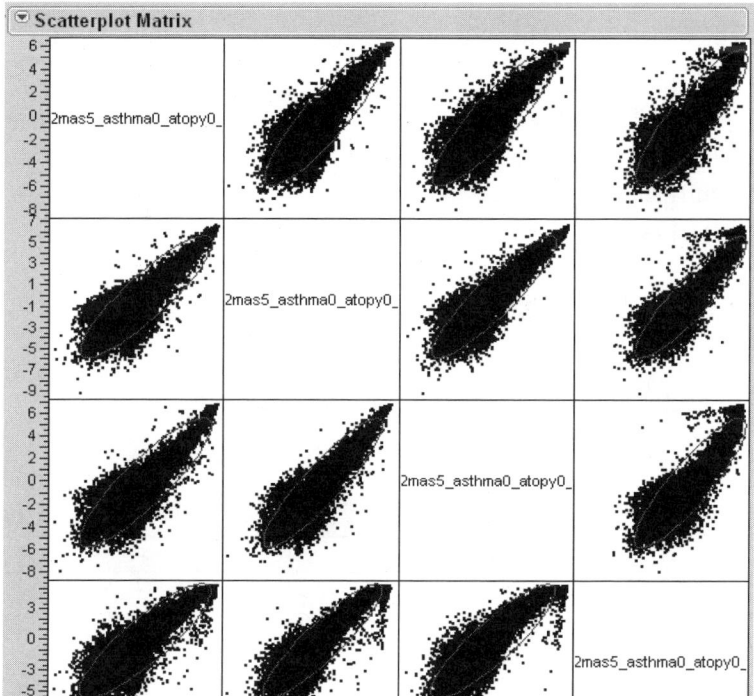

The data appear to have good quality, with correlations around 0.9 and a touch of saturation effects at the high end of certain arrays. The gray ellipses in Display 11.2, more easily seen on a color monitor, represent 95% Gaussian contours. These ellipses help to delineate the degree to which the LOG2MAS5 statistics have a higher degree of noise toward the low end.

Use the horizontal scroll bar to view the scatterplot matrices from the other groups. Move the mouse pointer over data points to view their labels, or click on the data points to highlight them in the data table and all the graphs simultaneously. You can also use **Rows → Colors** to color the points that are currently selected.

Finally, you can perform many additional JMP analyses from here. Feel free to click on items along the top menu bar to obtain a feel for the breadth of analyses possible. Selecting **Analyze → Distribution** is a good place to start. Note the data table has 22,266 rows (corresponding to the genes on the U133 chips) and columns corresponding to gene identifiers and annotation as well as separate columns for each of the 29 arrays containing the LOG2MAS5 statistics. Certain analyses will be too computationally intensive to perform directly in JMP because of the size of the data and the fact that JMP holds all of the data and results in memory. We will perform one such analysis next by first doing the processing on a SAS server and then displaying only the results in JMP.

11.4.4 Specifications for Mixed Model Analysis

Return to SAS Microarray Solution, and select the pre-loaded mixed model analysis. This kind of analysis addresses several research objectives simultaneously by fitting a linear statistical model to the data for each gene separately and then displaying results in a dynamically interactive fashion. The results include the following:

- analysis of variance on the treatment groups
- selection of significant genes based on customized hypothesis tests and popular multiple testing criteria
- mean expression profiles of significant genes
- hierarchical clustering of significant genes
- principal components display of significant genes
- links to relevant annotation and Web sites.

The input parameters are shown as four successive screen shots (Display 11.3A–D).

Display 11.3 Example input for mixed model analysis, shown across four screens.

A) B)

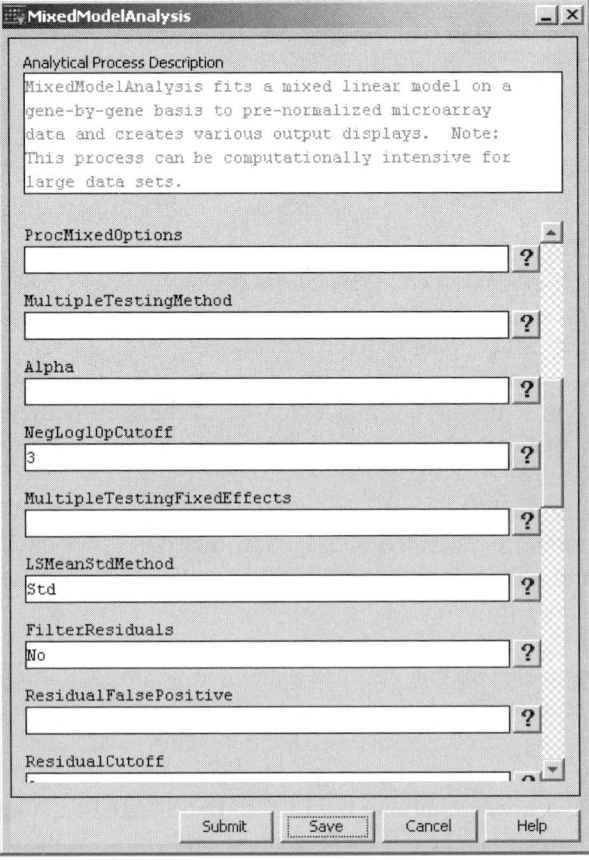

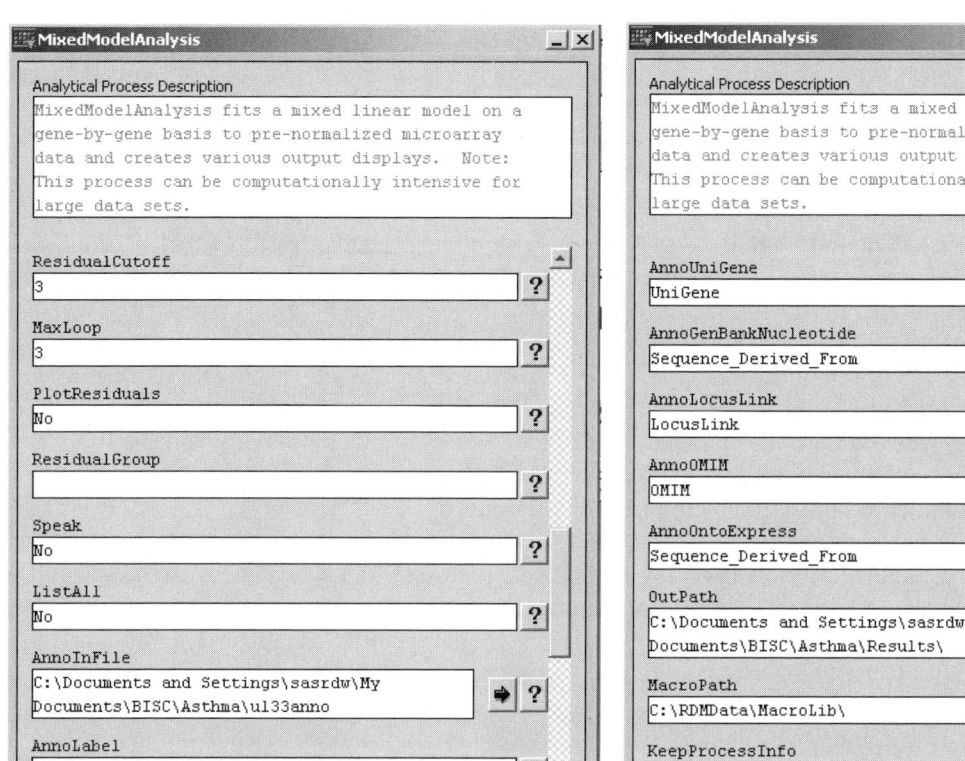

The central piece of SAS modeling code that is executed for each gene is as follows:

```
proc mixed;
   class trt;
   model log2mas5 = trt / outp=generesiduals;
   lsmeans trt;
   estimate "Asthma 1 minus 0"
      trt -1 -1 1 1 0;
   estimate "Atopy 1 minus 0"
      trt -1 1 -1 1 0;
   estimate "Asthma 2 minus 0 Atopy 1"
      trt 0 -1 0 0 1;
   estimate "Asthma 2 minus 1 Atopy 1"
      trt 0 0 0 -1 1;
run;
```

This code sets up a one-way ANOVA model of LOG2MAS5 on TRT. The LSMEANS statement requests mean profiles for each of the five treatment groups. The four ESTIMATE statements create custom hypothesis tests across the five levels.

11.4.5 Submit the Analysis

Upon completion of the analysis, a whole series of graphs will appear in JMP. Let's work through them now one by one. After each step, minimize the current window or press CTRL-B to send it to the back so that the next one is viewable.

The R-Squared Values window displays statistics of the percent of variability explained for each of the 22,266 linear model fits (Display 11.4). The results range from 0 to 70 percent, indicating a large amount of unexplained variability. The R-squared statistics, along with numerous other results from all of the model fits, are contained in the JMP table named MIXEDRESULTS.

The Variance Component Estimates window (Display 11.5) shows the distribution of the estimates of the residual variance from each of the linear model fits. Note this statistic is directly related to R-squared, since R-squared = (Total Variance minus Residual Variance)/Total Variance. Studying the variance estimates of individual genes can produce interesting insights. For example, by moving your mouse pointer over the top-most points in the box-plot portion of the graph, you can see that the genes with the most unexplained variability are those producing ribosomal proteins.

The Volcano Plots for Estimates window (Display 11.6) displays results for each of the hypothesis tests from the previously specified ESTIMATE statements. These plot the log base 2 fold change of the tested quantity (e.g., Asthma 1 minus 0) on the x-axis versus that quantity's statistical significance on the y-axis (negative log10 *p*-value, so that significant genes appear at the top of the graphs).

You can move your mouse pointer over points in these graphs to see their labels, or click on them to select them in the data table and all other associated graphs. SHIFT-click to select several at once, then from the top menu bar, click **Rows → Label/Unlabel** to display labels on the graphs. Click in an empty part of the graph to deselect all points. Right-click inside any of the graphs to select several more options (e.g., to change marker size and drawing mode).

Display 11.4 R-Squared Values window.

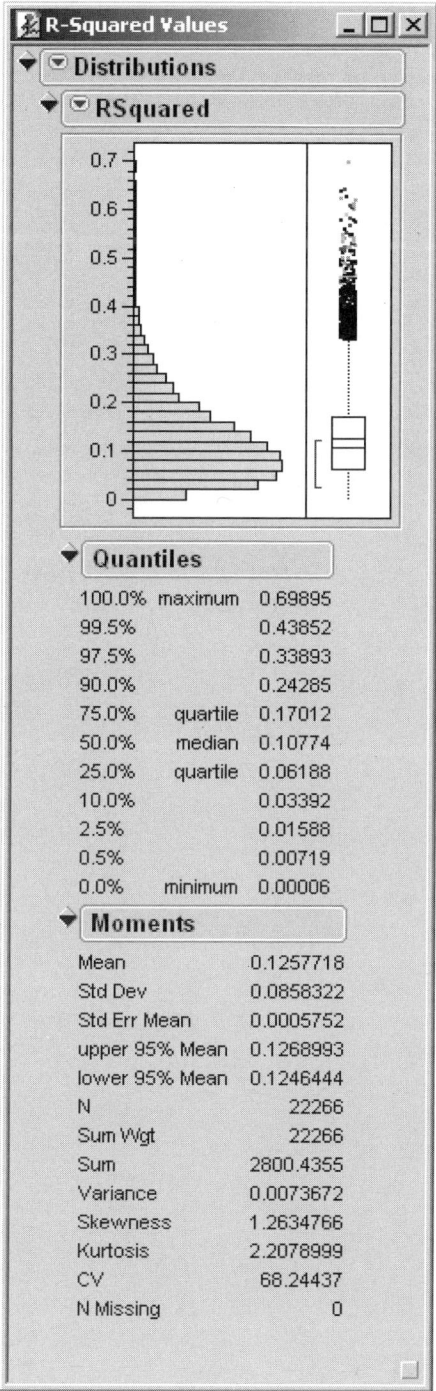

Display 11.5 Variance Component Estimates window.

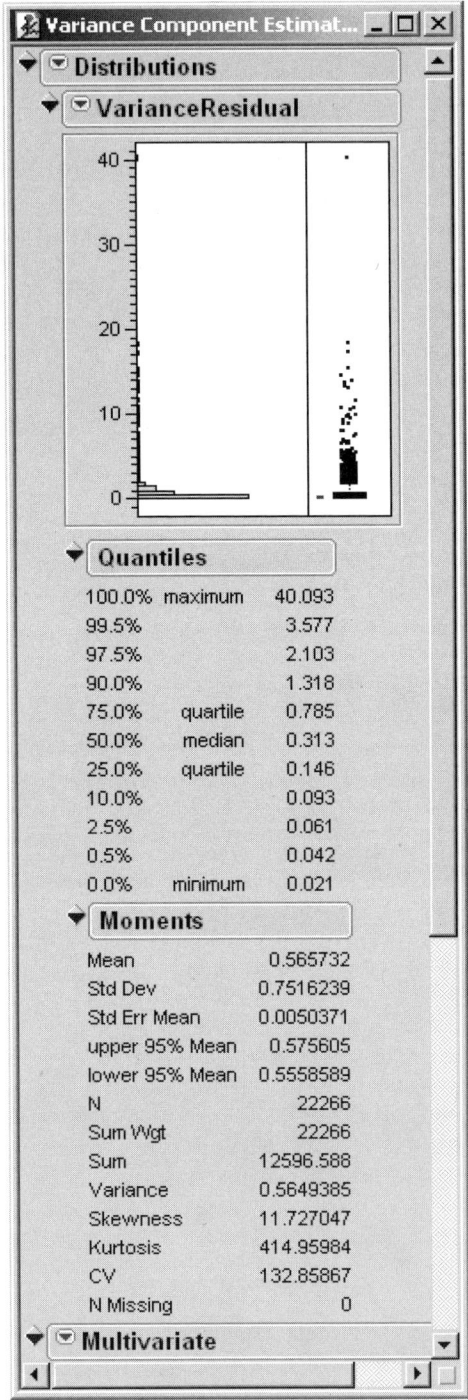

Display 11.6 Volcano Plots for Estimates window.

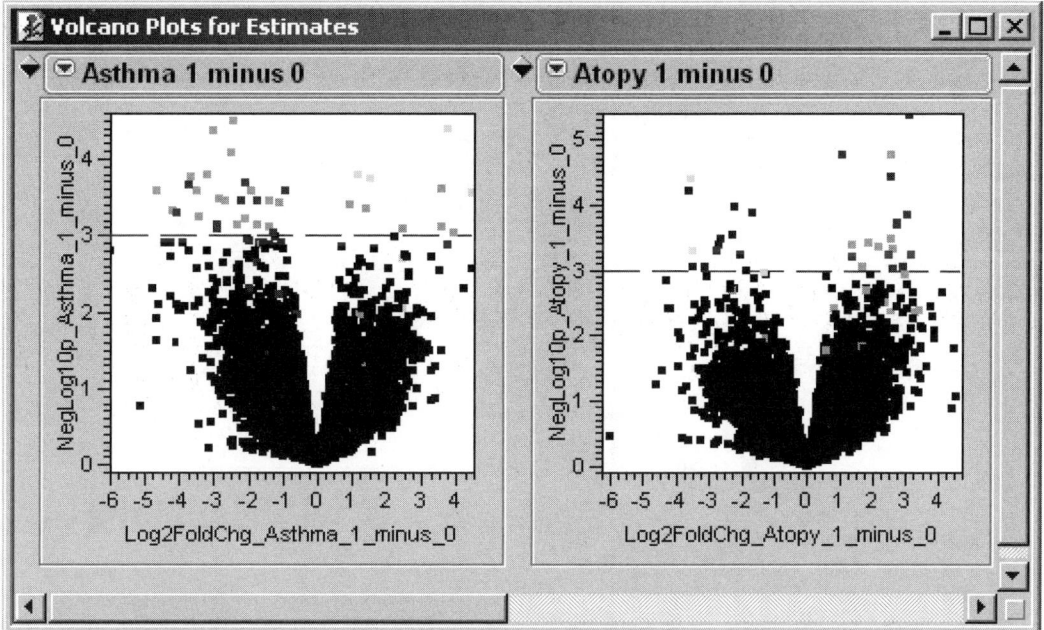

The dashed horizontal reference line (at $-\log 10(.001) = 3$) in each plot is constructed from a *p*-value cutoff of 0.001 chosen to provide a false-positive rate of 1 in every 1000 null hypothesis tests. Of the 22,266 genes, 74 exceed this cutoff in at least one of the hypothesis tests, and the JMP script creates a subtable of these genes from MixedResults called Significant Genes to be used for subsequent graphs.

The Parallel Plot of Least Squares Means window (Display 11.7) displays expression profiles for the 74 significant genes. You can click on profiles to select them in the data table and all other associated graphs. The Parallel Plot of Standardized Least Squares Means are the same profiles standardized to have mean 0 and variance 1 (Display 11.7). These are better suited for expression pattern matching, and are more closely related to the results in the volcano plots.

Display 11.7 Parallel plots for least squares means.

Note that many of the genes have peaks that occur in the asthma1_atopy0 group (in the middle of the plots). This is the treatment group that has only two chips, and so these results should be viewed with some caution because of the minimal replication.

The Significant Genes—Hierarchical Clustering window (Display 11.8) shows a two-way clustering analysis of the standardized least squares means for the significant genes. The heat map and dendrogram branches are clickable for selecting interesting subsets of these genes. The coloring in the other plots is established in this plot, and you can click and slide the cross-hairs to change the color groups. Click on the small red triangle to the left of "Hierarchical Clustering" to select several other useful options.

Display 11.8 Window displaying hierarchical clustering results.

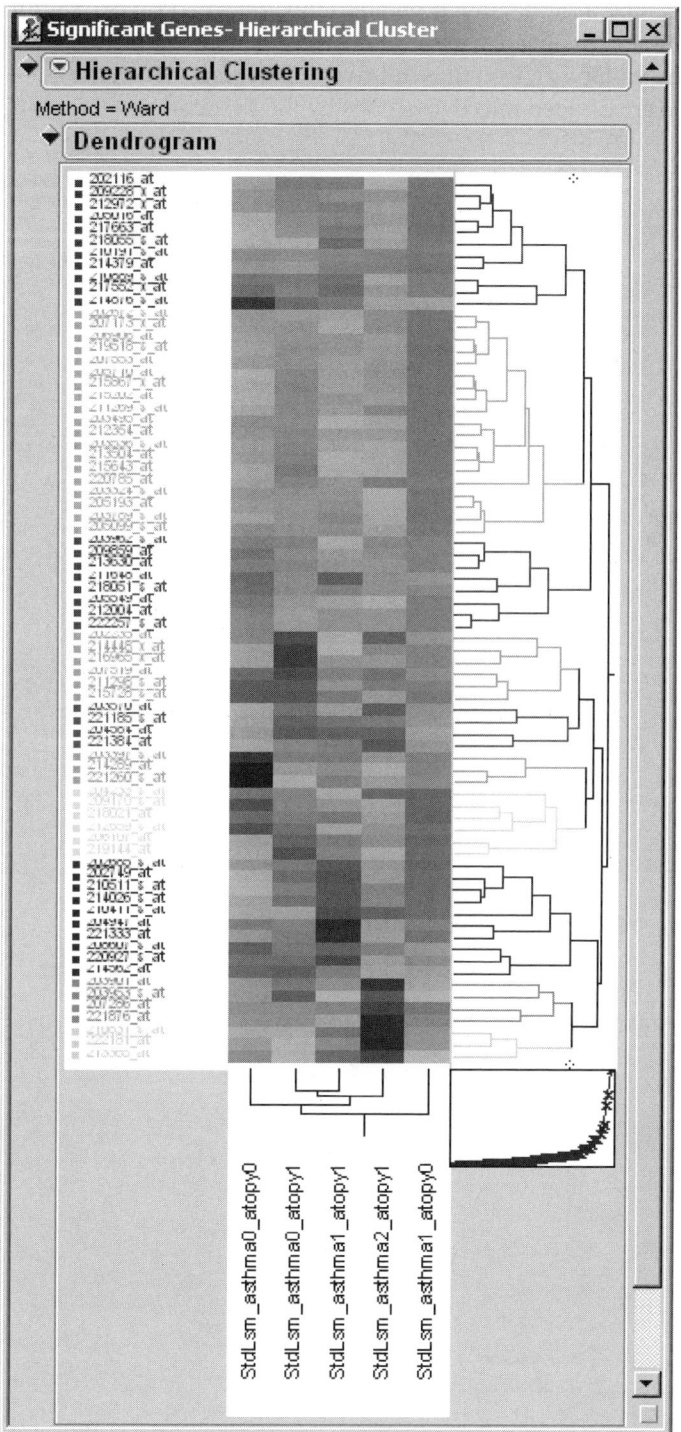

The heat map provides a colored view (on your computer) of the standardized least squares mean profiles. Several interesting patterns are apparent, and every gene has a significant difference somewhere in its row. The asthma1_atopy0 column has noticeable differences from the other four and is clustered to the far right. Again, since this group is derived from only two samples, some caution should be exercised in interpreting results.

The Principal Components of Standardized Least Squares Means window (Display 11.9) displays the two-dimensional projection of the expression profiles that has maximal separation in variance. Points closer to the edge of the invisible bounding ellipse are better explained by these first two principal components, and here the first two components largely capture the variability from the full five dimensions. Also, the clusters from the hierarchical clustering analysis are divided into two groups by the first component.

Display 11.9 Principal components display.

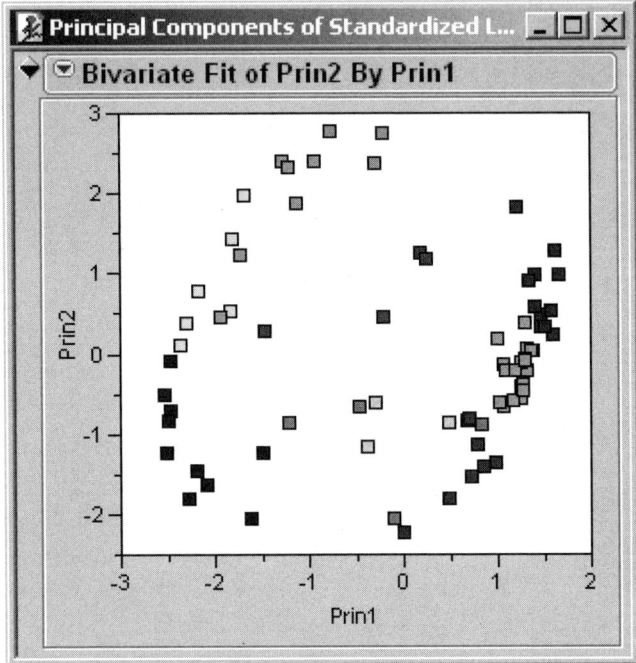

The Web Search dialog box enables you to open selected public Web pages on any genes you have selected (Display 11.10).

Display 11.10 Window that provides access to Web searches.

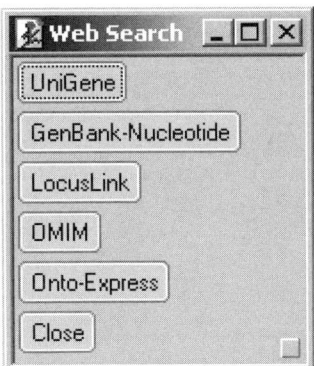

To illustrate one use of this facility, first select **Window** on the top menu bar, and select the JMP table entitled "Significant Genes." Use the horizontal scroll bar to bring the column "Title" into view. This column gives the names of the genes. Use the vertical scroll bar on the list to visually scan through the list of 74 significant genes. Let's assume row 43 of this table catches your interest because it contains "interleukin 1 receptor antagonist." Click on the row indicator column to highlight this entire row. Next, click **Window → Parallel Plot of Standardized Least Squares Means**, producing the image in Display 11.11.

Note that the expression profile for this gene is highlighted (with a thick light gray line in Display 11.11), and that expression of this gene is higher in asthmatic subjects.

Display 11.11 Least squares mean plot highlighting one gene.

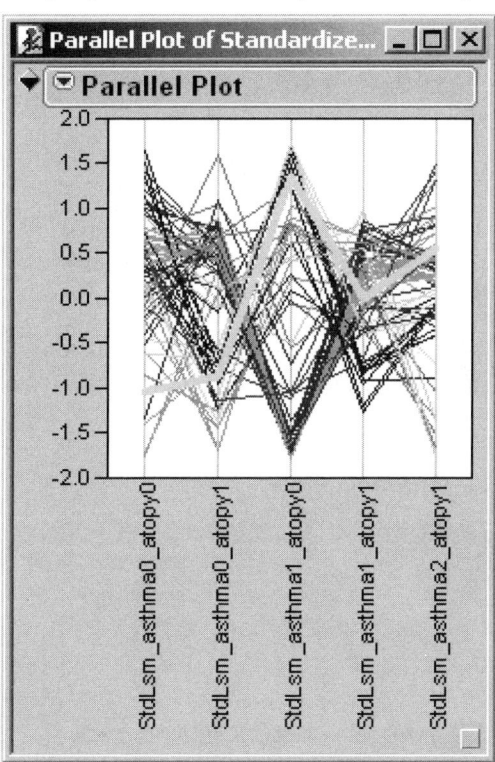

Display 11.12 NCBI Web access to specific genes.

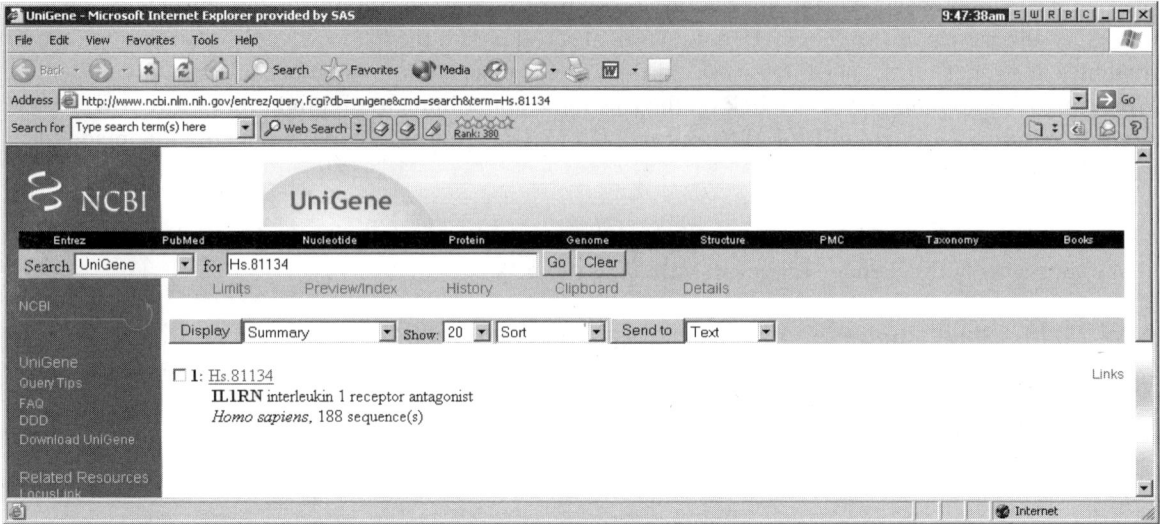

To easily access public resources on this gene, click **Window → Web Search** to bring the JMP Web search window back into view (Display 11.10). Click the **UniGene** button, and the NCBI Web page for this gene will appear in a new Web browser window (Display 11.12).

Click the **Hs.81134** link in this window to begin one exploration path. For another, return to the JMP Web Search window and click **GenBank – Nucleotide**, bringing up the image in Display 11.13. A third path for LocusLink brings up the window in Display 11.14.

Display 11.13 NCBI Web access to nucleotide sequence information.

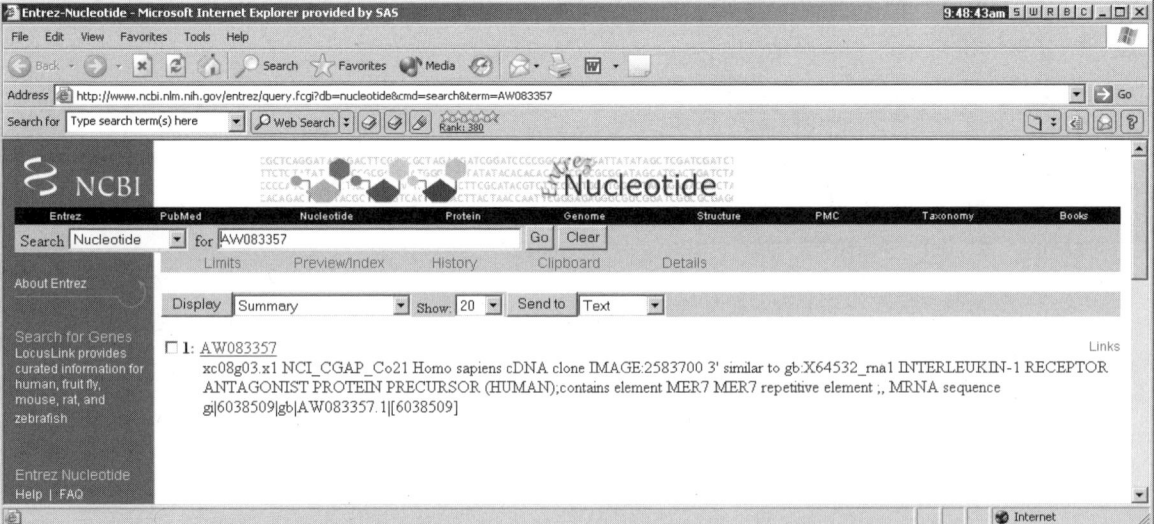

Display 11.14 NCBI Web access to gene level information.

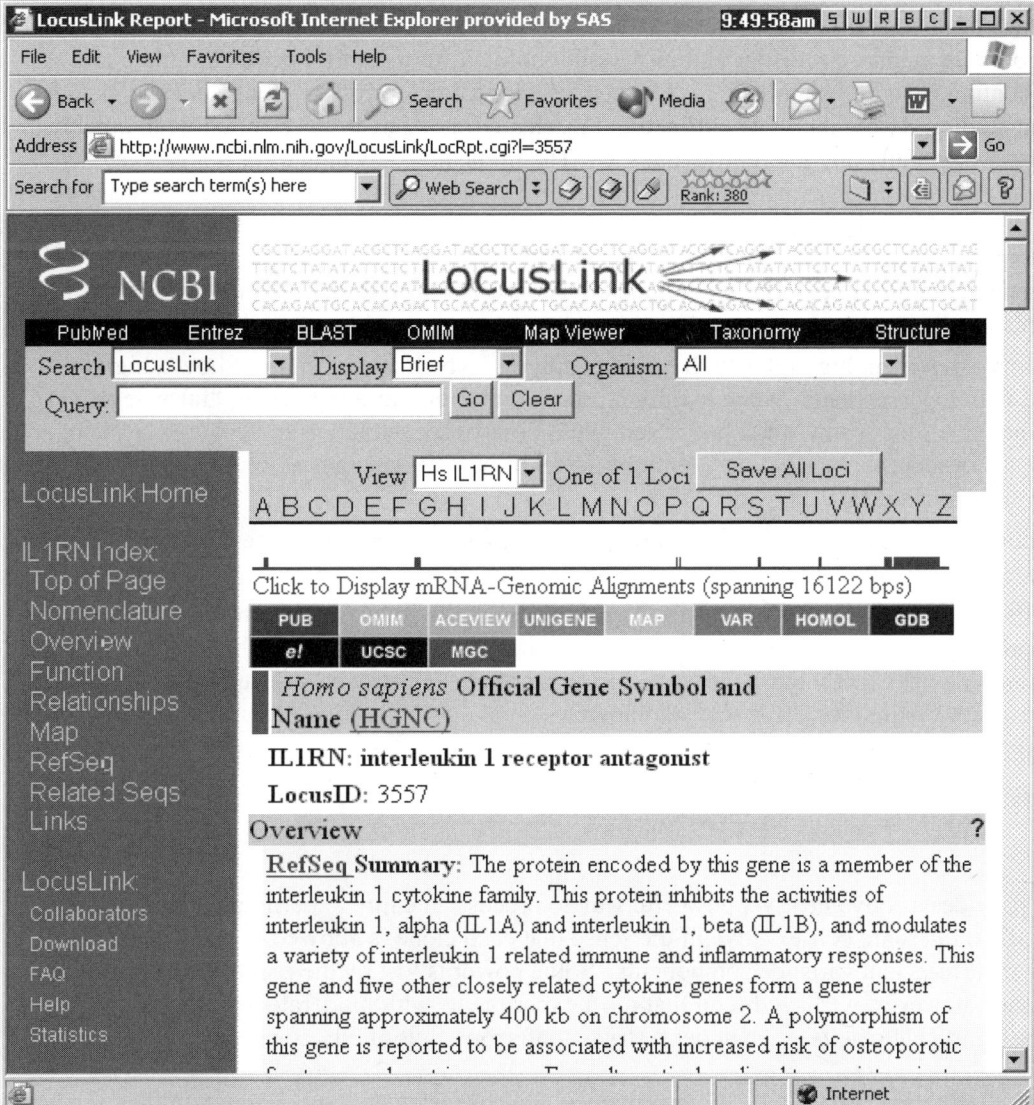

Additional buttons are available for OMIM and OntoExpress. Note it is easy to add your own buttons by adding to the JMP script driving all of these analyses. (This script is viewable in the MIXEDMODELANALYSIS window.) Additional annotation columns (e.g., SwissProt, Ensembl, and GO) are already available for use in the SIGNIFICANT GENES and MIXEDRESULTS JMP tables. The annotation for this example for the Affymetrix U133 GeneChip was loaded from netaffx.com. Continue exploring your favorite genes by selecting them from the table or graphs and then clicking the desired Web search button. If you have more than one gene selected, a separate Web page will appear for each gene. The OntoExpress facility allows you to submit an entire list.

11.5 Discussion

Statistical analysis of microarray data using mixed models is just the first phase of data analysis. Actually, many authors continue to ignore statistics and move straight to clustering and other data mining procedures, or continue to adopt simplistic fold-change criteria for determining which genes to include in their subsequent studies. Using the steps outlined in this chapter, it is nevertheless possible to quickly and efficiently analyze essentially any microarray data set. The immediate advantages of doing so are as follows:

- Gene expression values are computed relative to all of the sources of variance in the model: if two treatments are confounded (say, sex and alcohol consumption), then fitting a mixed model accounts for the identified sources of bias.

- Statistical assessment highlights both "false positives," which are genes that show a high fold change with low repeatability due to outlier arrays, and even more important, "false negatives," which are genes that show small but nevertheless highly repeatable differences. As a rule of thumb, six or more replicates give plenty of power to observe changes as small as 1.3-fold even at the 10^{-5} significance level.

- Interaction effects can be studied, allowing the investigator to focus on genes that only show a response to one treatment in the context of another treatment.

- All sorts of experimental designs can be analyzed, generally involving variants of the common loop and split-plot designs. Balance and completeness are desirable, but unbalanced designs can be processed with exactly the same code, and appropriate adjustments are automatically made with respect to estimation efficiency and power.

- Reference samples are unnecessary in two-color experiments, which means that the waste of expensive resources can be avoided, and that experiments performed by different groups on the same array platform can be analyzed together.

The preceding provides a convenient yet powerful way to process an entire experiment simultaneously. The clone-specific models are conservative in the sense that each clone is allowed to tell its own story in terms of treatment effects and variance components. Some power is lost by the fact that no pooling is done across clones; however, we have found that most experiments with moderate replication produce plenty of highly significant results, even under a strict Bonferroni multiplicity correction.

If you have a small design or very subtle effects, there is a fairly easy way to pool across clones and gain substantial power. You can do this by selecting prototypical values for the variance components based on results across all of the genes, and then add a PARMS statement to the gene-specific code that specifies these values along with the HOLD= option to hold them constant for each clone. Along with this, you can use the DDF= option in the MODEL statement to specify increased degrees of freedom for the fixed effects tests.

The basic procedures outlined here have been automated in SAS Microarray Solution software. SAS Microarray Solution consists of three major components: a SAS warehouse in which individual arrays and merged data sets are stored ready for analysis; a Java interface that facilitates writing of SAS code based on a small set of templates, forces the user to move through various data quality control steps, and makes available the full range of SAS procedures; and scripted JMP output that presents the results in a standard set of graphical formats (including volcano plots of significance against fold change, and hierarchical clusters) and also facilitates further data mining. This output is also linked through a Web server to public genome databases, allowing immediate access to the world of information relating to individual genes.

We have not discussed in any detail any of the numerous statistical issues that arise in the context of gene expression profiling. Powerful as this approach is, we recognize there are many steps that will not be familiar to molecular biologists or those new to SAS. Consequently, our final recommendation is that as much as possible, biologists and statisticians work together on data analysis side by side: pooled resources always increase power!

11.6 References

Chee, M., et al. 1996. Accessing genetic information with high-density DNA arrays. *Science* 274:610-614.

Chipping Forecast, The. 1999. *Nature Genetics Supplement* 21:1-60.

Chipping Forecast II, The. 2002. *Nature Genetics Supplement* 32:461-552.

Chu, T.-M., B. Weir, and R. Wolfinger. 2002. A systematic statistical linear modeling approach to oligonucleotide array experiments. *Mathematical Biosciences* 176:35-51.

Churchill, G. A. 2002. Fundamentals of experimental design for cDNA microarrays. *Nature Genetics* 32:490-495.

Durbin, B. P., J. S. Hardin, D. M. Dawkins, and D. M. Rocke. 2002. A variance-stabilizing transformation for gene expression microarray data. *Bioinformatics Suppl.* 18:S105-110.

Efron, B., and R. Tibshirani. 2002. Empirical Bayes methods and false discovery rates for microarrays. *Genet. Epidemiol.* 23:70-86.

Eisen, M. B., P. T. Spellman, P. O. Brown, and D. Botstein. 1998. Cluster analysis and display of genome-wide expression patterns. *Proc. Natl. Acad. Sci. USA* 95:14863-14868.

Friedman, N., M. Linial, I. Nachman, and D. Peer. 2000. Using Bayesian networks to analyze expression data. *J. Compu. Biol.* 7:601-620.

Jin, W., R. Riley, R. D. Wolfinger, K. P. White, G. Passador-Gurgel, and G. Gibson. 2001. Contributions of sex, genotype, and age to transcriptional variance in *Drosophila melanogaster*. *Nature Genetics* 29:389-395.

Kerr, M. K., and G. A. Churchill. 2001. Statistical design and the analysis of gene expression microarray data. *Genetical Research* 77:123-128.

Kerr, M. K., M. Martin, and G. A. Churchill. 2000. Analysis of variance for gene expression microarray data. *J. Compu. Biol.* 7:819-837.

Quackenbush, J. 2002. Microarray data normalization and transformation. *Nature Genetics* 32: 96-501.

Schadt, E. E., S. A. Monks, T. A. Drake, A. J. Lusis, N. Che, et al. 2003. Genetics of gene expression surveyed in maize, mouse, and man. *Nature* 422:297-302.

Schena, M., et al. 1996. Parallel human genome analysis: Microarray-based gene expression monitoring of 1000 genes. *Proc. Natl. Acad. Sci. USA* 93:10614-10619.

Slonim, D. K. 2002. From patterns to pathways: Gene expression data analysis comes of age. *Nature Genetics* 32:502-507.

Storey, J. D., and R. Tibshirani. 2003. Statistical significance for genomewide studies. Submitted. Available from *http://stat-www.berkeley.edu/~storey/papers/fdringenomics.pdf*.

Tusher, V. G., R. Tibshirani, and G. Chu. 2001. Significance analysis of microarrays applied to the ionizing radiation response. *Proc. Natl. Acad. Sci. USA* 98:5116-5121.

Wolfinger, R. D., G. Gibson, E. Wolfinger, L. Bennett, H. Hamadeh, P. Bushel, C. Afshari, and R. S. Paules. 2001. Assessing gene significance from cDNA microarray expression data via mixed models. *J. Compu. Biol.* 8:625-637.

Yang, Y.-H., M. J. Buckley, and T. P. Speed. 2001. Analysis of cDNA microarray images. *Briefings in Bioinformatics* 2:341-349.

Yang, Y.-H., S. Dudoit, P. Luu, D. Lin, V. Peng, J. Ngai, and T. P. Speed. 2002. Normalization for cDNA microarray data: A robust composite method addressing single and multiple slide systematic variation. *Nucleic Acids Research* 30:e15.

Yang, Y.-H., and T. Speed. 2002. Design issues for cDNA microarray experiments. *Nature Reviews Genetics* 3:579-588.

Yvert, G., R. B. Brem, J. Whittle, J. M. Akey, E. Foss, E. N. Smith, R. Mackelprang, and L. Kruglyak. 2003. Trans-acting regulatory variation in *Saccharomyces cerevisiae* and the role of transcription factors. *Nature Genetics* 35:57-64.

Additional Reading

Allen, O. B. 1983. A guide to the analysis of growth curve data with special reference to SAS. *Comp. Biomed. Research* 16:101-115.

Eckholdt, H. 2001. Growing genetic cows with SAS: Using low cost parallel processing with Linux – MS/NT networks to search genomes. *Proceedings of the Twenty-sixth Annual SAS Users Group International Conference*, Long Beach, CA.

Holland, J. B. 1998. EPISTACY: A SAS program for detecting two-locus epistatic interactions using genetic marker information. *Journal of Heredity* 89:374-375.

Hussein, M. A., A. Bjornstad, and A. H. Aastveit. 2000. SASG x ESTAB: A SAS program for computing genotype x environment stability statistics. *Agronomy Journal* 92:454-459.

Kang, M. S. 1985. SAS program for calculating stability-variance parameters. *Journal of Heredity* 76:142-143.

Kang, M. S. 1989. A new SAS program for calculating stability-variance parameters. *Journal of Heredity* 80:415.

Lessem, J. L., and S. S. Cherny. 2001. DeFries-Fulker multiple regression analysis of sibship QTL data: A SAS macro. *Bioinformatics* 17:371-372.

Linda, S. B. 1993. GRIFFING: A SAS macro implementing Griffing's analysis of diallel crossing systems. *Hort. Science* 28:61.

Liu, J. S., T. C. Wehner, and S. B. Donaghy. 1997. SASGENE: A SAS computer program for genetic analysis of gene segregation and linkage. *Journal of Heredity* 88:253-254.

Lu, H. Y. 1995. PC-SAS program for estimating Huhn's nonparametric stability statistics. *Agronomy Journal* 87:888-891.

Magari, R., and M. S. Kang. 1997. SAS-STABLE: Stability analyses of balanced and unbalanced data. *Agronomy Journal* 89:929-932.

May, W. L., and W. D. Johnson. 1997. A SAS macro for constructing simultaneous confidence intervals for multinomial proportions. *Computer Methods and Programs in Biomedicine* 53:153-162.

Mumm, R. H., and J. W. Dudley. 1995. A PC SAS computer program to generate a dissimilarity matrix for cluster analysis. *Crop Science* 35:925-927.

Piepho, H. P. 1999. Stability analysis using the SAS system. *Agronomy Journal* 91:154-160.

Saenz-Romero, C., E. V. Nordheim, R. R. Guries, and P. M. Crump. 2001. A case study of a provenance/progeny test using trend analysis with correlated errors and SAS PROC MIXED. *Silvae Genetica* 50:127-135.

Smiley, R. D., S. N. Hicks, L. G. Stinnett, E. E. Howell, and A. M. Saxton. 2002. Bisubstrate kinetics using SAS computer software. *Analytical Biochemistry* 301:153-156.

Sneller, C. H. 1994. SAS programs for calculating coefficient of parentage. *Crop Science* 34:1679-1680.

Thillainathan, M., and G. C. J. Fernandez. 2001. SAS applications for Tai's stability analysis and AMMI model in genotype x environmental interaction (GEI) effects. *Journal of Heredity* 92:367-371.

Wu, H. X., and A. C. Matheson. 2000. Analysis of half-diallel mating design with missing crosses: Theory and SAS program for testing and estimating GCA and SCA fixed effects. *Silvae Genetica* 49:130-137.

Wu, H. X., and A. C. Matheson. 2001. Analyses of half-diallel mating designs with missing crosses: Theory and SAS program for testing and estimating GCA and SCA variance components. *Silvae Genetica* 50:265-271.

Xiang, B., and B. Li. 2001. A new mixed analytical method for genetic analysis of diallel data. *Canadian J. Forest Research* 31:2252-2259.

Xu, S. 1993. INDUPDAT: A SAS/IML program for selection index updating. *Journal of Heredity* 84:316.

Zhang, Y. D., and M. S. Kang. 1997. DIALLEL-SAS: A SAS program for Griffing's diallel analyses. *Agronomy Journal* 89:176-182.

Index

A

ACF (autocorrelation function) 139, 141
additive covariance 12, 19
additive genetic correlation 12, 19, 40
additive variance 11, 16
ADJRSQ option, REG procedure 107
ADJUST option, MODEL statement (MIXED) 126
adjusted R-square 102, 107
Affymetrix GeneChip data (example) 260–276
AGRE (Autism Genetic Resource Exchange) 220, 222
allele frequencies
    See gene frequencies
ALLELE procedure 182
    HAPLO= option 188
    MAXDIST= option 188
    multiple-locus gene frequencies 188–190
    NDATA= option 186
    testing for Hardy-Weinberg proportions 186–187
    WHERE statement 187
alleles, fixation of (inbreeding)
    See inbreeding
alleles IBD
    See QTL mapping, sib-pair regression analysis for
allelic diversity 180
analysis of variance
    See ANOVA
ancestry, common
    See inbreeding
ANOVA (analysis of variance) 12
    Affymetrix data 260–276
    GEI analysis 70, 72–76
    GEI analysis, generalized distributions 87
    gene expression profiling 260–276
    marker-trait association tests 192–193
    MET (multi-environment trials) 83
    unbalanced data sets 13, 14
approximate critical value of QTL detection 209
AR= option, NOEST statement (ARIMA) 144
AR(1) model 125, 131–133
ARH(1) model, fitting 136–138
ARIMA procedure 141
    IDENTIFY statement 141
    NOEST statement 144
ARMA (autoregressive moving average) models 138–139, 141
artificial vs. natural selection 55
    See also genetic selection
Asthma-Atopy microarray data set (example) 260–276
ASYCOV option, MIXED procedure 160
aunt–nephew matings 44
Autism Genetic Resource Exchange) 220, 222
autocorrelation function (ACF) 139, 141
autoregressive moving average (ARMA) models 138–139, 141
autoregressive structure (AR) model 125
    MIXED procedure 131–133
autoregressive structure models, heterogeneous 136–138
AVERAGE option, INBREED procedure 47

B

BACK option, NOEST statement (ARIMA) 144
balanced data sets, ANOVA vs. REML 13, 14
Bayes inference 149–176
    example of (PEDIGREE data) 153–157
    generalized linear mixed models (GLMMs) 162–168
    linear mixed models 149–157
    MCMC vs. 169–174
    normally distributed data 157–160
    population-averaged 165
    REML 158–160
Bayesian mapping 225–250
    genetic map construction 238–240
    QTL analysis 240–249
    QTL analysis, example of 242–248
    three-point linkage analysis 231–238
    two-point linkage analysis 226–231
beef animal lactation
    See lactation modeling
best linear unbiased prediction
    See BLUP selection
BLUP selection 56, 65–66
    GEI analysis 71, 82–83, 86
body weight data (example) 42–43
BOUNDS statement, NLIN procedure 111
Box-Jenkins methodology 139
BRASSICA data set (example) 242–248

breeding value
    See BLUP selection
Brody growth equation 101, 117
BUFFALOES data set (example) 125–133

C

calf growth data (example), fitting with NLIN procedure 117–123
canalization 69
canonical link functions 86–90
CAPABILITY procedure 229, 236–237
cattle
    See lactation modeling
cDNA microarray analysis 251–252
    ANOVA-based analysis of Affymetrix data (example) 260–276
    normalization of data 252–254
Chianina calf growth 117–123
    fitting with NLIN procedure 117–123
classical quantitative genetics 2, 9
    empirical Bayes approaches to mixed model inference 149–176
    GEI analysis 25, 69–96
    genetic parameter estimation 35–54
    genetic selection 55–67
    growth and lactation curves 97–147
    REML estimation of genetic variances and covariances 11–34
clonal design, two environments (example) 23–27
cloning 44
coancestry, parental
    See inbreeding
code samples 2
common ancestry
    See inbreeding
compound symmetry (CS), fitting 125
    MIXED procedure 126–129
conditional density 150
CONTRAST statement 52
corn data (example) 49–54
CORR= option, %SELECT1 macro 58
CORRH2= option, %SELECT1 macro 58
cousin matings 44
covariance
    between parents 47
    dominance covariance 19
    environmental covariance 12, 19
    genetic covariance 12
    maternal-effect 19

    offspring and both parents (mid-parent) 39
    offspring on parents 36
    phenotypic 12
covariance function models 97–98
COVARIANCE option, INBREED procedure 34
covariances, estimating by REML using MIXED procedure 19–34
    clonal design, two environments (example) 23–27
    diallel (example) 27–33
    NC2 design (example) 33
    nested half-sib design, two traits (example) 19–23
COVB option, MODEL statement (MIXED) 157
cows
    See lactation modeling
cross heterosis 52–54
crossbreeding systems, rotational 52–54
crossing systems 52
crossover GEI 70
CS (compound symmetry), fitting 125
    MIXED procedure 126–129
culling, independent
    See independent culling
culling levels 60–62
cutoff values for independent culling 60

D

dairy breed lamb data (example) 121–123
    test day (TD) models, MIXED procedure for 133–138
dairy milk
    See lactation modeling
data, reading 4–5
data for examples 2
data input of pedigree 45
data normalization, microarrays 252–254
DATA step 4
DATALINES statement 4
DDF= option, MODEL statement (MIXED) 276
DDFM= option, MODEL statement (MIXED) 126
degree of resemblance 36
    See also genetic parameter estimation
DESIRED= option, %SINDEX macro 64
diallel 12, 27–33
    heterosis estimation 52–54

DIVISOR option, ESTIMATE statement (GLM) 52
dominance covariance 19
dominance variance 11
downloading example data and programs 2
Drosophila melanogaster data (example) 23–27, 39
Durbin-Watson statistic 105
DW option, MODEL statement (MIXED) 105

E

economic weights 60
effective population size 48–49, 171
EM algorithm for estimating haplotype frequencies 190–191
empirical Bayes inference 149–176
    example of (PEDIGREE data) 153–157
    generalized linear mixed models (GLMMs) 162–168
    linear mixed models 149–157
    MCMC vs. 169–174
    normally distributed data 157–160
environmental covariance 12, 19
environmental variance 11, 16
environments in multi-environment trials 70
EPD (expected progeny difference) 65–66
epistatic variance 11
EQCONS= option, PARMS statement (MIXED) 163
eQTL (expression QTL) 252
ESTIMATE statement (GLM)
    diallel (example) 52
    for heterosis estimation 51–52, 86
evolution (natural selection) 55
EW= option, %SINDEX macro 64
EWES data set (example) 139–146
example data 2
expected heterozygosity 180
expected progeny difference (EPD) 65–66
expression QTL (eQTL) 252
extended sib-pair regression analysis, QTL mapping 202–223
    inferring proportion of genes IBD 204–208
    maximum likelihood (MML) estimation 208
    single marker analysis and interval mapping 209–222
    statistical model 203

F

F-tests 12–15
families, marker-trait associations 195–199
FAMILY= option, %SELECT1 macro 57–58
fixation of alleles
    See inbreeding
fixed effects (GLMMs), Bayes inference on 162–166
    MML estimates 166–168
fixed GEI model 74–75
fixed linear-bilinear models for GEI analysis 76–82
%FIXEDBIPLOT macro 77
fold differences 259
FOUNDERS data set (example), marker-trait association 192–195
frequencies, gene
    See gene frequencies
full-sib matings 44

G

G= option, %SINDEX macro 64
gain from genetic selection 41–43
    See also genetic gain
GAM procedure 90
gamma distribution model, GEI analysis 87
gamma function, incomplete
    See Wood's incomplete gamma function
GAWSNPS data set (example) 185
    marker-trait association tests 195–199
GEI analysis 25, 69–96
    See also MET for GEI analysis
    ANOVA model with fixed GEI 72–76
    BLUP selection 71, 82–83, 86
    crossover GEI 70
    fixed GEI model 74–75
    gamma distribution model 87
    GENMOD procedure 71, 86–90
    GLMs (generalized mixed models) 86–90
    GREG and SREG models 80–82
    homoscedasticity in 71
    LBMs (linear-bilinear models) with fixed GEI 76–82
    LMMs (linear mixed models) 71, 82–86
    LRTs (likelihood ratio tests) 77–78
    negative binomial distribution model 87
    NLINMIXED procedure 71
    non-crossover 70
    Poisson distribution model 87

GEI analysis (*continued*)
   principal component analysis 76
   regression analysis 75–76
   REML 83
   SREGH model 81
   SSGA (smoothing spline genotype analysis) 71, 90–94
   univariate regression approach 75–76
gene expression profiling 251–278
   ANOVA-based analysis 260–276
   normalization of microarray data 252–254
   two-color microarray data 254–260
gene frequencies 179–200
   Hardy-Weinberg proportions 180
   Hardy-Weinberg proportions, testing for 184–187
   marker-trait association tests 191–199
   multiple-locus gene frequencies 188–191
   QTL detection 208–209
   single-locus gene frequencies 180–184
generalized linear mixed models
   See GLMMs
generalized linear models (GLMs) 160
   GEI analysis 86–90
generation interval 56
genes, searching for 272–273
genes IBD
   See QTL mapping, sib-pair regression analysis for
genetic correlation (additive) 12, 19, 40
genetic covariance 12
genetic gain 41–43
   BLUP selection 56, 65–66
   BLUP selection, GEI analysis 71
   expressing 60
   genetic selection index 62–65
   independent culling 59–62
   single-trait genetic selection 56–59
genetic linkage maps
   three-point analysis 231–238
   two-point analysis 226–231
genetic map construction, Bayesian approach for 238–240
genetic merit, estimating 65–66
genetic parameter estimation 35–54
   See also covariances, estimating by REML
   See also variances, estimating by REML
   hybrid vigor (heterosis) 49–54
   inbreeding 44–49
   inbreeding, heterosis estimation 49–54
   realized heritability 41–43
   relationship coefficients 44
genetic selection 55–67
   BLUP selection 56, 65–66
   BLUP selection, GEI analysis 71
   gain from 41–43
   independent culling 59–62
   selection index 62–65
   single-trait selection 56–59
genetic variance 11
genetics, molecular
   See molecular genetics
genetics, quantitative
   See quantitative genetics
GENMOD procedure, GEI analysis 71, 86–90
genotype-by-environment interaction analysis
   See GEI analysis
genotype rank changes 70
genotype regression (GREG) models for GEI analysis 80–82
%GETBLUPS macro 66
GLABELS= option, %SINDEX macro 64
%GLBM macro 81
GLIMM, empirical Bayes inference in 166–168
%GLIMMIX macro 88, 163–166
   MCMC procedures vs. 169–174
   MML estimates 166–168
GLM procedure
   ESTIMATE statement, heterosis estimation 51–52
   for regression models 40–41
GLMMs (generalized linear mixed models) 160–168
   Bayes inference 162–168
   distinguishing from LMMs 166–167
   inverse link functions 160–161
   link functions 160–161
   logistic link function 161
   MML estimates 166–168
   probit link GLMMs 161
   random effects 162–168
GLMs (generalized mixed models) 160
   GEI analysis 86–90
Gompertz equation 101
   LAMB data set (example) 121–123
grandparent–grandoffspring matings 44
GREG models for GEI analysis 80–82
GROUP= option, REPEATED statement (MIXED) 25, 81
growth curve equations 100–101
   fitting with NLIN procedure 117–123
   fitting with REG procedure 101–110

growth curve equations, test day models 123–125
    MIXED procedure for 133–138
    predicting individual data 138–146
GROWTH data set (example) 117–123

H

H2= option, %SINDEX macro 64
half-sib design
    See nested half-sib design
half-sib matings 44–46
HAPLO= option, ALLELE procedure 188
haplotype frequencies, estimating 188–191
HAPLOTYPE procedure 190–191
    marker-trait association, unrelated individuals 193–195
    OUT= option 193
haplotype trend regression (HTR) 193–195
Hardy-Weinberg proportions 180
    testing for 184–187
heritability, estimating
    inbreeding 44–49
    inbreeding, heterosis estimation 49–54
    realized heritability 41–43
    Wood function parameters 108
heritability estimation using regression of offspring on parent 35–41
heterogeneous autoregressive structure (ARH) models 136–138
heterosis estimation 49–54, 86
heterozygosity, increasing 49–50, 180
hierarchical models 151–153
    example of (PEDIGREE data) 153–157
HOLD= option, PARMS statement (MIXED) 276
homoscedasticity in GEI analysis 71
homozygosity
    See heritability, estimating
    See inbreeding
HTR (haplotype trend regression) method 193–195
HWE (Hardy-Weinberg equilibrium) 180
    testing for 184–187
hybrid seed corn data (example) 49–54
hybrid vigor 49–54
hyperparameters 151

I

IDENTIFY statement, ARIMA procedure 141

IML procedure
    QTL mapping with 210, 214–216
    three-point linkage analysis 233–234
    two-point linkage analysis 228–230
IMPORT procedure 4–5
INBREED procedure 34, 45–49
    linear mixed model inference, example of 153–154
    MATING statement 47, 48
    MATRIX option 34, 47
    OUTCOV= option 34
    VAR statement 47
inbreeding 44–49
    effective population size 48–49
    heterosis 49–54
%INCLUDE statement 5
incomplete gamma function
    See Wood's incomplete gamma function
%INDCULL macro 60–62
independent culling 59–62
    cutoff values 60
    selection index vs. 62
indirect single-trait genetic selection 58
individual heterosis 50
individual single-trait genetic selection 56
inferring proportion of genes IBD 204–208
    multipoint estimation 207–208, 212–214
INPUT statement 4
inputting pedigree data 45
interval mapping of QTL 209–222
    program demo 220–222
    program execution 216–220
    program implementation 210–216
inverse link functions, GLMMs 160–161
Italian water buffaloes (example) 125–133

J

JMP analysis 263
joint posterior density 150

K

Kenward-Roger adjustment 126

L

lactation modeling 98–101
    fitting with NLIN procedure 110–123
    fitting with REG procedure 101–110
    linear phase 98–99

lactation modeling (*continued*)
  mechanistic models for curves 111
lactation modeling, test day models 123–125
  MIXED procedure for 125–133
  predicting individual data 138–146
LAMB data set (example) 121–123
    test day (TD) models, MIXED procedure for 133–138
LBMs (linear-bilinear models) with fixed GEI 76–82
  GREG and SREG models 80–82
LD (linkage disequilibrium) 179, 188–190
LDATA= option, REPEATED statement (MIXED) 34
LEAD option, NOEST statement (ARIMA) 144
least squares regression analysis 36
likelihood ratio tests (LRTs) 17
  GEI analysis 77–78
  QTL detection 209
linear-bilinear models (LBMs) with fixed GEI 76–82
  GREG and SREG models 80–82
linear mixed models
  See LMMs
linear phase, lactation models 98–99
linear regression 35–41
link functions
  canonical 86–90
  GLMMs 160–161
linkage disequilibrium (LD) 179, 188–190
linkage maps
  three-point analysis 231–238
  two-point analysis 226–231
LMMs (linear mixed models)
  approach to GEI analysis 71
  distinguishing from GLMMs 166–167
  GEI analysis 82–86
  generalized (GLMMs) 160–168
  test day models 126–129
  test day models, MIXED procedure for 133–138
LMMs (linear mixed models), Bayes approaches to 149–176
  example of (PEDIGREE data) 153–157
  generalized linear mixed models (GLMMs) 162–168
  linear mixed models 149–157
  MCMC vs. 169–174
  normally distributed data 157–160
LMMs (linear mixed models), for lactation and growth curve modeling
    See growth curve equations
    See lactation modeling
logistic link function, GLMMs 161
logistic model 100–101, 119–120
loop designs, microarray data 253
low body weight data (example) 42–43
LOWERB option, PARMS statement (MIXED) 31–32
LRTs (likelihood ratio tests) 17
  GEI analysis 77–78
  QTL detection 209
LSMEANS statement 73

M

MA= option, NOEST statement (ARIMA) 144
macros 5–6
mapping, Bayes
  See Bayesian mapping
mapping of QTL
  See QTL mapping, sib-pair regression analysis for
marginal densities 150
marker-trait association tests 191–199
  family data 195–199
  unrelated individuals 192–195
Markov chain Monte Carlo (MCMC) methods
  See MCMC procedures
maternal-effect covariance 19
maternal-effect variance 14, 16
maternal heterosis 50
MATING statement, INBREED procedure 47, 48
MATRIX option, INBREED procedure 34, 47
MAX keyword, DESIRED option (%SINDEX) 64
MAXDIST= option, ALLELE procedure 188
maximum likelihood (MML) estimates 166–168, 208
MCMC (Markov chain Monte Carlo) procedures 226
  Bayesian QTL analysis 240–249
  Bayesian QTL analysis, example of 242–248
  empirical Bayes vs. 169–174
  genetic map construction 238–240
  %GLIMMIX macro vs. 169–174
  three-point linkage analysis 231–238
  two-point linkage analysis 228–230
mechanistic models for lactation curves 111

MET (multi-environment trials) for GEI analysis 69–70, 72, 83
  canonical link functions 86–87
  environments in 70
  fixed GEI model approach 74–75
method of moments 12–13
Metropolis-Hastings algorithm 232, 239
  Bayesian QTL analysis 243–245
MEYER data set (example)
  generalized linear mixed model inference 163–166
  linear mixed model inference 153–157
  MCMC procedures vs. empirical Bayes inference 169–174
  MML estimates 167–168
microarray analysis 251–252
  ANOVA-based analysis of Affymetrix data (example) 260–276
  for QTL mapping 252
  loop designs 253
  normalization of data 252–254
  split-plot designs 253
  two-color data 254–260
milk production
  See lactation modeling
mixed models, gene expression profiling 251–278
  ANOVA-based analysis 260–276
  normalization of microarray data 252–254
  two-color microarray data 254–260
mixed models, linear
  See GLMMs
  See LMMs
MIXED procedure
  See also MODEL statement
  See also PARMS statement
  See also RANDOM statement
  See also REPEATED statement
  AR model 131–133
  Bayes inference, MCMC procedures vs. 169–174
  canonical link functions 88–90
  ESTIMATE statement (GLM) for heterosis estimation 51–52
  fitting compound symmetry 126–129
  fitting UN structure 125–126
  GEI analysis 71, 82–86
  GEI analysis, ANOVA model for 72–76
  hierarchical models 151–153
  hierarchical models, example of 153–157
  lactation modeling, TD models 125–133
  LMMs, TD models 133–138
  QTL mapping 210, 216
  regression line calculation 37
  regression models 40–41
  REP variation, controlling 42
  repeated measures design 124–125
  test-day models 125–138
  test-day models, growth curve equations 133–138
  two-color microarray data 254–260
  TYPE=ML option 17
MIXED procedure, estimating variances and covariances 11–34
  clonal design, two environments (example) 23–27
  diallel (example) 27–33
  NC2 design (example) 33
  nested half-sib design, one trait (example) 13–19
  nested half-sib design, two traits (example) 19–23
MLEs of genotype frequencies
  See gene frequencies
MML (maximum likelihood) estimates 166–168, 208
MODEL statement, MIXED procedure 20
  ADJUST option 126
  COVB option 157
  DDF= option 276
  DDFM= option 126
  DW option 105
  NOINT option 156
  SOLUTION option 16
molecular genetics 3, 177
  Bayesian mapping 225–250
  gene expression profiling 251–278
  gene frequencies 179–200
  QTL mapping by sib-pair regression analysis 202–223
moments, method of 12–13
MPM module 212–214
MU= option, NOEST statement (ARIMA) 144
multi-environment trials
  See MET for GEI analysis
multiple-locus gene frequencies 188–191
multiple-trait genetic selection
  See genetic selection
multipoint estimation of proportion of genes IBD 207–208, 212–214
multivariate normal distribution of traits 60

N

narrow-sense heritability 11, 16
natural vs. artificial selection 55
    See also genetic selection
NC2 design 12, 27, 33
NDATA= option, ALLELE procedure 186
negative binomial distribution model, GEI
        analysis 87
nested half-sib design 12–27
    disadvantages of 27
    one trait (example) 13–19
    one trait, two environments (example) 23–27
    two traits (example) 19–23
NLIN procedure
    fitting growth curve equations 117–123
    fitting incomplete gamma function 110–123
    lactation and growth curve modeling 110–123
NLINMIXED procedure, GEI analysis 71
NOEST statement, ARIMA procedure 144
NOINT option, MODEL statement (MIXED) 156
NOITER option, PARMS statement (MIXED) 154
non-crossover GEI 70
non-linear models of Wood equation 110–123
NOPROFILE option, PARMS statement (MIXED) 154
normalization of microarray data 252–254
normally distributed data, Bayes inference for 157–160
norms of reaction 69
North Carolina Design 2 (NC2) 12, 27, 33
NSIB= option, %SELECT1 macro 57–58
nuisance parameters 151

O

observed heterozygosity 180
optimal culling levels, determining 60–62
ordinary likelihood function 17
OUT= option, HAPLOTYPE procedure 193
OUTCOV= option, INBREED procedure 34
OUTEST statement, REG procedure 107
overdispersion 161–162

P

P= option, %SINDEX macro 64

Parallel Plot of Least Squares Means window 269–270
parametric empirical Bayes inference 158
parent–offspring matings 44
parental coancestry
    See inbreeding
parental selection
    See genetic selection
parental variance 36
PARITIES data set (example)
    fitting with NLIN procedure 110–123
    fitting with REG procedure 101–110
PARMS statement, MIXED procedure 23, 31–32
    EQCONS= option 163
    HOLD= option 276
    LOWERB option 31–32
    NOITER and NOPROFILE options 154
paternal heterosis 50
%PC1PC2 macro 81
PCA (principal component analysis), GEI
        analysis 76
PDIFF option, LSMEANS statement 73
%PDMIX800 macro 73
peanut yield data (example) 72
    ANOVA model with fixed GEI 72–76
    LBMs (linear-bilinear models) with fixed GEI 77–82
    LMMs (linear mixed models) for GEI analysis 81
Pearson chi-square statistic, HWE 184–187
pedigree data, inputting 45
PEDIGREE data set (example) 153–157
phenotypic covariance 12
phenotypic measures, index weights on 62–65
phenotypic plasticity 69
phenotypic variance 11, 16
PIC (polymorphism information content) 180
PLABELS= option, %SINDEX macro 64
Poisson distribution model, GEI analysis 87
polymorphism information content (PIC) 180
population-averaged Bayes inference 165
population size, effective 48–49, 171
posterior inference regarding recombination rate 235
predicting individual test data 138–146
prediction of genetic gain
    See genetic gain
    See genetic selection
principal component analysis (PCA), GEI
        analysis 76

Principal Components of Standardized Least Squares Means window 272
PROBCHI function 17
probit link GLMMs 161
program code, obtaining 2
proportions of genes IBD, inferring 204–208
    multipoint estimation 207–208, 212–214
PSD= option, %SINDEX macro 64

Q

%QTDT macro 195–198
QTL analysis, Bayesian approaches for 240–249
    example of (*Brassica napus*) 242–248
QTL mapping
    IML procedure 210, 214–216
    microarray analysis for 252
    MIXED procedure 210, 216
    random model approach 202–203
    scanning increment 210–211
    test statistic profile 209
    variance component analysis to 202–203
QTL mapping, sib-pair regression analysis for 202–223
    inferring proportion of genes IBD 204–208
    maximum likelihood (MML) estimation 208
    single marker analysis and interval mapping 209–222
    statistical model 203
%QTLCRITIC macro 202, 210, 219
%QTLIBD macro 202, 210, 218
    scanning increment 210, 211
%QTLPLOT macro 202, 210, 219
%QTLSCAN macro 202, 210
%QTLSCAN1 macro 210, 219, 220, 222
%QTLSCAN2 macro 210
quantitative genetics 1
quantitative genetics, classical 2, 9
    empirical Bayes approaches to mixed model inference 149–176
    GEI analysis 25, 69–96
    genetic parameter estimation 35–54
    genetic selection 55–67
    growth and lactation curves 97–147
    REML estimation of genetic variances and covariances 11–34
quantitative genetics, molecular 3, 177
    Bayesian mapping 225–250
    gene expression profiling 251–278
    gene frequencies 179–200
    QTL mapping by sib-pair regression analysis 202–223
quantitative trait loci
    See QTL analysis, Bayesian approaches for
    See QTL mapping, microarray analysis for
    See QTL mapping, sib-pair regression analysis for
quantitative traits, GEI of
    See GEI analysis

R

R-Squared Values window 266–267
random effects (GLMMs), Bayes inference on 162–168
random model approach to QTL mapping 202–203
random regression models 97–98, 110–123
    fitting with REG procedure 101–110
RANDOM statement, MIXED procedure
    repeated measures design 124–125
    SIRE effect 17
    SOLUTION option 156, 168
    TYPE=CS option 27
    TYPE=LIN option 29–32
    TYPE=TOEP(1) option 17
    TYPE=UN option 19–20, 22, 27
    TYPE=UNC option 22
    TYPE=UNR option 22, 25
RANGAM function 241
reading data 4–5
realized heritability 41–43
recombination rates
    posterior inference 235
    three-point linkage analysis 231–238
    two-point linkage analysis 227–230
REG procedure
    fitting incomplete gamma function 101–110
    lactation and growth curve modeling 101–110
    OUTEST statement 107
regression analysis 35–41
    GEI analysis 75–76
    GLM procedure 40–41
    GREG and SREG models, GEI analysis 80–82
    least squares 36
    line calculation 37
    MIXED procedure 40–41
    QTL mapping by sib-pair analysis 202–223
    random models 97–98, 110–123

regression analysis (*continued*)
    random models, fitting with REG procedure 101–110
relationship equation and coefficients 44–49
relatives, single-trait genetic selection on 57
REML (restricted maximum likelihood) 13
    as GLM parameter 87
    empirical Bayes inference 158–160
    GEI analysis 83
REML estimation of variances and covariances 11–34
    clonal design, two environments (example) 23–27
    diallel (example) 27–33
    NC2 design (example) 33
    nested half-sib design, one trait (example) 13–19
    nested half-sib design, two traits (example) 19–23
    unbalanced data sets 13, 14
RENAME= data set option 186
REP variation, controlling 42
repeated measures 97
    lactation and growth curve modeling 123–125
    MIXED procedure 124–125
REPEATED statement, MIXED procedure 19–20
    GROUP= option 25, 81
    LDATA= option 34
resemblance, degree of 36
    See also genetic parameter estimation
residual variance 111
residuals, processing 37
restricted likelihood function 16–17
restricted maximum likelihood (REML)
    See REML
    See REML estimation of variances and covariances
RFI values 253
rotational crossbreeding systems 52–54
%RUNMIXED macro 84–85

S

Sarda dairy breed lambs (example) 121–123
    test day models, MIXED procedure for 133–138
    time series prediction of test day data 139–146
SAS Microarray Solution 1.0.1 260–276
    gene expression profiling 260–276
    Parallel Plot of Least Squares Means window 269–270
    Principal Components of Standardized Least Squares Means window 272
    R-Squared Values window 266–267
    searching for genes 272–273
    Significant Genes—Hierarchical Clustering window 270–271
    Variance Component Estimates window 266, 268
    Web Search dialog box 272–273
scanning increment, QTL mapping 210–211
SCC (somatic cell count) 37–41
searching for genes 272–273
seasonal ARMA models for EWES data (example) 141
seed beetles (example)
    nested half-sib design, two traits 19–23
    nested half-sib design with one trait 13–19
seed corn data (example) 49–54
%SELECT1 macro 56
selection experiments 41
    See also genetic selection
selection index 62–65
SELECTON option, %SINDEX macro 64
self-accelerating/decelerating phases 98–99
    See also lactation modeling
selfing 44
SHEEP data set (example) 112–116
sheep milk
    See lactation modeling
sib-pair regression analysis, QTL mapping 202–223
    inferring proportion of genes IBD 204–208
    maximum likelihood (MML) estimation 208
    single marker analysis and interval mapping 209–222
    statistical model 203
sibling tests of linkage and association 198–199
%SIBQTDT macro 198–199
SIBR= option, %SELECT1 macro 57–58
SIBT= option, %SELECT1 macro 57–59
Significant Genes—Hierarchical Clustering window 270–271
%SINDEX macro 62–65
single-locus gene frequencies 180–184
single marker analysis, QTL mapping 209–222
    program demo 220–222
    program execution 216–220
    program implementation 210–216

single-trait genetic selection 56–59
singular value decomposition (SVD) 76
SIRE effect 17
sites regression (SREG) models for GEI analysis 80–82
smoothing spline genotype analysis (SSGA) 71, 90–94
SOLUTION option
    MODEL statement (MIXED) 16
    RANDOM statement (MIXED) 156, 168
somatic cell count (SCC) 37–41
SPECTRA procedure 140–146
spline analysis 71, 90–94
split-plot designs, microarray data 253
SREG models for GEI analysis 80–82
SREGH model for GEI analysis 81
SSGA (smoothing spline genotype analysis) 71, 90–94
STMTS= option, %GLIMMIX macro 163
stochastic MCMC procedures, empirical Bayes vs. 169–174
SVD (singular value decomposition) 76

T

TD (test day) models 123
    growth curve equations 123–125
    lactation modeling 123–125
    linear mixed models 126–129, 133–138
    MIXED procedure for 125–138
    predicting individual data 138–146
    repeated measures theory 123–125
    time series prediction 138–146
terminal crosses 52
test day data
    See TD models
test statistic profile, QTL mapping 209
three-point linkage analysis 231–238
time series prediction of test day data 138–139
    SPECTRA procedure for 140–146
%TPLOT macro 189–190, 196
transmitting ability
    See BLUP selection
Tribolium data (example) 42–43
two-color microarray data, mixed model analysis 254–260
two-point linkage analysis 226–231
TYPE= options, RANDOM statement (MIXED)
    ML option 17
    TYPE=CS 27
    TYPE=LIN 29–32
    TYPE=TOEP(1) 17
    TYPE=UN 19–20, 22, 27
    TYPE=UNC 22
    TYPE=UNR 22, 25
    UNR option 22

U

unbalanced data sets, ANOVA vs. REML 13, 14
uncle–niece matings 44
UNIVARIATE procedure 37
univariate regression approach to GEI analysis 75–76
unrelated individuals, marker-trait associations 192–195
unrestricted likelihood function 17
unstructured (UN) structure, fitting 124
    MIXED procedure for 125–126

V

VAR statement, INBREED procedure 47
variance, analysis of
    See ANOVA
variance
    dominance variance 11
    environmental variance 11, 16
    epistatic variance 11
    genetic variance 11
    maternal-effect 14, 16
    offspring on mid-parent average 39
    parental 36
    phenotypic 11, 16
    residual variance 111
variance component analysis to QTL mapping 202–203
Variance Component Estimates window 266, 268
variances, estimating by REML using MIXED procedure 11–34
    clonal design, two environments (example) 23–27
    diallel (example) 27–33
    NC2 design (example) 33
    nested half-sib design, one trait (example) 13–19
volcano plots 258–259, 266, 269

W

water buffaloes (example) 125–133
Web Search dialog box 272–273
WHERE statement, ALLELE procedure 187
Wood's incomplete gamma function 99–100
    fitting with NLIN procedure 110–123
    fitting with REG procedure 101–110
    heritability of parameters 108
    non-linear models 110–123
    technical meaning of parameters 105–107

Z

z-tests 16

Call your local SAS office to order these books from
Books by Users Press

Advanced Log-Linear Models Using SAS®
by **Daniel Zelterman**................................ Order No. A57496

Analysis of Clinical Trials Using SAS®: A Practical Guide
by **Alex Dmitrienko, Walter Offen, Christy Chuang-Stein,**
and **Geert Molenbergs** .... Order No. A59390

Annotate: Simply the Basics
by **Art Carpenter**.. Order No. A57320

Applied Multivariate Statistics with SAS® Software, Second Edition
by **Ravindra Khattree**
and **Dayanand N. Naik** .... Order No. A56903

Applied Statistics and the SAS® Programming Language, Fourth Edition
by **Ronald P. Cody**
and **Jeffrey K. Smith**................................ Order No. A55984

An Array of Challenges — Test Your SAS® Skills
by **Robert Virgile**....................................... Order No. A55625

Carpenter's Complete Guide to the SAS® Macro Language, Second Edition
by **Art Carpenter**....................................... Order No. A59224

The Cartoon Guide to Statistics
by **Larry Gonick**
and **Woollcott Smith**................................ Order No. A55153

Categorical Data Analysis Using the SAS® System, Second Edition
by **Maura E. Stokes, Charles S. Davis,**
and **Gary G. Koch** .... Order No. A57998

Cody's Data Cleaning Techniques Using SAS® Software
by **Ron Cody** ....Order No. A57198

Common Statistical Methods for Clinical Research with SAS® Examples, Second Edition
by **Glenn A. Walker**................................. Order No. A58086

Debugging SAS® Programs: A Handbook of Tools and Techniques
by **Michele M. Burlew** .... Order No. A57743

Efficiency: Improving the Performance of Your SAS® Applications
by **Robert Virgile**....................................... Order No. A55960

Genetic Analysis of Complex Traits Using SAS®
Edited by **Arnold M. Saxton** .... Order No. A59454

A Handbook of Statistical Analyses Using SAS®, Second Edition
by **B.S. Everitt**
and **G. Der** .... Order No. A58679

Health Care Data and the SAS® System
by **Marge Scerbo, Craig Dickstein,**
and **Alan Wilson**....................................... Order No. A57638

The How-To Book for SAS/GRAPH® Software
by **Thomas Miron** .... Order No. A55203

In the Know ... SAS® Tips and Techniques From Around the Globe
by **Phil Mason** .... Order No. A55513

Instant ODS: Style Templates for the Output Delivery System
by **Bernadette Johnson** .... Order No. A58824

Integrating Results through Meta-Analytic Review Using SAS® Software
by **Morgan C. Wang**
and **Brad J. Bushman** .... Order No. A55810

Learning SAS® in the Computer Lab, Second Edition
by **Rebecca J. Elliott** .... Order No. A57739

The Little SAS® Book: A Primer
by **Lora D. Delwiche**
and **Susan J. Slaughter** .... Order No. A55200

The Little SAS® Book: A Primer, Second Edition
by **Lora D. Delwiche**
and **Susan J. Slaughter** .... Order No. A56649
(*updated to include Version 7 features*)

The Little SAS® Book: A Primer, Third Edition
by **Lora D. Delwiche**
and **Susan J. Slaughter** .... Order No. A59216
(*updated to include SAS 9.1 features*)

Logistic Regression Using the SAS® System: Theory and Application
by **Paul D. Allison** .... Order No. A55770

Longitudinal Data and SAS®: A Programmer's Guide
by **Ron Cody** .... Order No. A58176

Maps Made Easy Using SAS®
by **Mike Zdeb** .... Order No. A57495

Models for Discrete Data
by **Daniel Zelterman** .... Order No. A57521

support.sas.com/pubs

Multiple Comparisons and Multiple Tests Using SAS®
Text and Workbook Set
(*books in this set also sold separately*)
by **Peter H. Westfall, Randall D. Tobias,
Dror Rom, Russell D. Wolfinger,**
and **Yosef Hochberg** .... Order No. A58274

Multiple-Plot Displays: Simplified with Macros
by **Perry Watts** .... Order No. A58314

*Multivariate Data Reduction and Discrimination with
SAS® Software*
by **Ravindra Khattree**
and **Dayanand N. Naik** .... Order No. A56902

Output Delivery System: The Basics
by **Lauren E. Haworth** .... Order No. A58087

*Painless Windows: A Handbook for SAS® Users,
Third Edition*
by **Jodie Gilmore** .... Order No. A58783
(*updated to include Version 8 and SAS 9.1 features*)

PROC TABULATE by Example
by **Lauren E. Haworth** .... Order No. A56514

Professional SAS® Programming Shortcuts
by **Rick Aster** .... Order No. A59353

Quick Results with SAS/GRAPH® Software
by **Arthur L. Carpenter**
and **Charles E. Shipp**................................ Order No. A55127

Quick Results with the Output Delivery System
by **Sunil K. Gupta** ....Order No. A58458

Quick Start to Data Analysis with SAS®
by **Frank C. Dilorio**
and **Kenneth A. Hardy** .... Order No. A55550

Reading External Data Files Using SAS®: Examples Handbook
by **Michele M. Burlew** .... Order No. A58369

*Regression and ANOVA: An Integrated Approach Using
SAS® Software*
by **Keith E. Muller**
and **Bethel A. Fetterman**......................... Order No. A57559

SAS® Applications Programming: A Gentle Introduction
by **Frank C. Dilorio** .... Order No. A56193

SAS® for Forecasting Time Series, Second Edition
by **John C. Brocklebank,**
and **David A. Dickey** .... Order No. A57275

SAS® for Linear Models, Fourth Edition
by **Ramon C. Littell, Walter W. Stroup,**
and **Rudolf J. Freund**............................... Order No. A56655

*SAS® for Monte Carlo Studies: A Guide for Quantitative
Researchers*
by **Xitao Fan, Ákos Felsővályi, Stephen A. Sivo,**
and **Sean C. Keenan**................................ Order No. A57323

SAS® Functions by Example
by **Ron Cody** .... Order No. A59343

SAS® Macro Programming Made Easy
by **Michele M. Burlew** .... Order No. A56516

SAS® Programming by Example
by **Ron Cody**
and **Ray Pass** .... Order No. A55126

*SAS® Programming for Researchers and Social Scientists,
Second Edition*
by **Paul E. Spector**.................................. Order No. A58784

*SAS® Survival Analysis Techniques for Medical Research,
Second Edition*
by **Alan B. Cantor** ....Order No. A58416

*SAS® System for Elementary Statistical Analysis,
Second Edition*
by **Sandra D. Schlotzhauer**
and **Ramon C. Littell**............................... Order No. A55172

SAS® System for Mixed Models
by **Ramon C. Littell, George A. Milliken, Walter W. Stroup,**
and **Russell D. Wolfinger** .... Order No. A55235

SAS® System for Regression, Third Edition
by **Rudolf J. Freund**
and **Ramon C. Littell**............................... Order No. A57313

SAS® System for Statistical Graphics, First Edition
by **Michael Friendly** .... Order No. A56143

The SAS® Workbook and Solutions Set
(*books in this set also sold separately*)
by **Ron Cody** .... Order No. A55594

*Selecting Statistical Techniques for Social Science Data:
A Guide for SAS® Users*
by **Frank M. Andrews, Laura Klem, Patrick M. O'Malley,
Willard L. Rodgers, Kathleen B. Welch,**
and **Terrence N. Davidson** .... Order No. A55854

Statistical Quality Control Using the SAS® System
by **Dennis W. King** .... Order No. A55232

*A Step-by-Step Approach to Using the SAS® System
for Factor Analysis and Structural Equation Modeling*
by **Larry Hatcher**..................................... Order No. A55129

*A Step-by-Step Approach to Using the SAS® System
for Univariate and Multivariate Statistics*
by **Larry Hatcher**
and **Edward Stepanski**............................ Order No. A55072

Step-by-Step Basic Statistics Using SAS®: Student Guide
and *Exercises*
(*books in this set also sold separately*)
by **Larry Hatcher**..................................... Order No. A57541

*Survival Analysis Using the SAS® System:
A Practical Guide*
by **Paul D. Allison** .... Order No. A55233

*Tuning SAS® Applications in the OS/390 and z/OS
Environments, Second Edition*
by **Michael A. Raithel** .... Order No. A58172

support.sas.com/pubs

*Univariate and Multivariate General Linear Models:
Theory and Applications Using SAS® Software*
by **Neil H. Timm**
and **Tammy A. Mieczkowski** .... Order No. A55809

Using SAS® in Financial Research
by **Ekkehart Boehmer, John Paul Broussard,**
and **Juha-Pekka Kallunki** .... Order No. A57601

Using the SAS® Windowing Environment: A Quick Tutorial
by **Larry Hatcher** .... Order No. A57201

Visualizing Categorical Data
by **Michael Friendly** .... Order No. A56571

Web Development with SAS® by Example
by **Frederick Pratter** .... Order No. A58694

Your Guide to Survey Research Using the SAS® System
by **Archer Gravely** .... Order No. A55688

JMP® Books

JMP® Start Statistics, Third Edition
by **John Sall, Ann Lehman,**
and **Lee Creighton** .... Order No. A58166

Regression Using JMP®
by **Rudolf J. Freund, Ramon C. Littell,**
and **Lee Creighton** .... Order No. A58789

support.sas.com/pubs